AI and the Bomb

AI and the Bomb

Nuclear Strategy and Risk in the Digital Age

James Johnson

OXFORD
UNIVERSITY PRESS

OXFORD
UNIVERSITY PRESS

Great Clarendon Street, Oxford, ox2 6dp,
United Kingdom

Oxford University Press is a department of the University of Oxford.
It furthers the University's objective of excellence in research, scholarship,
and education by publishing worldwide. Oxford is a registered trade mark of
Oxford University Press in the UK and in certain other countries

Published in the United States of America by Oxford University Press
198 Madison Avenue, New York, NY 10016, United States of America

British Library Cataloguing in Publication Data
Data available

Library of Congress Control Number: 2022948348

ISBN 978–0–19–285818–4

DOI: 10.1093/oso/9780192858184.001.0001

Printed and bound by
CPI Group (UK) Ltd, Croydon, CR0 4YY

Acknowledgments

I have incurred many debts in writing this book, which I cannot possibly repay. What I can do, however, is acknowledge them and express my sincere thanks. Many colleagues read, commented, and in various ways contributed to the book at its various stages. I am especially grateful to James Acton, John Amble, Greg Austin, John Borrie, Lyndon Burford, Jeffrey Cummings, Jeffrey Ding, Mona Dreicer, Sam Dunin, Andrew Futter, Erik Gartzke, Andrea Gilli, Rose Gottemoeller, Rebecca Hersman, Michael Horowitz, Patrick Howell, Keir Lieber, Jon Lindsay, Giacomo Persi Paoli, Kenneth Payne, Tom Plant, Daryl Press, Bill Potter, Benoit Pelopidas, Adam Quinn, Andrew Reddy, Brad Roberts, Mick Ryan, Daniel Salisbury, John Shanahan, Michael Smith, Wes Spain, Reuben Steff, Oliver Turner, Chris Twomey, Tristen Volpe, Tom Young, and Benjamin Zala. The book greatly benefits from their comments and criticisms on the draft manuscript and separate papers from which the book draws. My appreciation also goes to the many experts who challenged my ideas and sharpened my arguments on the presentations I have given at various international forums during the development of this book.

I have also enjoyed the generous support of several institutions that I would like to acknowledge, including: the James Martin Center for Non-Proliferation Studies; the Project on Nuclear Issues at the Center for Strategic and International Studies and the Royal United Services Institute; the Modern War Institute at West Point; the Vienna Center for Disarmament and Non-Proliferation; the Center for Global Security Research at Lawrence Livermore Laboratory; the US Naval War College, the UK Deterrence & Assurance Academic Alliance; the International Institute for Strategic Studies, and the Towards a Third Nuclear Age Project at the University of Leicester. I would also like to express my thanks for the encouragement, friendship, and support of my colleagues in the Department of Politics and International Relations at the University of Aberdeen.

My appreciation also to the excellent team at Oxford University Press for their professionality, guidance, and support. Not to mention the anonymous reviewers, whose comments and suggestions kept me honest and improved the book in many ways. Finally, thanks to my amazing wife, Cindy, for her unstinting support, patience, love, and encouragement. This book is dedicated to her.

Contents

List of figures and tables

Figures

Table

List of abbreviations

A2/AD	anti-access and area denial
AGI	artificial general intelligence
APT	advanced persistent threat
ASAT	anti-satellite weapons
ATR	automatic target recognition
C2	command and control
C3I	command, control, communications, and intelligence
DARPA	Defense Advanced Research Projects Agency (US DoD)
DL	deep learning
DoD	Department of Defense (United States)
GANs	generative adversarial networks
ICBM	intercontinental ballistic missile
ISR	intelligence, surveillance, and reconnaissance
LAWS	lethal autonomous weapon systems
MIRV	multiple independent targetable re-entry vehicles
ML	machine learning
NC3	nuclear command, control, and communications
NPT	Non-Proliferation Treaty
PLA	People's Liberation Army
R&D	Research and Development
RMA	revolution in military affairs
SSBN	nuclear-powered ballistic missile submarine
UAVs	unmanned aerial vehicles
UNIDIR	United Nations Institute for Disarmament Research
UUVs	unmanned underwater vehicles
USVs	unmanned surface vehicles

Introduction
Artificial intelligence and nuclear weapons

On the morning of December 12, 2025, political leaders in Beijing and Washington authorized a nuclear exchange in the Taiwan Straits. In the immediate aftermath of the deadly confrontation—which lasted only a matter of hours, killing millions and injuring many more—leaders on both sides were dumbfounded about what caused the "flash war."[1] Independent investigators into the 2025 "flash war" expressed sanguinity that neither side deployed AI-powered "fully autonomous" weapons, nor intentionally violated the law of armed conflict—or the principles of proportionality and distinction that apply to nuclear weapons in international law.[2] Therefore, both states acted in a contemporaneous and *bona fide* belief that they acted in self-defense—that is, lawfully under *jus ad bellum* use of military force.[3]

In an election dominated by the island's volatile relations with Communist China in 2024, President Tsai Ing-wen—in another major snub to Beijing—pulled off a sweeping victory, securing her third term for the pro-independence Democrats. As the mid-2020s dawned, tensions across the Straits continued to sour, as both sides—held hostage to hardline politicians and hawkish military generals—maintained uncompromising positions, jettisoning diplomatic gestures and inflamed by escalatory rhetoric, fake news, and campaigns of mis/disinformation. At the same time, both China and the US deployed artificial intelligence (AI) technology to support battlefield awareness, intelligence, surveillance, and reconnaissance (ISR), and early-warning and other decision-support tools to predict and suggest tactical responses to enemy actions in real-time.

By late 2025, the rapid improvements in the fidelity, speed, and predictive capabilities of commercially produced dual-use AI applications[4] persuaded great military powers not only to feed data-hungry machine learning (ML) to enhance tactical and operational maneuvers, but increasingly to inform strategic decisions. Impressed by the early adoption and fielding by Russia, Turkey, and Israeli of AI tools to support autonomous drone swarms to outmaneuver and crush counterterrorist incursions on their borders, China

AI and the Bomb. James Johnson, Oxford University Press. © James Johnson (2023).
DOI: 10.1093/oso/9780192858184.003.0001

synthesized the latest iterations of dual-use AI, despite sacrificing rigorous testing and evaluation in the race for first-mover advantage.[5]

With Chinese military incursions—aircraft flyovers, island blockade drills, and drone surveillance operations—in the Taiwan Straits marking a dramatic escalation in tensions, leaders in China and the US demanded the immediate fielding of the latest strategic AI to gain the maximum asymmetric advantage in scale, speed, and lethality. These state-of-the-art strategic AI systems were trained on a combination of historical combat scenarios, experimental wargames, game-theoretic rational decision-making, intelligence data, and learning from previous versions of themselves to generate novel and unorthodox strategic recommendations—a level of sophistication which often confounded designers and operators alike.[6] As the incendiary rhetoric playing out on social media—exacerbated by disinformation campaigns and cyber intrusions on command and control (C2) networks—reached a fever pitch on both sides, a chorus of voices expounded the immediacy of a forced unification of Taiwan by China.

Buoyed by the escalatory situation unraveling in the Pacific—and with testing and evaluation processes incomplete—the US decided to bring forward the fielding of its prototype autonomous AI-powered "Strategic Prediction & Recommendation System" (SPRS)—supporting decision-making in non-lethal activities such as logistics, cyber, space assurance, and energy management. China, fearful of losing the asymmetric upper hand, fielded a similar decision-making support system, "Strategic & Intelligence Advisory System" (SIAS), to ensure its autonomous preparedness for any ensuing crisis, while similarly ensuring it could not authorize autonomous lethal action.

On June 14, 2025, the following events transpired. At 06:30 local time, a Taiwanese coast guard patrol boat collided with and sank a Chinese autonomous sea-surface vehicle conducting an intelligence recon mission within Taiwan's territorial waters. On the previous day, President Tsai hosted a senior delegation of US congressional staff and White House officials in Taipei on a high-profile diplomatic visit. This controversial visit caused outrage in China, sparking a furious—and state abetted—response from Chinese netizens who called for a swift and aggressive military response.

By 06:50, the cascading effect that following—turbo-charged by AI-enabled bots, deepfakes, and false-flag operations—far exceeded Beijing's pre-defined threshold, and thus capacity to contain. By 07:15, these information operations coincided with a spike in cyber intrusions targeting US Indo-Pacific Command and Taiwanese military systems and defensive maneuvers of orbital Chinese counter space assets, automated People's Liberation Army (PLA) logistics systems were activating, and there was suspicious movement of the PLA's nuclear road-mobile transporter erector launchers. At 07:20, US SPRS assessed this behavior as an impending major national security threat and recommended an elevated deterrence posture and a powerful demonstration of force. The White House authorized an autonomous strategic bomber flyover in the Taiwan Straits at 07:25.

In response, at 07:35, China's SIAS notified Beijing of an increased communication loading between US Indo-Pacific Command and critical command and communication nodes at the Pentagon. By 07:40, SIAS raised the threat level for a pre-emptive US strike in the Pacific to defend Taiwan, attack Chinese-held territory in the South China Seas, and contain China. At 07:45, SIAS advised Chinese leaders to use conventional counterforce weapons, including cyber, anti-satellite, hypersonic weapons, and other smart precision missile technology (i.e., "fire and forget" munitions) to achieve early escalation dominance through a limited pre-emptive strike against critical US Pacific assets.

At 07:50, Chinese military leaders, fearful of an imminent disarming US strike and increasingly reliant on the assessments of SIAS, authorized the attack—which SIAS had already anticipated and thus planned and prepared for. By 07:55, SPRS alerted Washington of the imminent attack—US C2 were reeling from cyber and anti-satellite weapons attacks, and Chinese hypersonic missiles would likely impact US bases in Guam in sixty seconds—and recommended an immediate limited nuclear strike on China's mainland to compel Beijing to cease its offensive. SPRS judged that US missile defenses in the region could successfully intercept the bulk of the Chinese theater (tactical) nuclear countervail—predicting that China would only authorize a response in kind and avoid a counter-value nuclear strike on the continental United States. SPRS proved correct. After a limited US-China atomic exchange in the Pacific, leaving millions of people dead and tens of millions injured, both sides agreed to cease hostilities.

In the aftermath, both sides attempted to retroactively reconstruct a detailed analysis of decisions made by SPRS and SIAS. However, the designers of the ML algorithms underlying SPRS and SIAS reported that it was not possible to explain the decision rationale and reasoning of the AI behind every subset decision. Besides, because of the various time, encryption, and privacy constraints imposed by the end military and business users, it was impossible to keep retroactive back-testing logs and protocols. Did AI technology cause the 2025 "flash war"?

The emerging AI-nuclear nexus

> "A problem well-stated is a problem half solved."
>
> —Charles Kettering[7]

We are now in an era of rapid disruptive technological change, especially in AI. "AI technology"[8] is already being infused into military machines, and global armed forces are well advanced in their planning, research and development, and, in some cases, deployment of AI-enabled capabilities. Therefore, the embryonic journey to reorient military forces to prepare for the future digitized battlefield is no longer merely purview speculation or science fiction. While much of the recent discussion has focused on specific technical issues and uncertainties involved as militaries develop AI applications at the tactical and operational level of war, strategic issues have only been lightly researched. This book addresses this gap. It examines the intersection between technological change, strategic thinking, and nuclear risk—or the "AI-nuclear dilemma."

The book explores the impact of AI technology on the critical linkages between the gritty reality of war at a tactical level and strategic affairs—especially nuclear doctrine and strategic planning. In his rebuttal of the tendency of strategic theorists to bifurcate technologically advanced weaponry into tactical and strategic conceptual buckets, Colin Gray notes that all weapons are tactical in their employment and strategic in their effects.[9] An overarching question considered in the book is whether the gradual adoption of AI technology—in the context of the broader digital information ecosystem—will increase or decrease that state's (offensively or defensively motivated) resort to nuclear use.

The book's central thesis is two-fold. First, rapid advances in AI—and a broader class of "emerging technology"[10]—are transforming how we should conceptualize and thus manage and control technology and the risk of nuclear use. Second, while remaining cognizant of the potential benefits of innovation, complexity, information distortion, and decision-making compression associated with the burgeoning digital information age, they may also exacerbate existing tensions between adversaries and create new mechanisms and novel threats which increase the risk of nuclear accidents and inadvertent escalation.

To unpack these challenges, the book applies Cold War-era cornerstone nuclear strategic theorizing and concepts (nuclear deterrence, strategic stability, the security dilemma, inadvertent escalation, and accidental "catalytic" nuclear war) to examine the impact of introducing AI technology into the nuclear enterprise. Given the paucity of real-world examples of how AI may affect crises, deterrence, and escalation, bringing conceptual tools from the past to bear is critical. This theoretical framework moves beyond the hidebound precepts and assumptions—associated with bipolarity, symmetrical state-centric world politics, and classical rational-based deterrence—that still dominate the literature. The book argues that this prevailing wisdom is misguided in a world of asymmetric nuclear-armed dyads, nihilistic non-state actors, and non-human actors—or intelligent machines.

Drawing on insights from political psychology, cognitive neuroscience, and strategic studies, the book advances an innovative theoretical framework to consider AI technology and nuclear risk. Connecting the book's central research threads are human cognitive-psychological explanations, dilemmas, and puzzles. The book draws out the prominent political psychological (primarily perceptual and cognitive bias) features of the Cold War-era concepts chosen and offers a novel theoretical explanation for why these matter for AI applications and nuclear strategic thinking. Thus, the book propounds an innovative reconceptualization of Cold War-era nuclear concepts considering the impact of technological change on nuclear risk and an improved understanding of human psychology.

It is challenging to make theoretical predictions *a priori* about the relationship between the development of AI technology and the risk of nuclear war. This is ultimately an empirical question. Because of the lack of empirical datasets in the nuclear domain, it is very difficult to talk with confidence about the levels of risk probably associated with nuclear use (accidental or deliberate). How can we know for sure, for instance, the number of nuclear accidents and near misses, and determine where, when, and the frequency with which these incidents occurred?[11] Because much of the discussion about AI's impact on nuclear deterrence, stability, escalation, crisis scenarios, and so on is necessarily speculative, counterfactual thinking can help policymakers who engage in forward-looking scenario planning but at the same time seek historical explanations.[12]

As scholar Richard Ned Lebow notes, the use of counterfactual thinking (or "what-if's") allows for scholars to account for the critical role of luck, loss of control, accidents, and overconfidence as possible causes of escalation, and the validity of other "plausible worlds" that counterfactuals can reveal.[13] According to Lebow, for instance, "if the Cuban missile crisis had led to war—conventional or nuclear—historians would have constructed a causal chain leading ineluctably to this outcome."[14] In a similar vein, Pentagon insider Richard Danzig opined that while predictions about technology and the future of war are usually wrong, it is better to be circumspect about the nature of future war and be prepared to respond to unpredictable and uncertain conditions when prediction fails.[15] Chapters 3 to 5 use counterfactual scenarios to illuminate some of the possible real-world implications of AI technology for mishaps, misperception, and overconfidence in the ability to control nuclear weapons, as a potential cause of escalation to nuclear use.

There is an opportunity, therefore, to establish a theoretical baseline on the AI-nuclear strategic nexus problem-set. It is not the purpose of this book to make predictions or speculate on the timescale of AI technological progress, but rather, from AI's current state, to explore a range of possible scenarios and their strategic effects. In this way, the book will stimulate thinking about concrete (short-to-medium term) vexing policymaking issues, including: trade-offs in human-machine collaboration and automating escalation (i.e., retaining human control of nuclear weapons while not relinquishing the potential benefits of AI-augmentation); modern deterrence (i.e., regional asymmetric threats short of nuclear war and systemic global stability between great powers); and designing nuclear structures, postures, and philosophies of use in the digital age. How should militaries be using AI and applying it to military systems to ensure mutual deterrence and strategic stability?

The remainder of this chapter has two goals. First, it defines military-use AI (or "military AI").[16] It offers a nuanced overview of the current state of AI technology and the potential impact of dramatic advances in this technology (e.g., ML, computer vision, speech recognition, natural language processing, and autonomous technology) on military systems. This section provides an introductory primer to AI in a military context for non-technical audiences.[17] Because of the rapidly developing nature of this field, this primer can only provide a snapshot in time. However, the underlying AI-related technical concepts and analysis described in this section will likely remain applicable in the near term. How, if at all, does AI differ from other emerging technology? How can we conceptualize AI and technological change in the context of nuclear weapons?

Second, it highlights the developmental trajectory of AI technology and the associated risks of these trends as they relate to the nuclear enterprise. This section describes how AI technology is already being researched, developed, and, in some cases, deployed into the broader nuclear deterrence architecture (e.g., early-warning, ISR, C2, weapon delivery systems, and conventional weapons) to enhance the nuclear deterrence architecture in ways that could impact nuclear risk and strategic planning. In doing so, it demystifies AI in a broader military context, thus debunking several misperceptions and misrepresentations surrounding AI—and the key enabling technologies associated with AI, including, *inter alia*, big-data analytics, autonomy, quantum computing, robotics, miniaturization, additive manufacturing, directed energy, and hypersonic weapons. The main objective of this chapter is to establish a technical baseline that informs the book's theoretical framework for considering AI technology and nuclear risk.

Military-use AI primer: What is AI and what can it do?

AI research began as early as the 1950s as a broad concept concerned with the science and engineering of making intelligent machines.[18] In the decades that followed, AI research went through several development phases—from early exploitations in the 1950s and 1960s, the "AI summer" during the 1970s, through to the early 1980s, and the "AI Winter" from the 1980s. Each of these phases failed to live up to its initial, and often over-hyped, expectations—particularly when *intelligence* has been confused with *utility*.[19] Since the early 2010s, an explosion of interest in the field (or the "AI renaissance") has occurred, due to the convergence of four critical enabling developments:[20] the exponential growth in computing processing power and cloud computing; expanded datasets (especially "big-data" sources);[21] advances in the implementation of ML techniques and algorithms (especially deep "neural networks");[22] and the rapid expansion of commercial interest and investment in AI technology.[23] Notwithstanding the advances in the algorithms used to construct ML systems, AI experts generally agree that the last decade's progress has had far more to do with increased data availability and computing power than improvements in algorithms per se.[24]

AI is concerned with machines that emulate capabilities usually associated with human intelligence, such as language, reasoning, learning, heuristics, and observation. Today, all practical (i.e., technically feasible) AI applications fall into the "narrow" category (or "weak AI"), or, less so, into the category of artificial general intelligence (AGI)—also referred to as "strong AI" or "superintelligence."[25] Narrow AI has been broadly used in a wide range of civilian and military tasks since the 1960s,[26] and involves statistical algorithms (mostly based on ML techniques) that learn procedures by analyzing large training datasets designed to approximate and replicate human cognitive tasks (see Chapter 1).[27] Narrow AI is the category of AI to which this book refers when it assesses the impact of AI technology in a military context.

Most experts agree that the development of AGI is at least several decades away, if feasible at all.[28] While the potential of AGI research is high, the anticipated exponential gains in the ability of AI systems to provide solutions to problems today are limited in scope. Moreover, these narrow-purpose applications do not necessarily translate well to more complex, holistic, and open-ended environments (i.e., modern battlefields), which exist simultaneously in the virtual (cyber/non-kinetic) and physical (or kinetic) plains.[29] However, that is not to say that the conversation on AGI and its potential impact should be entirely eschewed (see Chapters 1 and 2). If AGI does emerge, then ethical, legal, and normative frameworks will need to be devised to anticipate the implications for what would be a potentially pivotal moment in the course of human history.[30] To complicate matters further, the distinction between narrow and general AI might prove less of an absolute (or binary) measure. Thus, research on narrow AI applications, such as game playing, medical diagnosis, and travel logistics often results in incremental progress on general-purpose AI—moving researchers closer to AGI.[31]

AI has generally been viewed as a sub-field of computer science, focused on solving computationally complex problems through search, heuristics, and probability. More broadly, AI also draws heavily from mathematics, human psychology, biology, philosophy, linguistics, psychology, and neuroscience (see Figure 0.1).[32] Because of the divergent risks involved and development

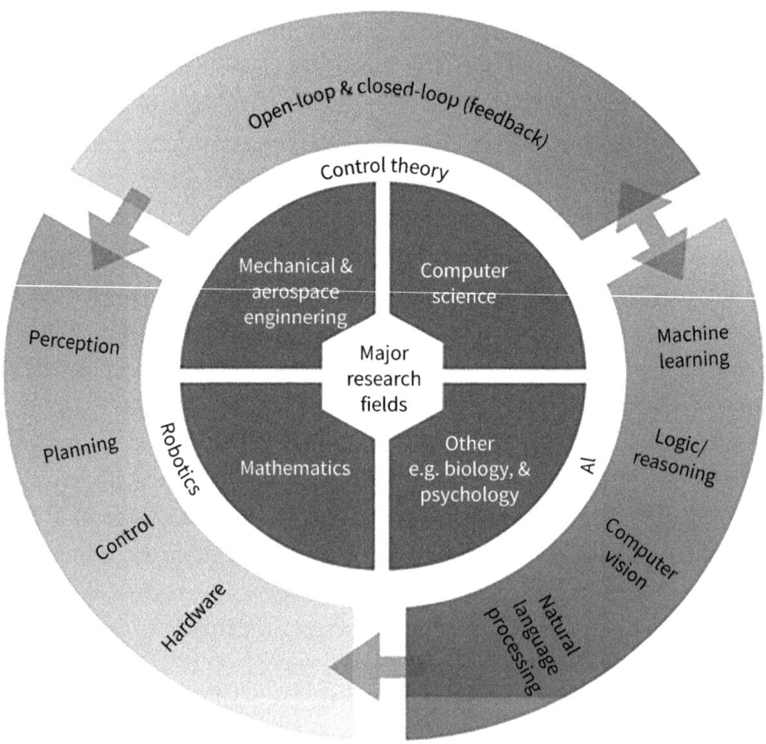

Figure 0.1 Major research fields and disciplines associated with AI
Source: James Johnson, *Artificial Intelligence & the Future of Warfare: USA, China, and Strategic Stability* (Manchester: Manchester University Press, 2021), p. 19.

timeframes in the two distinct types of AI, the discussion in this book is careful not to conflate them.[33] Given the diverse approaches to

research in AI, there is no universally accepted definition of AI,[34] which is confusing when the generic term "artificial intelligence" is used to make grandiose claims about its revolutionary impact on military affairs—or revolution in military affairs (RMA).[35] Moreover, if AI is defined too narrowly or too broadly, we risk understating the potential scope of AI capabilities; or, juxtaposed, fail to specify the unique capacity that AI-powered applications might have. A recent US congressional report defines AI as follows:

> Any artificial system that performs tasks under varying and unpredictable circumstances, *without significant human oversight*, or can learn from their experience and improve their performance may solve tasks requiring human-like perception, cognition, planning, learning, communication, or physical action (emphasis added).[36]

In a similar vein, the US DoD defines AI as:

> The ability of machines to *perform tasks that typically require human intelligence*—for example, recognizing patterns, learning from experience, drawing conclusions, making predictions, or taking action—whether digitally or as the smart software behind autonomous physical systems (emphasis added).[37]

AI can be best understood as a universal term for improving the performance of automated systems to solve a wide variety of complex tasks, including:[38] *perception* (sensors, computer vision, audio, and image processing); *reasoning and decision-making* (problem-solving, searching, planning, and reasoning); *learning and knowledge representation* (ML, deep networks, and modeling);

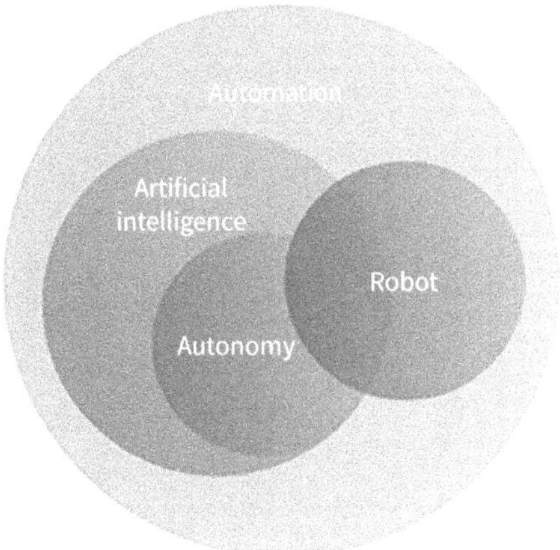

Figure 0.2 The linkages between AI and autonomy

Source: James Johnson, *Artificial Intelligence & the Future of Warfare: USA, China, and Strategic Stability* (Manchester: Manchester University Press, 2021), p. 20.

communication (language processing); *automatic* (or autonomous) systems and robotics (see Figure 0.2); and *human-AI collaboration* (humans define the systems' purpose, goals, and context).[39] As a potential enabler and force multiplier of a portfolio of capabilities, therefore, military AI is more akin to electricity, radios, radar, and ISR support systems than a "weapon" per se.[40]

Machine learning is not alchemy or magic

ML is an approach to software engineering developed during the 1980s and 1990s, based on computational systems that can "learn" and "teach"[41] themselves through a variety of techniques, such as neural networks, memory-based learning, case-based reasoning, decision trees, supervised learning, reinforcement learning, unsupervised learning, and, more recently, generative adversarial networks—which examine how systems that leverage ML algorithms might be tricked or defeated. [42] Consequently, the need for cumbersome human hand-coded programming has been dramatically reduced.[43] From the fringes of AI until the 1990s, advances in ML algorithms with more sophisticated connections (i.e., statistics and control engineering) emerged as one of the most prominent AI methods (see Figure 0.3). In recent years, a subset of ML, deep learning (DL), has become the avant-garde AI software engineering approach, transforming raw data into abstract representations for a range of complex tasks, such as image recognition, sensor data, and simulated interactions (e.g., game playing).[44] The strength of DL is its ability to build complex concepts from simpler representations.[45]

Alongside the development of AI and ML, a new ecosystem of AI subfields and enablers have evolved, including: image recognition, machine

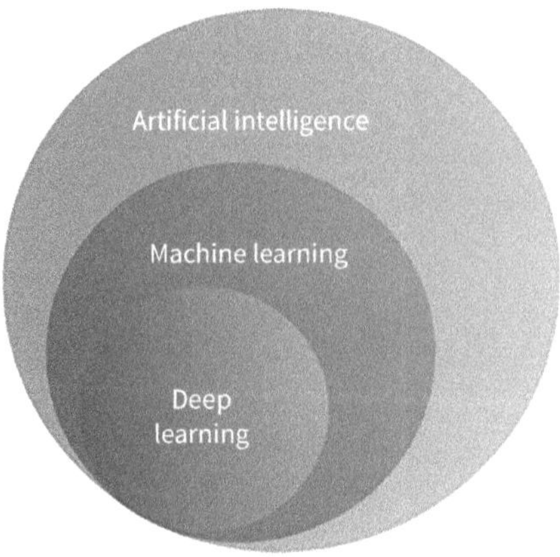

Figure 0.3 Hierarchical relationship: ML is a subset of AI, and DL is a subset of machine learning

Source: James Johnson, *Artificial Intelligence & the Future of Warfare: USA, China, and Strategic Stability* (Manchester: Manchester University Press, 2021), p. 22.

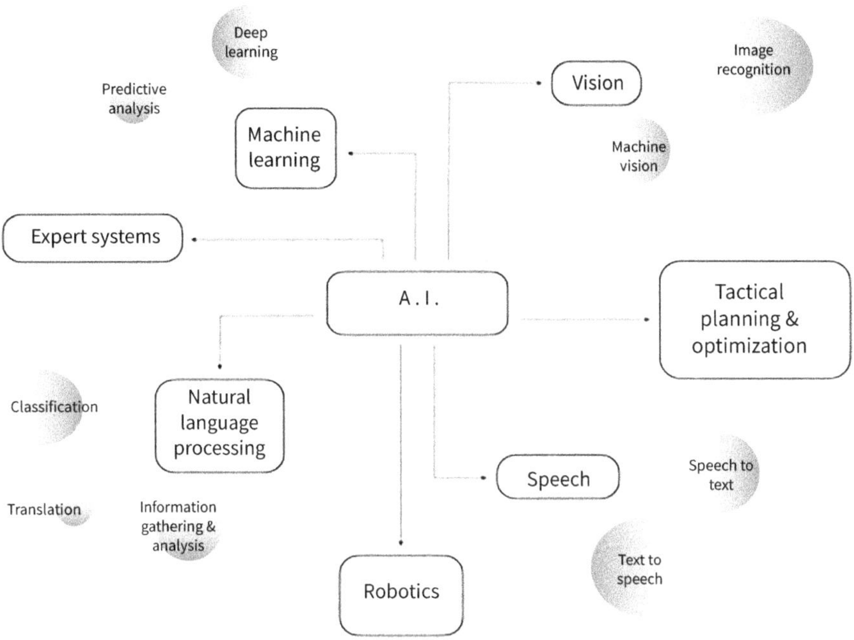

Figure 0.4 The emerging "AI ecosystem"
Source: James Johnson, *Artificial Intelligence & the Future of Warfare: USA, China, and Strategic Stability* (Manchester: Manchester University Press, 2021), p. 23.

vision, predictive analysis and planning, reasoning and representation, natural language representation and processing, robotics, and data classification (see Figure 0.4).[46] In combination, these techniques have the potential to enable a broad spectrum of increasingly autonomous applications, *inter alia*: big-data mining and analytics; AI voice assistants; language and voice recognition aids; structured query language data-basing; autonomous weapons and autonomous vehicles; and information gathering and analysis, to name a few. One of the critical advantages of ML is that human engineers no longer need to explicitly define the problem to be resolved in a particular operating environment.[47] For example, ML image recognition systems can be used to express mathematically the differences between images, which human hard-coders struggle to do.

ML's recent success can be mainly attributed to the rapid increase in computing power and the availability of vast datasets to train ML algorithms. Today, AI-ML techniques are routinely used in many everyday applications, including: empowering navigation maps for ridesharing software; by banks to detect fraudulent and suspicious transactions; making recommendations to customers on shopping and entertainment websites; to support virtual personal assistants that use voice recognition software to offer their users content; and to enable improvements in medical diagnosis and scans.[48]

While advances in ML-enabled AI applications for IRS performance—used in conjunction with human intelligence analysis—could improve the ability of militaries to locate, track, and target an adversary's nuclear assets,[49] four major technical bottlenecks remain unresolved: brittleness; a dearth of "quality" data; automated detection weaknesses; and the so-called "black box" (or "explainability") problem-set.[50]

Brittleness problem

Today, AI suffers from several technical shortcomings that should prompt prudence and restraint in the early implementation of AI in a military context—and other safety-critical settings such as transportations and medicine. AI systems are brittle in situations when an algorithm is unable to adapt to or generalize conditions beyond a narrow set of assumptions. Consequently, an AI's computer vision algorithms are unable to recognize an object due to minor perturbations or changes to an environment.[51] Because deep reinforcement learning computer vision technology is still relatively nascent, problems have been uncovered which demonstrate the vulnerability of these systems to uncertainty and manipulation.[52] In the case of autonomous cars, for example, AI-ML computer vision cannot easily cope with volatile weather conditions. For instance, road lane markings partially covered with snow, or a tree branch partially obscuring a traffic sign—which would be self-evident to a human—cannot be fathomed by a computer vision algorithm because the edges no longer match the system's internal model.[53] Figure 0.5 illustrates how an image can be deconstructed into its edges (i.e., through mathematical computations to identify transitions between dark and light colors); while humans see a tiger, an AI algorithm "sees" sets of lines in various clusters.[54]

ML algorithms are unable to effectively leverage "top-down" reasoning (or "System 2" thinking) to make inferences from experience and abstract reasoning when perception is driven by cognition expectations, which is an essential element in safety-critical systems where confusion, uncertainty, and imperfect information require adaptions to novel situations (see Chapter 5).[55] This reasoning underpins people's common sense and allows us to learn complex tasks, such as driving, with very little practice or foundational learning. By contrast, "bottom-up" reasoning (or "System 1" thinking) occurs when information is taken in at the perceptual sensor level (eyes, ears, sense of touch) to construct a model of the world to perform narrow tasks

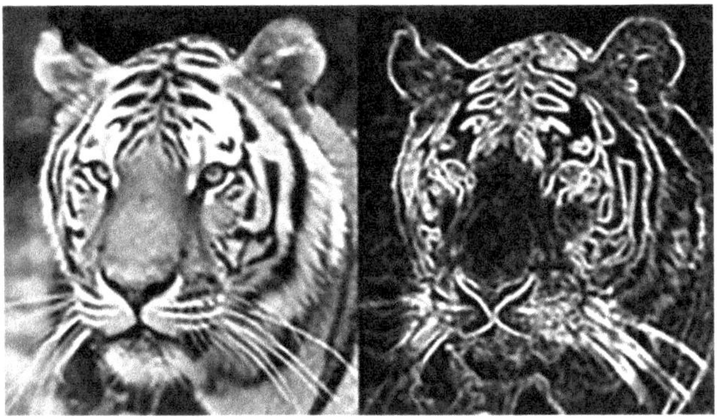

Figure 0.5 Edge decomposition illustration
Source: https://commons.wikimedia.org/wiki/File:Find_edges.jpg, courtesy of Wessam Bahnassi [Public domain].

such as object and speech recognition and natural language processing, which AIs perform well.[56]

People's ability to use judgment and intuition to infer relationships from partial information (i.e., "top-down" reasoning) is illustrated by the Kanizsa triangle visual illusion (see Figure 0.6). AI-ML algorithms lack the requisite sense of causality that is critical for understanding what to do in novel *a priori* situations and have thus been unable to either recognize or replicate visual illusions.[57] While advancements have recently been made in computer vision—and in the DL algorithms that power these systems—the perceptual models of the real-world constructed by these systems are inherently brittle. The real world characterized by uncertainty (i.e., partial observability), data scarcity, ambiguous and nuanced goals, and disparate timescales of decision-making and so on, creates technological gaps, which has important implications for future collaboration with humans (see Chapter 3).

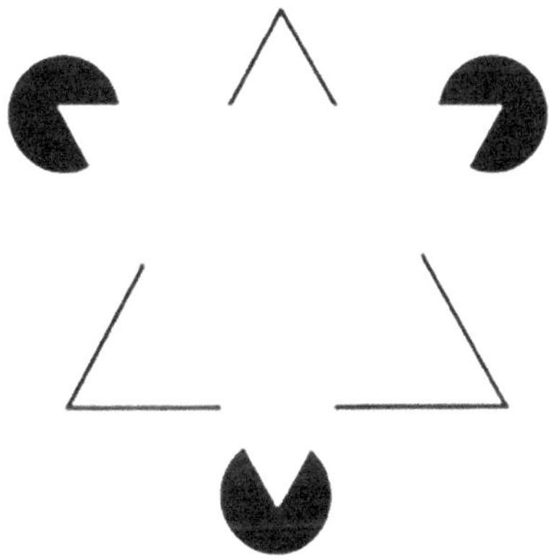

Figure 0.6 The Kanizsa triangle visual illusion
Source: https://commons.wikimedia.org/wiki/File:Kanizsa_triangle.svg., courtesy of Fibonacci [Public domain].

Consequently, AI cannot effectively and reliably diagnose errors (e.g., sampling errors or intentional manipulation) from complex datasets and the esoteric mathematics underlying AI algorithms.[58] Moreover, AI systems cannot handle novel situations reliably; AI relies on *a posteriori* knowledge (i.e., "skills-based" knowledge) to make inferences and inform decision-making. Failure to execute a particular task, especially if biased results are generated, would likely diminish the level of trust placed in these applications.[59] In a recent study by AI researchers at MIT Lincoln Laboratory, researchers used the card game Hanabi—a game of full cooperation and limited information—to explore whether a state-of-the-art deep-learning reinforcement learning program that surpassed human abilities could become a reliable and trusted coworker. The study found that high performance by the AI did not translate into a good collaboration experience. Instead, players complained about the program's poor understanding of human subjective preferences and a resultant lack of trust.[60]

A key future challenge for the AI-ML community is how to design algorithmic architectures and training frameworks that incorporate human inductive biases and *a priori* qualities such as reasoning, intuition, shared knowledge, and understanding social cues.[61] With this goal in mind, researchers from IBM, MIT, and Harvard University recently collaborated on a Defense Advanced Research Projects Agency-sponsored "Common Sense AI" dataset for benchmarking AI intuition.[62] Inspired by development studies of infants, this was the first study of its kind to use human psychology to accelerate the development of AI systems that exhibit common sense—to learn core intuitive human psychology while maintaining their autonomy. The researchers constructed two ML approaches to test real-world scenarios, thus establishing a valuable baseline for these ML models to learn from humans while maintaining their autonomy.

Dearth of quality of data—"garbage in garbage out"

ML systems depend on vast amounts of high-quality pre-labeled datasets (with both positive and negative examples) and trails to learn from, which humans are able to generalize with far less experience.[63] ML algorithms operate based on correlation, not causation—algorithms' use incoming data inputs and well defined outputs to identify patterns of correlation.[64] AI-ML is, therefore, only as good as the *quantity and quality* of the data it is trained on (i.e., input data linked to associated outcomes) before, and supplied with during, operations—or "garbage in garbage out."[65] The performance of ML systems can be improved by scaling them up with higher volumes of data and increased computation power. For example, OpenAI's language processing model GPT-3, with 175 billion parameters—which is still much less than the number of synapses in the human brain—generates significantly more accurate text than its processor, GPT-2, with only 1.5 billion parameters.[66] Despite these significant milestones, there remain fundamental deficiencies of current ML approaches that cannot be resolved by scaling or computational power alone.[67]

No matter how sophisticated the datasets are, however, they are unable to replicate real-world situations perfectly. Each situation includes some irreducible error (or an error due to variance) because of incomplete and imprecise measurements and estimates. According to the former US DoD's Joint Artificial Intelligence Center director Lt. General Jack Shanahan, "if you train [ML systems] against a very clean, gold-standard dataset, it will *not work in real-world conditions*" (emphasis added).[68] Further, relatively easy, and cheap, efforts (or "adversarial AI") to fool these systems would likely render even the most sophisticated systems redundant (see Chapter 5).[69]

As a corollary, an ML system that memorizes data within a particular training set may fail when exposed to new and unfamiliar data. Because there is an infinite number of engineering options, it is impossible to tailor an algorithm for all eventualities. Thus, as new data emerges, a near-perfect algorithm will likely quickly become redundant. There is minimal room for error in a military context in balancing the pros and cons of deciding how much data to supply ML systems to learn from—or "bias variance trade-off."[70] Even if a method was developed that performs flawlessly on the data fed to it, there is no guarantee that the system will perform in the same way on images it receives subsequently. As a result, less sophisticated AI systems will exhibit more significant levels of bias but with lower accuracy.

In other words, there is no way to know for sure how ML-infused autonomous weapon systems, such as AI-enhanced conventional counterforce capabilities, would function in the field (see Chapter 4). AI-enhanced capabilities will be prone to errors and accidents because of the critical feedback from testing, validation, prototyping, and live testing (associated with the development of kinetic weapon systems).[71] Because there is a naturally recursive relationship in how AI systems interact with humans and information, the likelihood of errors may be both amplified and reduced (see Chapter 2).

Three technical limitations contribute to this data shortage problem.[72] First, in the military sphere, AI systems have relatively few images to train on (e.g., mobile road and rail-based missile launchers). This data imbalance will cause an AI system to maximize its accuracy by substituting images with a greater abundance of data, resulting in false positives.[73] In other words, to maximize its accuracy in classifying images, ML algorithms are incentivized to produce false negatives, such as misclassifying a regular truck as a mobile missile launcher. Thus, without distinguishing between true and false, *all moving targets* could be considered viable targets by AI systems.

Second, AI is at a very early stage of "concept learning."[74] Images generally depict reality only poorly. Whereas humans can deduce—using common sense—the function of an object from its external characteristics, AI struggles with this seemingly simple task. In situations where an object's form explicitly tells us about its function (i.e., language processing, speech, and handwriting recognition), this is less of an issue, and narrow AI generally performs well. However, in situations where an object's appearance does not offer this kind of information, AI's ability to induce or infer is limited. Thus, the ability of ML algorithms to reason is far inferior to human conjecture and criticism.[75] For example, in a military context, AI would struggle to differentiate the military function of a vehicle or platform. This problem is compounded by the shortage of quality datasets and the likelihood of AI adversarial efforts poised to exploit this shortcoming.

Third, ML becomes exponentially more difficult as the number of features, pixels, and dimensionality increases.[76] Greater levels of resolution and dimensional complexity—requiring more memory and time for ML algorithms to learn—could mean that images become increasingly difficult for AI to differentiate. For example, a rail-based mobile missile launcher might appear to AI as a cargo train car, or a military aircraft as a commercial airliner. In short, similar objects will become increasingly dissimilar to the AI, while images of different and unrelated objects will become increasingly indistinguishable.[77]

Automated image-detection limitations

ML's ability to autonomously detect and cue precision-guided munitions is limited, particularly in cluttered and complex operational environments. This automated image recognition and detection weakness are mainly caused by AI's inability to effectively mimic human vision and cognition, which is notoriously opaque and irrational.[78] This limitation could, for example, reduce a commander's confidence in the effectiveness of AI-augmented counterforce capabilities, with potentially significant implications for nuclear deterrence and escalation (see Chapters 2 and 3). Besides this, strategic competitors will continue to pursue countermeasures (e.g., camouflage, deception, decoys, and concealment) to protect their strategic forces against these advances.

AI's "black box" problem-set

"What I cannot make, I do not understand."

—Richard Feynman[79]

ML algorithms today are inherently opaque—especially those built on neural networks—creating "black box" computational mechanisms that human engineers frequently struggle to fathom.[80] This "black box" problem-set may cause unpredictability, for instance algorithms reacting in unexpected ways to datasets used during the training phase, which could have severe consequences at a strategic level—not least complicating the problem of attributing responsibility and accountability in the event of an accident or error in the safety-critical nuclear domain (see Chapter 5). For instance, in a military context it may not be possible to determine with any precision the weighting given by an algorithm to troop movements, weapons stockpiling, human intelligence, social media, and geospatial opensource information, or other historical data that an AI uses to reach a particular outcome or recommendation.[81]

While it is conceptually possible for AIs to reveal the information on which their prediction or recommendation is derived, because of the vast quantity of potential data an AI can feed from, the complexity of the weighting, the relationships between data points, and the continuous adaptions made by algorithms in dynamic environments, building explainable and transparent algorithms—that is, algorithms that can list the factors considered and explicate the weighting given to each factor—that humans can comprehend will be extremely challenging.[82] In sum, algorithmic brittleness, a dearth of quality datasets for ML algorithms to learn from, automated detection technical limitations, the likelihood of adversarial countermeasures and exploitation, and the opacity of ML algorithms will significantly reduce the *a priori* knowledge AI systems can obtain from a situation—particularly in complex adversarial environments.

To mitigate the potentially destabilizing effects of deploying either poorly conceptualized, immature, or accidental-prone AIs into the safety-critical nuclear domain, therefore, decision-makers need to better understand what AI is (and what it is not), its limitations, and how best to implement AI in a military context.[83] Furthermore, while AI augmentation of humans in safety-critical domains is already within reach, this technical landmark

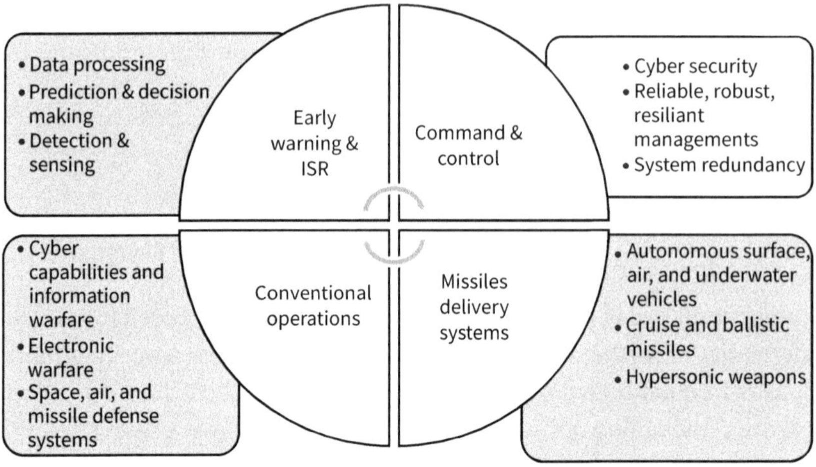

Figure 0.7 AI applications and the nuclear deterrence architecture

should not be mistaken for the ability of AI to replace humans and go under-supervised in the operation of these systems.[84] In short, AI-enabled systems should not be operating in safety-critical systems without significant human oversight.

AI nuclear deterrence architecture

This section assesses how advances in AI technology are being researched and developed, and, in some cases, are deployed and operational in the context of the broader nuclear deterrence architecture (see Figure 0.7): early-warning and ISR; C2; nuclear weapon delivery systems; and non-nuclear operations.[85] It also considers how and to what degree AI-augmentation marks a departure from automation in the nuclear enterprise, which goes back several decades.[86] How transformative are these developments?

Intelligence, surveillance, and reconnaissance (ISR)

Early-warning and ISR

AI-ML might, in three ways, quantitatively enhance existing early-warning and ISR systems.[87] First, ML, in conjunction with cloud computing, unmanned aerial vehicles (UAVs),[88] and big-data analytics, could be used to enable mobile ISR platforms to be deployed in geographically long ranges, and in complex, dangerous environments (e.g., contested anti-access/area-denial zones,[89] urban counterinsurgency, or deep-sea) to process real-time data (i.e., signals and objects), and alert commanders to potentially suspicious or threatening situations (e.g., military drills and suspicious troop or mobile missile launcher movements).

Second, ML algorithms could be used to gather, mine, and analyze large volumes of intelligence (open-source and classified) sources to detect correlations in heterogeneous—and possibly contradictory, compromised, or otherwise manipulated—datasets. Third, and related, ML algorithmic processed intelligence may be used to support commanders to anticipate—and thus more rapidly pre-empt—an adversary's preparations for a nuclear strike (e.g., the manufacturer, deployment, or shift in alert status of nuclear weapons).[90] In short, AI technology could offer human commanders operating in complex and dynamic environments vastly improved situational awareness and decision-making tools, allowing for more time to make informed decisions with potentially stabilizing effects.

Nuclear command and control (C2)

In contrast to ISR and early-warning, the impact of AI technology is unlikely to have a significant qualitative impact on nuclear C2—which for several decades has synthesized automation but not autonomy (see Chapter 5). As we have seen, the ML algorithms that underlie complex autonomous systems today are too unpredictable, vulnerable (i.e., to adversarial cyber-attacks), unexplainable (the "black-box" problem), and brittle to be used unsupervised in safety-critical domains.[91] For now, there is a broad consensus amongst nuclear experts and nuclear-armed states that, even if the technology permitted,[92] AI decision-making which directly impacts nuclear C2 functions (i.e., missile launch decisions), should not be pre-delegated to AIs.[93] Whether this fragile consensus can withstand mounting first-mover advantage temptations in a multipolar nuclear order (see Chapter 4), or human commanders—predisposed to anthropomorphize subjects, cognitive offloading, and automation bias—viewing AIs as a panacea for the cognitive fallibilities of human analysis and decision-making is an open question (see Chapters 1 and 2).[94] The question, therefore, is perhaps less whether nuclear-armed states will adopt AI technology into the nuclear enterprise, but rather by whom, when, and to what degree it will be adopted. The book considers these questions and the implications for nuclear strategy and risk.

Irrespective of whether AIs are directly fused with Nuclear Command, Control, and Communications (NC3) systems to fulfill strategic decision-making functions, however, the effects of AI technology may yet have significant (and less understood) quantitative implications for the NC3 ecosystem—and thus nuclear deterrence. For example, AI-enhanced cybersecurity applications may enhance the robustness of NC3 cyber defenses against adversarial attacks; long-endurance AI-augmented UAVs might replace the role of signal rockets to support airborne communications in situations where satellites are comprised or destroyed; and more efficient and resilient AI tools may improve the stockpiling and logistical management of personnel and nuclear arsenals, thus potentially reducing the risk of accidents involving nuclear weapons (see Chapter 5).

Nuclear and non-nuclear missile delivery systems

AI technology will likely affect the nuclear weapon delivery systems in several ways. First, ML algorithms may be used to improve the accuracy, navigation (i.e., pre-programed guidance parameters), autonomy (i.e., "fire-and-forget" functionality) of missiles, and precision—mainly used in conjunction with hypersonic glide vehicles. For example, China's DF-ZF maneuverable hypersonic glide vehicle is a dual-capable (nuclear and conventionally armed) prototype with autonomous functionality.[95] Second, it could improve the resilience and survivability of nuclear launch platforms against adversary countermeasures such as electronic warfare jamming or cyber-attacks—that is, autonomous AI-enhancements would remove the existing vulnerabilities of communications and data links between launch vehicles and operators. Third, the extended endurance of AI-augmented unmanned platforms (i.e., unmanned underwater vehicles (UUVs) and unmanned combat aerial vehicles)[96] used in extended ISR missions—that cannot be operated remotely—can potentially increase their ability to survive countermeasures and reduce states' fear of a nuclear decapitation, especially in asymmetric dyads.[97] AI and autonomy might strengthen states' second-strike capability—and thus deterrence—and even support escalation management during a crisis or conflict (see Chapter 4).[98]

Conventional counterforce operations

AI technology could be used to enhance a range of conventional capabilities, with potentially significant strategic implications—especially strategic non-nuclear weapons used in conventional counterforce operations. First, ML could increase the onboard intelligence of manned and unmanned fighter aircraft, thus increasing their capacity to penetrate enemy defenses using conventional high-precision munitions. Moreover, increased levels of AI-enabled autonomy might allow unmanned drones—possibly in swarms—to operate in environments hitherto considered inaccessible or too dangerous for manned systems (e.g., anti-access and area denial (A2/AD) zones, or deep-water and outer space environments).[99] The 2021 Azerbaijani-Armenian war demonstrated, for instance, how smaller states can integrate new weapon systems to amplify their battlefield effectiveness and lethality.[100]

Second, and related, AI-ML could substantially enhance missile, air, and space defense systems' ability to detect, track, target, and intercept. While AI technology has been integrated with automatic target recognition (ATR) to support defense systems since the 1970s, the speed of defense systems' target-identification—because of the limited database of target signatures that an ATR system uses to recognize its target—has progressed slowly.[101] Advances in AI-ML DL and generative adversarial networks (GANs) could alleviate this technical bottleneck.[102] Specifically, deep-learning techniques could enhance the ability of ATR systems to learn independently (or "unsupervised") to detect the differences between types of targets (i.e., military vs. civilian objects).[103] To support the increased fidelity of ATR systems, GANs technology could generate realistic synthetic data to train and test ATR systems. In addition, advances in AI technology could enable multi-vehicle coordinated autonomous drone swarms to conduct A2/AD operations—increasing the risk of offensive strikes, thus potentially bolstering deterrence against nuclear and conventional strikes. Besides, autonomous drone swarms might also be used defensively (e.g., decoys or flying mines) to buttress traditional air defenses.[104]

Third, recent advances in AI-ML have already begun to transform how existing automation and autonomy in the cyber domain operate. AI technology is also changing how (both offensive and defensive) cyber capabilities are designed and operated.[105] On the one hand, AI might reduce a military's vulnerability to cyber-attacks and electronic warfare operations. AI cyber-defensive tools (e.g., GANs technology) and anti-jamming capabilities—designed, for example, to recognize changes to patterns of behavior and anomalies in a network and automatically identify malware or software code vulnerabilities—could protect NC3 systems against cyber intrusions or jamming operations.[106]

On the other hand, advances in AI-ML (notably an increase in the speed, stealth, and anonymity of cyber warfare) might enable identifying an adversary's "zero-day vulnerabilities"—that is, undetected or unaddressed software vulnerabilities.[107] While there is no definitive evidence in publicly available sources that North Korea has applied ML and autonomy in the nuclear domain, reports indicate that AI technology is already being used to facilitate identification by North Korea of zero-day vulnerabilities in South Korean and US computer systems.[108] An adversary might also use malware to take control, manipulate, or fool the behavior and pattern recognition systems of autonomous systems such as the DoD's Project Maven—for example, using GANs to generate synthetic and realistic-looking data poses a threat to both ML and rules-based forms of attack detection.[109] In short, AI technology in the nuclear domain will likely be a double-edged sword: strengthening the NC3 systems while expanding the pathways and tools available to adversaries to conduct cyber-attacks and electronic warfare operations against these systems (e.g., "left of launch" or "false-flag" operations enhanced with ML-enhanced cyber and anti-jamming tools, see Chapter 5).[110]

In addition, AI-ML applications could also be used to (directly or indirectly) manipulate the information ecosystem in which strategic decisions involving nuclear weapons take place. Cheaper and more sophisticated AI-ML tools to conduct information mis/disinformation (e.g., fake news, rumors, and propaganda) or other forms of manipulation will make it inexorably easier for actors (especially third-party and non-state) to achieve their nefarious goals asymmetrically (see Chapters 4 and 5).[111] For example, GANs might be used to create fake (video, audio, or text) orders designed to trick or spoof nuclear weapon operators—monitoring AI-infused ISR early-warning systems—into launching a nuclear weapon (a false positive) or not responding to an attack (a false negative).[112] To be sure, these risks would likely be amplified with the increasing automation of NC3 systems by nuclear-armed states to gather and analyze intelligence—that is, creating new potential mechanisms (e.g., data poisoning or spoofing) to undermine or manipulate the information fed into ISR and early-warning systems. The potential implications of these developments for deterrence, inadvertent escalation, and accidental nuclear war are explored in the remainder of the book.

Finally, advances in AI technology could contribute to the physical security of nuclear weapons, particularly against threats posed by third-party and non-state actors (see Chapter 5). Autonomous vehicles (e.g., "anti-saboteur robots") could be used, for example, to protect states' nuclear forces, patrol the parameter of sensitive facilities, or form armed automated surveillance systems (e.g., South Korea's Super Aegis II robotic sentry weapon that includes a fully autonomous mode),[113] along vulnerable borders.[114] In addition, AI technology—coupled with other emerging technologies—could be harnessed to provide novel solutions to support nuclear risk and non-proliferation efforts; for example, removing the need for "boots on the ground" inspectors in sensitive facilities to support non-interference mechanisms for arms control verification agreements (see Conclusion below).

Book plan

The remainder of the book is organized into five chapters. Chapter 1 examines the potential stabilizing and, especially, destabilizing effects of employing AI-enabled—both nuclear and non-nuclear—capabilities against nuclear-armed powers. It highlights the multifaceted possible intersections of AI technology with nuclear stability. How might AI affect the risk of nuclear war? And how could intelligent machines escalate a situation, and how easily might they be manipulated? The gap that currently exists between the expectations and fears of policymakers and the public surrounding the intersection of AI and other emerging technologies with nuclear weapons—caused in part by the hyperbolic depictions of AI in popular culture and science fiction—is diminishing predictability, shared understanding, and trust, and obscuring sober debate on these critical topics. Specifically, balancing the delicate trade-off between harnessing the axiomatic operational and qualitative tactical benefits of leveraging AI while managing the risks posed to strategic stability (first-strike stability, crisis stability, and arms-race stability) and nuclear security—or the "AI-nuclear dilemma." This chapter addresses this gap.

The chapter argues that AI's impact on states' nuclear strategy will depend as much (or perhaps more so) on adversaries' perceptions of its capabilities as on what a particular AI-powered application can do. Much like previous transformative technologies, uncertainty and misperception about the perceived utility—pace, proliferation, diffusion, and potential impact of deterrence and decision-making—of AI applications will likely have a significantly outsized impact on strategic stability. AI's impact on strategic stability and, in turn, the risk of nuclear war, will hinge on the fear perceived by others and the manipulation of this fear in competitive dyads. The chapter considers the ways military AI might exacerbate tension between nuclear-armed powers (especially the US and China), thus complicating crisis stability. A core finding is that the extent to which military AI poses risks to strategic stability will largely depend on this technology's pace and scope to facilitate new ways to improve the delivery of and defense against nuclear weapons and strategic non-nuclear weapons.

Chapter 2 builds on the rich canon of work on nuclear deterrence theory and practice, and highlights some of the variegated and contradictory—especially human cognitive psychological—effects of AI technology for nuclear deterrence. It considers the potential effects of introducing non-human agents into a crisis between nuclear powers' effect on deterrence, escalation, and crisis stability. How might nuclear deterrence be affected by the proliferation of AI and autonomous systems? And how will the introduction of intelligent machines affect human-to-human (and human-to-machine) deterrence? Is the answer for AI and humans to combine their respective talents as a synergetic unit? A growing number of great military powers invest political capital and financial resources in developing AI technology and AI-enhanced autonomous weapon systems in deriving the maximum potential military benefits these systems offer.

The chapter's main objective is to examine the impact of the adoption of these technologies on nuclear deterrence. The ubiquity of these new classes of advanced capabilities—and the incentives for militaries to adopt them—on the future battlefield is fast becoming a foregone certainty. The chapter considers how these capabilities—and various configurations—may weaken or enhance deterrence and how states with different political systems, philosophies of use, and deterrence policies will likely respond to these dynamics. It argues that AI technology could raise the risks of misperception, arms races, and potential escalation to nuclear use through miscalculation, thus undermining deterrence. In short, conventional theories and assumptions associated with classical deterrence have questionable applicability in the digital age.

Chapter 3 revisits Cold War-era thinking about inadvertent escalation—notably, Barry Posen's *Inadvertent Escalation*[115]—to consider how AI technological augmentation of specific advanced conventional weapons, against adversaries, and through different mechanisms—the security dilemma, cognitive bias and heuristics, military doctrine, strategic culture, and the Clausewitzian "fog of war"—could affect inadvertent escalation risk. The chapter argues that existing notions of inadvertent nuclear escalation, and escalation dominance more broadly, need to be reconsidered in the digital age. How might AI be incorporated into nuclear and conventional operations in ways that affect escalation risks? How might specific applications and configurations of AI amplify existing risks of inadvertent escalation?

The chapter argues that co-mingling (or entangling) nuclear and non-nuclear weapons will likely exacerbate the escalation risks of a new generation of AI-enhanced conventional capabilities. Specifically, the multifaceted possible intersections of AI technology with a range of advanced conventional weapons (primarily conventional counterforce) can compromise nuclear weapons, thus amplifying the destabilizing effects of these weapons at a strategic level. The chapter describes how AI technology might be incorporated into nuclear and conventional operations in ways that affect inadvertent escalation risks in strategically competitive dyads—especially China and the US. It also assesses the divergences emerging in how different political systems conceptualize the use of AI to enhance their military power. The chapter uses new and lightly researched empirical evidence, which indicates that both China and the US are pursuing these capabilities in ways that support the chapter's theoretical premises.

How might dual-use disruptive AI technology influence the security dilemma? Chapter 4 revisits the classical "security dilemma theory" and the broader spiral models found in classical Cold-War era security international relations scholarship, considering the implications of advances in dual-use AI technology—and the role of emerging security technologies generally—for strategic competitive dyads. The chapter examines the intersection between the security dilemma, emerging technology, human psychology, and great power—especially US-China—strategic rivalry. Addressing a gap in the literature, the chapter not only establishes a theoretical framework for exploring dual-use AI technology and the security dilemma (or the "AI-security dilemma") but also illuminates the potential dynamics and dangers of this rapidly evolving international security phenomena—that is, self-reinforcing spirals of distrust, arms racing, and inadvertent and unnecessary war.

The chapter considers the AI offense-defense balance and how psychological factors might influence the distinguishability of this balance. Drawing on multidisciplinary literature, the chapter demonstrates the importance of human cognitive-psychology and security dilemma theorizing to understand better the nature and potential dangers posed by the "AI security dilemma" within competitive dyads—that is, new pathways arms races and potential inadvertent escalation through misperception. The chapter's findings have theoretical and policy implications, especially apposite given the elevation and prominence of dual-use AI technology into strategically competitive dyads.

Chapter 5 revisits the Cold War-era concept of "catalytic nuclear war"—in which the collateral effects of third-party or non-state actor actions provoke a nuclear war between the two nuclear-armed powers. It argues that the catalyzing chain of reaction and counter-retaliation dynamics set in motion by a non-state or third-party actor's deliberate action is fast becoming a more plausible scenario in the digital era. The chapter explains why the notion of a "catalytic nuclear war" should be revisited considering recent technological change and an improved understanding of human psychology. Specifically, the human propensity to make fast, intuitive, reflexive, and heuristic judgments (known as "System 1" thinking) is exacerbated when information overload and unfamiliar technologies are a more prominent feature of decision-making.

How can militaries maintain effective command and control of nuclear forces in a rapidly evolving, uncertain, and complex conflict environment? The chapter is not a treatise on the motives or goals of non-state actors for committing such acts. Instead, it examines the evolving means at the disposal of actors to actualize their nefarious goals in the digital age. The chapter considers the various factors that may aggravate escalation pathways during a crisis. It concludes that AI technology is creating new—and exacerbating old—escalation pathways that risk catalyzing accidental nuclear confrontation between nuclear-armed powers, particularly under irrational (or sub-rational) conditions.

The concluding chapter brings together the key arguments and findings from the book and considers possible nuclear risk reduction measures and policy recommendations for managing the intersection of AI technology with nuclear weapons. Going beyond these admonitions will not be easy. This section elucidates the essential dynamics (e.g., normative, behavioral, and broader stakeholder cooperation) and risk reduction mechanisms (e.g., arms control, verification, confidence building, and strategic dialogue) that decision-makers need to consider if we are to mitigate or prevent the worst-case scenario effects, dangers, and risks of our AI-nuclear future. That is, before the pace of technological change outpaces (or surpasses) strategic affairs.

Notes

1. The UN Institute for Disarmament Research coined the term "flash war" to describe compressed decision-making cycles—incentivized by increased levels of autonomy—between decision and action. UNIDIR Resources, "No. 6 The Weaponization of Increasingly Autonomous Technologies: Concerns, Characteristics, and Definitional Approaches—a Primer," *United Nations Institute for Disarmament Research* (2017), http://www.unidir.org/files/publications/pdfs/the-weaponization-of-increasingly-autonomoustechnologies-concerns-characteristics-and-definitionalapproaches-en-689.pdf.
2. Under the applicable international laws, any use of a nuclear weapon would be judged *jus ad bellum* (international law governing the interstate use of force) and *jus in bello* (international law applicable in armed conflict). See Gro Nystuen, Annie Golden Bersage, and Stuart Casey-Maslen, eds., *Nuclear Weapons under International Law* (Cambridge: Cambridge University Press, 2014), p. 5.
3. Ibid. p. 8.

4. Alyssa Schroer, "34 Artificial Intelligence Companies Building a Smarter Tomorrow," *BuiltIn*, May 1, 2019, https://builtin.com/artificial-intelligence/ai-companies-roundup.
5. A 2021 UN report on Libya stated that a Turkish Kargu-2 drone—which has both autonomous and manual functionality—was used for the first time to attack humans autonomously using the drone's AI capabilities. In the same year, in what is thought to be the world's first of drone swarming in combat, the Israeli Defense Force used "dozens" of drones in coordination with mortars and ground-based missiles to strike at Hamas combatants in the Gaza region. David Hambling, "Israel Used World's First AI-Guided Combat Drone Swarm in Gaza Attacks," *New Scientist*, June 30, 2021, https://www.newscientist.com/article/2282656israel-used-worlds-first-ai-guided-combat-drone-swarm-in-gaza-attacks; and Zachary Kallenborn, "If a Killer Robot Were Used, Would We Know?" *Bulletin of the Atomic Scientists*, June 4, 2021, https://thebulletin.org/2021/06/if-a-killer-robot-were-used-would-we-know/.
6. Demis Hassabis, "The Power of Self-Learning Systems," *Video. United States: Center for Brains, Minds & Machines*, March 2019, https://cbmm.mit.edu/video/power-self-learning-systems.
7. Ray Poynter, "A Problem Well Stated is Half Solved," *Potentiate*, January 10, 2022, https://www.potentiate.com/blog/a-problem-well-stated-is-half-solved.
8. The term "AI technology" used in the book refers to a broad category of technologies that are either enablers of AI capabilities and applications (e.g., ML and deep-learning neural networks, 5G networks, and quantum computing) or enabled/enhanced by AI systems (e.g., robotics and autonomy, GANs, big-data analytics, and the internet of things). See the "military-use AI primer" below for a fuller description.
9. Colin Gray, *Weapons for Strategic Effect: How Important Is Technology?* (Alabama: Center for Strategy and Technology, 2001).
10. The concept of "emerging technologies" is a loose and ill-defined one and often used interchangeably when discussing the impact of technologies such as cyberspace, social media, autonomous weapons, quantum computing, rail guns, directed energy weapons, 3D printing, and AI. Emerging technologies can mean different things to different people and are analytically very distinct; for example, cyber capabilities can enable offensive capabilities to penetrate and compromise an adversary's systems, while social media can permit attackers to influence perceptions, create false memes, or otherwise disrupt or upend prevailing norms. AI can amplify and enhance the potency and efficacy of these effects. This distinction also implies that there will be variation among third-party non-state actors in their relative capabilities to achieve their malevolent objectives in the context of nuclear weapons, which the chapter highlights.
11. For example, it took several decades to discover the critical role that misperceptions played in the 1962 Cuban missile crisis. Patricia Lewis, Heather Williams, Benoît Pelopidas, and Sasan Aghlani, *Too Close for Comfort: Cases of Near Nuclear Use and Options for Policy* (London: Chatham House, 2014), p. 30.
12. "Counterfactual thinking" in this context refers to a "possible," "parallel," or "alternate" world in which key features of the real world are not present or take on different values. A counterfactual involves "the mental construction of a course of events which is altered through modification in one or more 'conditions.'" Edward A. Shils and Henry A. Finch, ed. and trans., *Max Weber: The Methodology of the Social Sciences* (New York: Free Press, [1905] 1949), p. 173.

13. Richard Ned Lebow, "Counterfactuals and Security Studies," *Security Studies* 24, no. 3 (2015): pp. 403–412.
14. Ibid., p. 406.
15. Richard Danzig, "Surviving on a Diet of Poisoned Fruit: Reducing the National Security Risks of America's Cyber Dependencies," *Center for a New American Security*, July 21, 2014, https://www.cnas.org/publications/reports/surviving-on-a-diet-of-poisonedfruit-reducing-the-national-security-risks-of-americas-cyber-dependencies.
16. This section is derived in part from James Johnson, *Artificial Intelligence & the Future of Warfare: USA, China, and Strategic Stability* (Manchester: Manchester University Press, 2021), Ch. 1.
17. Stuart Russell and Peter Norvig, *Artificial Intelligence: A Modern Approach*, 3rd ed. (Harlow: Pearson Education, 2014); and Stuart Russell, *Human Compatible* (New York: Viking Press, 2019).
18. Andy Pearl, "Homage to John McCarthy, the Father of Artificial Intelligence (AI)," *Artificial Solutions*, June 2, 2017, www.artificial-solutions.com/blog/homage-to-john-mccarthy-the-father-of-artificialintelligence.
19. Nils J. Nilsson, *The Quest for Artificial Intelligence: A History of Ideas and Achievements* (New York: Cambridge University Press, 2010).
20. Executive Office of the President, National Science and Technology Council, Committee on Technology, *Preparing for the Future of Artificial Intelligence*, October 12, 2016, https://obamawhitehouse.archives.gov/sites/default/files/whitehouse_files/microsites/ostp/NSTC/preparing_for_the_future_of_ai.pdf, p. 6.
21. Though many AI systems rely on large amounts of data, AI does not necessarily entail the volume, velocity, and variety usually associated with big-data analytics. Thus, while big-data is not new, and does not necessarily incorporate AI, AI techniques will likely supercharge the mining of data-rich sources. E. Gray et al., "Small Big Data: Using Multiple Datasets to Explore Unfolding Social and Economic Change," *Big Data & Society* 2, no. 1 (2015): pp. 1–6.
22. ML is a concept that encompasses a wide variety of techniques designed to identify patterns in, and "learn" and make predictions from, datasets. Successful "learning" depends on having access to vast pools of reliable data about past behavior and successful outcomes. The "neural network" approach to AI represents only a small segment of the improvements in AI techniques. AI also includes, for example, language processing, knowledge representation, and inferential reasoning, which is being actualized by the rapid advancements in software, hardware, data collection, and data storage. See, for example, Jürgen Schmidhuber, "Deep Learning in Neural Networks: An Overview," *Neural Networks* 61 (January 2015): pp. 85–117; Dzmitry Bahdanau, Kyunghyun Cho, and Yoshua Bengio, "Neural Machine Translation by Jointly Learning to Align and Translate," (2014) arXiv:1409.0473; and Yoshua Bengio, Yann Lecun, and Geoffrey Hinton, "Deep Learning for AI," *Communications of the ACM* 64, no. 7 (July 2021): pp. 58–65.
23. US technology companies reportedly invested an estimated US$20 to US$30 billion in narrow AI algorithms in 2016, and this amount is expected to reach $126 billion by 2025. At the same time, global AI-generated revenue is projected to climb from US$643.7 million in 2016 to US$36.8 billion in 2025, a factor of almost sixty times greater. Daniel S. Hoadley and Lucas J. Nathan, "Artificial Intelligence and National Security,"

Congressional Research Report for Congress, November 2, 2020, https://fas.org/sgp/crs/natsec/R45178.pdf, p. 2.

24. See Ben Buchanan, "The AI Triad and What it Means for National Security Strategy," Center for Security and Emerging Technology, August 2020, https://cset.georgetown.edu/wp-content/uploads/CSET-AI-Triad-Report.pdf.
25. Nick Bostrom, *Superintelligence: Paths, Dangers, Strategies* (Oxford: Oxford University Press, 2014).
26. See Russell and Norvig, *Artificial Intelligence*; and Nilsson, *The Quest for Artificial Intelligence*.
27. ML algorithms generate statistical models to accomplish a specified task in situations they have not previously encountered during the "learning" process. Most recent advances in AI applications have occurred with non-symbolic ML approaches—computationally intensive, linear algebra, and statistical methods where input and knowledge representation are numerical, e.g., image pixels and audio frequencies. See Russell and Norvig, *Artificial Intelligence*.
28. AI "narrow" breakthroughs have led to speculation about the arrival of AGI. Stuart Armstrong, Kaj Sotala, and Seán S. ÓhÉigeartaigh, "The Errors, Insights, and Lessons of Famous AI Predictions—And What They Mean for the Future," *Journal of Experimental & Theoretical Artificial Intelligence* 26, no. 3 (2014): pp. 317–342.
29. Technology for Global Security (T4GS), "AI and Human Decision-Making: AI and the Battlefield," *T4GSReports* (November 28, 2018), www.tech4gs.org/ai-and-human-decisionmaking.html.
30. See Amandeep Singh Gill, "Artificial Intelligence and International Security: The Long View," *Ethics and International Affairs* 33, no. 2 (2019): pp. 169–179.
31. Russell, *Human Compatible*.
32. For an excellent overview of AI and its various subfields, see Margaret A. Boden, *AI: Its Nature and Future* (Oxford: Oxford University Press, 2016); and David Vernon, *Artificial Cognitive Systems: A Primer* (Cambridge, MA. MIT Press, 2014).
33. "Narrow" AI has inherent limitations and lacks a capacity for will and intent. Therefore, used in a military context, it is a tool to enhance or enable weapon systems, not an independent actor. See Lawrence Lewis, "AI and Autonomy in War," (Washington, DC: Center for Naval Analysis, August 2018), www.cna.org/CNA_files/PDF/Understanding-Risks.pdf. p. 17.
34. Historical definitions of AI can be grouped as follows: systems that think like humans; systems that act like humans, and systems that act and reason. See Andrew W. Moore, "AI and National Security in 2017," Presentation at AI and Global Security Summit, Washington, DC, November 1, 2017; and Andrew Ilachinski, *AI, Robots, and Swarms: Issues, Questions, and Recommended Studies* (Washington, DC: Center for Naval Analysis, January 2017), p. 6.
35. The term revolution in military affairs (RMA) was popularized in the 1990s and early 2000s, especially within the US defense community. An RMA is usually caused by technology but is not *revolutionary* unless it changes how states fight or instigate wars. See Knox MacGreger and Murray Williamson, *The Dynamics of Military Revolution* (Cambridge: Cambridge University Press, 2001), pp. 1,300–2,050; and Colin S. Gray, *Strategy for Chaos: Revolutions in Military Affairs and the Evidence of History* (London: Frank Cass, 2002). For Chinese views on the debate, see Fu Wanjuan, Yang Wenzhe, and Xu Chunlei,

"Intelligent Warfare: Where Does it Not Change?" *PLA Daily*, January 14, 2020, www.81.cn/jfjbmap/content/2020-01/14/content_252163.htm; and Lin Juanjuan, ZhangYuantao and Wang Wei, "Military Intelligence is Profoundly Affecting Future Operations," *Ministry of National Defense of the People's Republic of China*, September 10, 2019, www.mod.gov.cn/jmsd/201909/10/content_4850148.htm.

36. Hoadley and Lucas, "Artificial Intelligence and National Security," pp. 1–2.
37. US Department of Defense, "Summary of the 2018 Department of Defense Artificial Intelligence Strategy," https://media.defense.gov/2019/Feb/12/2002088963/-1/-1/1/SUMMARY-OF-DOD-AISTRATEGY.PDF, p. 5.
38. Ibid., p. 3; Moore, "AI and National Security in 2017."
39. "Autonomy" in the context of military applications can be defined as the condition or quality of being self-governing to achieve an assigned task, based on a system's situational awareness (integrated sensing, perceiving, and analyzing), planning, and decision-making. An autonomous weapon system (or lethal autonomous weapon system) is a weapon system that, once activated, can identify, target, and attack without the intervention of a human operator. A fully autonomous system is never, however, completely free from human interaction; either a system designer or an operator would retain the ability to program the system to function pursuant to specified parameters. See US Department of Defense, *Directive 3000.09, Autonomy in Weapon Systems,* www.esd.whs.mil/Portals/54/Documents/DD/issuances/DODd/300009p.pdf.
40. Michael C. Horowitz, "Artificial Intelligence, International Competition, and the Balance of Power," *Texas National Security Review* 1, no. 3 (2018): pp. 37–57.
41. The concept of "learning" and "teaching" in the context of ML refers to finding statistical relationships in past data, as opposed to anthropomorphic interpretations that confuse human learning with machine learning. See David Watson, "The Rhetoric and Reality of Anthropomorphism in Artificial Intelligence," *Minds and Machines* 29 (2019): pp. 417–440.
42. Matthew Hutson, "AI Researchers Allege that Machine Learning is Alchemy," *Science* (May 3, 2018), www.sciencemag.org/news/2018/05/ai-researchers-allege-machine-learning-alchemy.
43. A computational method known as "artificial neural networks" draws on knowledge of the human brain, statistics, and applied mathematics. Nilsson, *The Quest for Artificial Intelligence*, Chs. 28–29.
44. DL is an approach to ML whereby the system "learns" how to undertake a task in either a supervised, semi-supervised, or unsupervised manner. Ian Goodfellow, Yoshua Bengio, and Aaron Courville, *Deep Learning* (Cambridge, MA: MIT Press, 2016).
45. Several recent AI innovations continue to use the previous generation of hard-coded techniques such as Bayesian statistics, probabilistic relational models, and other evolutionary algorithms. Nilsson, *The Quest for Artificial Intelligence*, Chs. 28–29.
46. Ibid., p. 347.
47. In contrast, hand-coded programming often requires a significant amount of research on how the world works—or contextual analysis. Russell and Norvig, *Artificial Intelligence*, Ch. 18.
48. Lewis, *AI and Autonomy in War.*
49. See James Johnson, "'Catalytic Nuclear War' in the Age of Artificial Intelligence and Autonomy: Emerging Military Technology and Escalation Risk between Nuclear-Armed States," *Journal of Strategic Studies* (2021) doi: 10.1080/01402390.2020.1867541.

50. Juanjuan, Yuantao, and Wei, "Military Intelligence is Profoundly Affecting Future Operations."
51. See Nadine Carter, David Woods, and Charles Billings, "Automation Surprises," in *Handbook of Human Factors and Ergonomics*, 2nd ed., edited by G. Salvendy (New York: John Wiley & Sons Inc., 1997).
52. For example, AI researchers recently discovered that neural nets are not capturing accurate depth information in images, which may have important safety ramifications. See Tom van Dijk and Guido de Croon, "How do Neural Networks See Depth in Single Images?" (2019), https://arxiv.org/abs/1905.07005.
53. Tom Krishner, "5 Reasons Why Autonomous Cars Aren't Coming Anytime Soon," *AP News*, February 4, 2019, https://apnews.com/b67a0d6b6413406fb4121553cdf0b95a; and Roger K. Lewis, "Reality is Going to Stall for Some Time the Advent of Driverless Cars," *Washington Post*, August 2, 2019, https://www.washingtonpost.com/realestate/reality-is-going-to-stallfor-some-time-the-advent-of-driverless-cars/2019/08/01/343c9458-afa8-11e9-a0c9-6d2d7818f3da_story.html.
54. Mary L. Cummings, "Rethinking the Maturity of Artificial Intelligence in Safety-Critical Settings," *AI Magazine*, in press, https://hal.pratt.duke.edu/sites/hal.pratt.duke.edu/files/u39/2020-min.pdf. p. 2.
55. Bengio, Lecun, and Hinton, "Deep Learning for AI," pp. 58–65.
56. Using ML-DL for top-down (or "system 2") tasks is an area of research that is still in its infancy. Ibid., p. 63.
57. By contrast, a computer vision algorithm would learn that this image has three equally spaced Vs with three alternately spaced circles, each missing a one-sixth piece—whereas most people will see two triangles superimposed over one another. See Robert Max Williams and Roman V. Yampolskiy, "Optical Illusions Images Dataset," (2018), https://arxiv.org/abs/1810.00415.
58. AI systems are not merely collections of hardware and algorithms but, instead, supervisory control systems. These systems need to be observable, explainable, controllable, and predictable, and direct the attention of their human supervisors when required by a particular situation or operation. David Woods and Erik Hollnagel, *Joint Cognitive Systems: Patterns in Cognitive Systems Engineering* (London: Routledge, 2006), pp. 136–137.
59. Rob A. Osonde and William Welser IV, *An Intelligence in Our Image: The Risks of Bias and Errors in Artificial Intelligence* (Santa Monica, CA: RAND Corporation, 2017), www.rand.org/pubs/research_reports/RR1744.html.
60. Ben Dickson, "Reinforcement Learning Frustrates Humans in Teamplay, MIT Study Finds," *TechTalks*, November 1, 2021, https://bdtechtalks.com/2021/11/01/reinforcement-learning-hanabi/.
61. For recent research with this objective in mind, see Alison Gopnik et al., "A Theory of Causal Learning in Children: Causal Maps And Bayes Nets," *Psychological Review* 111, no. 1 (2004): pp. 3–32; Nan R. Ke et al., "Learning Neural Causal Models from Unknown Interventions," (2019), arXiv:1910.01075; and David Lopez-Paz et al., "Discovering Causal Signals in Images," in *Proceedings of the IEEE Conf. Computer Vision and Pattern Recognition* (2017): pp. 6,979–6,987.
62. Tianmin Shu et al., "AGENT: A Benchmark for Core Psychological Reasoning," *Proceedings of the 38th International Conference on Machine Learning, PMLR* 139 (2021), https://arxiv.org/pdf/2102.12321.pdf.

63. Recent studies have made important progress is clarifying how different neural net architectures fare in terms of this systematic generalization ability. Dzmitry Bahdanau et al., "Systematic Generalization: What is Required and Can it be Learned?" (2018), arXiv:1811.12889.
64. Anjanette H. Raymond et al., "Building a Better HAL 9000: Algorithms, the Market, and the Need to Prevent the Engraining of Bias," *Journal of Technology and Intellectual Property* 15 (2018): pp. 224–225.
65. Carl T. Bergstrom and Jevin D. West, *Calling Bullshit: The Art of Skepticism in a Data-Driven World* (New York: Random House, 2020), p. 183.
66. Will Heaven, "OpenAI's New Language Generator GPT-3 is Shockingly Good—and Completely Mindless," July 20, 2020, *MIT Technology Review*, https://www.technologyreview.com/2020/07/20/1005454/openai-machine-learning-language-generator-gpt-3-nlp/.
67. Bengio, Lecun, and Hinton, "Deep Learning for AI," pp. 63–64.
68. Lt. General John Shanahan, Director of the DoD Joint Artificial Intelligence Center, quoted in Sydney J. Freedberg Jr., "EXCLUSIVE Pentagon's AI Problem is 'Dirty' Data: Lt. Gen. Shanahan," *Breakingdefense*, November 13, 2019, https://breakingdefense.com/2019/11/exclusive-pentagons-ai-problem-is-dirty-data-lt-genShanahan/.
69. Ian Goodfellow, Patrick McDaniel, and Nicolas Papernot, "Making Machine Learning Robust against Adversarial Inputs," *Communications of the ACM* 61, no. 7 (2018): pp. 56–66; and Ian Goodfellow, Jonathon Shlens, and Christian Szegedy, "Explaining and Harnessing Adversarial Examples," December 20, 2014, *arXiv* preprint arXiv: 1412.6572.
70. Many algorithms currently used to categorize users or contents tend to face a bias/variance trade-off. Where human users interact with algorithmically generated content that they consider as misleading, formal definitions of bias (incorrectly categorized content) and variance (wide-ranging categorizations) are likely to be less relevant than the way in which the felt nature of the displayed content interacts with the knowledge, disposition, and belief systems of users—as variance decreases, bias usually increases. See Pedro Domingos, "A Few Useful Things to Know about Machine Learning," *Communications of the ACM* 55, no. 10 (2012): pp. 85–86.
71. Recent research in adversarial AI has discovered blind spots and a minimal understanding of the context in a fast-moving and complex environment—or "brittleness." See Goodfellow, Shlens, and Szegedy, "Explaining and Harnessing Adversarial Examples."
72. Joseph Johnson, "MAD in an AI Future?" Center for Global Security Research (Livermore, CA: Lawrence Livermore National Laboratory), pp. 4–6.
73. As "unsupervised" ML techniques mature, the reliance on data and labeling (i.e., images, videos, and text) to support AI system's training environments is expected to decrease. Unlike current "supervised" (or reinforcement) learning techniques that depend on labeling images to detect patterns, unsupervised ML is designed to create autonomous AI by rewarding agents (i.e., ML algorithms) for learning about the data they observe without a particular task in mind. See Alexander Graves and Kelly Clancy, "Unsupervised Learning: The Curious Pupil," *Deepmind*, June 25, 2019, https://deepmind.com/blog/article/unsupervised-learning; and Richard S. Sutton and Andrew G. Barto, *Reinforcement Learning: An Introduction*, 2nd ed. (Cambridge, MA: MIT Press, 2020).

74. When solving problems, humans learn high-level concepts with relatively little data, which apply to other problems. AI does not possess this broader knowledge or common sense. Goodfellow, Shlens, and Szegedy, "Explaining and Harnessing Adversarial Examples"; and Brenden M. Lake, Ruslan Salakhutdinov, and Joshua B. Tenenbaum, "Human-Level Concept Learning through Probabilistic Program Induction," *Science* 350, no. 6,266 (2015): pp. 1,332–1,338.
75. AI systems are limited to what they can infer from datasets because of the relatively few higher-level mathematical concepts from which computational learning theory is derived. David Deutsch, "Creative Blocks," *Aeon*, October 3, 2012, https://aeon.co/essays/how-close-are-we-to-creating-artificial-intelligence.https://aeon.co/essays/how-close-are-we-to-creating-artificial-intelligence.
76. Domingos, "A Few Useful Things to Know about Machine Learning," pp. 78–88.
77. Technical advances in decision trees, support vector machines, and DL have enabled a greater level of flexible discriminators into AI, but they still exhibit similar learning issues as the resolution and dimensions of images increases. For a study on the nature of human cognition, see Daniel Kahneman, *Thinking, Fast and Slow* (New York: Penguin, 2011).
78. Deutsch, "Creative Blocks."
79. Chris Lysy, "Richard Feynman: What I Cannot Create, I Do Not Understand," *Fresh Spectrum*, https://freshspectrum.com/richard-feynman-what-i-cannot-create-i-do-not-understand/.
80. Leslie Kaelbling, Michael L. Littman, and Anthony R. Cassandra, "Planning and Acting in Partially Observable Stochastic Domains," *Artificial Intelligence* 10, no. 3 (June 2017): pp. 967–968.
81. Matt Turek, *Explainable Artificial Intelligence (XAI), DARPA*, https://www.darpa.mil/program/explainable-artificial-intelligence.
82. Several AI researchers have posited that ML algorithms could be designed to monitor the performance of other algorithms, thus providing a potential solution. Matt Burgess, "Holding AI to Account: Will Algorithms Ever Be Free from Bias if They're Created by Humans?" *Wired*, January 11, 2016, https://www.wired.co.uk/article/creating-transparent-ai-algorithms-machine-learning.
83. See Carter, Woods, and Billings, "Automation Surprises."
84. To address the known gaps in the brittleness of AI, there has been increasing interest in the fusing of symbolic (or classical AI) and connectionist deep-neural network approaches (or brain-inspired AI), which mimics neural connections in the brain to AI. However, neither approach—either separately or combined—can approximate top-down reasoning to deal with significant uncertainty. Mary L. Cummings, "Rethinking the Maturity of Artificial Intelligence in Safety-Critical Settings," *AI Magazine*, in press, https://hal.pratt.duke.edu/sites/hal.pratt.duke.edu/files/u39/2020-min.pdfp, p. 9.
85. For example, "narrow" AI currently used in a military context to support defense systems has exhibited some degree of both automation and autonomy since the 1970s. See Ingvild Bode and Tom Watts, "Meaning-Less Human Control: Lessons from Air Defence Systems on Meaningful Human Control for the Debate on AWS," *University of Southern Denmark*, February 2021, https://dronewars.net/wpcontent/uploads/2021/02/DW-Control-WEB.pdf.

86. Vincent Boulanin, ed., *Artificial Intelligence, Strategic Stability and Nuclear Risk* (Stockholm: SIPRI Publications, June 2020), pp. 23–30.
87. Ibid. pp. 23–24.
88. Most nuclear-armed states are developing, or have already deployed, autonomous unmanned surface vehicles (USVs), under-water unmanned vehicles (UUVs), and unmanned aerial vehicles (UAVs), which may be used for a variety of ISR missions. See Paddy Walker, "Leadership Challenges from the Deployment of Lethal Autonomous Weapon Systems," *The RUSI Journal* (2021) doi: 10.1080/03071847.2021.1915702.
89. See Luis Simon, "Demystifying the A2/AD Buzz," *War on the Rocks*, January 4, 2017, https://warontherocks.com/2017/01/demystifying-the-a2ad-buzz/.
90. US Department of Defense, "Defense Science Board, Report of the Defense Science Board Summer Study on Autonomy," (Washington: DC US DOD, June 2016).
91. See Rafael Loss and Joseph Johnson, "Will Artificial Intelligence Imperil Nuclear Deterrence," *War on the Rocks*, September 19, 2019, https://warontherocks.com/2019/09/will-artificial-intelligence-imperil-nuclear-deterrence/; Arthur M. Holland, *Known Unknowns: Data Issues & Autonomous Military Systems* (Geneva: UNIDIR, 2021); and Johnson, *Artificial Intelligence & the Future of Warfare*, Ch. 1.
92. Recent advances in ML and AI have already led to significant qualitative improvements to a broad range of autonomous weapon systems, resolving several technical bottlenecks in existing military technology. Ibid., p. 27.
93. For a notable exception, see Adam Lowther and Curtis McGiffin, "AMERICA needs a 'Dead Hand,'" *War on the Rocks*, August 16, 2019, https://warontherocks.com/2019/08/america-needs-a-dead-hand/; and Matt Field, "Strangelove Redux: US Experts Propose Having AI Control Nuclear Weapons," *Bulletin of the Atomic Scientists*, August 30, 2019, https://thebulletin.org/2019/08/strangelove-redux-us-experts-propose-having-ai-control-nuclear-weapons/.
94. See Watson, "The Rhetoric and Reality of Anthropomorphism in Artificial Intelligence," pp. 417–440; and Linda J. Skitka, Kathleen L. Mosier, and Mark Burdick, "Does Automation Bias Decision-Making?" *International Journal of Human-Computer Studies* 51, no. 5 (1999): pp. 991–1006.
95. Lora Saalman, "China's Integration of Neural Networks into Hypersonic Glide Vehicles," in *AI, China, Russia, and the Global Order: Technological, Political, Global, and Creative Perspectives*, edited by N. D. Wright, White Paper (US Department of Defense and Joint Chiefs of Staff: Washington, DC, December 2018), pp. 153–160.
96. The ability to change or terminate autonomous UAV or UUV on ISR missions distinguishes them from missiles and torpedoes (e.g., ICBMs and submarine launched ballistic missiles), offering commandeers potentially new tools for managing escalation.
97. At least two nuclear-armed states (Russia and the US) are reportedly considering deploying UAVs or UUVs for nuclear weapon delivery operations. See Roland Oliphant, "Secret Russian Radioactive Doomsday Torpedo Leaked on Television," *Daily Telegraph*, November 15, 2015, https://www.telegraph.co.uk/news/worldnews/europe/russia/11991729/Secret-Russian-radioactive-doomsday-torpedo-leaked-on-television.html; and Dave Majumdar, "USAF Leader Confirms Manned Decision for New Bomber," *Flight International*, April 23, 2013, https://www.flightglobal.com/usaf-leader-confirms-manned-decision-for-new-bomber/109511.article.

98. See Michael C. Horowitz, Paul Scharre, and Alexander Velez-Green, "A Stable Nuclear Future? The Impact of Autonomous Systems and Artificial Intelligence," December 2019, *arXix*, https://arxiv.org/pdf/1912.05291.pdf.
99. Notable capabilities in this field include: Russia's extra-large UUV designed for ISR missions, Orca; the US *Sea Hunter*, an autonomous surface system capable of anti-submarine operations and mine countermeasures; the US Navy's Control Architecture for Robotic Agent Command and Sensing system, a software architecture that enables swarms of USVs and UAVs to conduct A2/AD operations autonomously. See Vincent Boulanin, Ian Davis, and Maaike Verbruggen, "The Convention on Certain Conventional Weapons and Lethal Autonomous Weapon Systems," *SIPRI Yearbook 2019* (London: Oxford University Press, 2019).
100. Azerbaijan's combination of drones and artillery weaponry effectively targeted Armenia's air defense units, blunting Armenia's ability to counter Azerbaijan's UAV attacks. Shaan Shaikh and Wes Rumbaugh, "The Air and Missile War in Nagorno-Karabakh: Lessons for the Future of Strike and Defense," *CSIS*, December 8, 2020, https://www.csis.org/analysis/air-and-missile-war-nagorno-karabakh-lessons-future-strike-and-defense.
101. See David Mindell, "Automation's Finest Hour: Radar and System Integration in World War II," in *Systems, Experts and Computers: The Systems Approach in Management and Engineering, World War II and After*, edited by A. C. Hughes and T. P. Hugues (Cambridge, MA: MIT Press, 2000), pp. 27–56, pp. 40–44.
102. GANs can augment existing datasets by generating new data incorporated into a classifier's training data. GAN-augmented datasets of spam and malware have been used, for example, by research teams to create classifiers that perform better than those trained only on real-world data. See Gray Stanton and Athirai A. Irissappane, "GANs for Semi-Supervised Opinion Spam Detection," *arXiv* (March 19, 2019), http://arxiv.org/abs/1903.08289; and Hojjat Aghakhani et al., "Detecting Deceptive Reviews Using Generative Adversarial Networks," *arXiv* (May 25, 2018), http://arxiv.org/abs/1805.10364.
103. Mark Berlin and Matt Young, "Automatic Target Recognition Systems," *Technology Today* 1 (2018): pp. 10–13.
104. See Matthew Hipple, "Bring on the Countermeasure Drones," *Proceedings* (US Naval Institute), February 2014, https://www.usni.org/magazines/proceedings/2014/february/bring-countermeasure-drones.
105. Johnson, *Artificial Intelligence & the Future of Warfare*, Ch. 7.
106. In 2016 the US Defense Advanced Research Projects Agency (DARPA) launched an open challenge to develop systems to identify and analyze new enemy signals on the fly. US Defense Advanced Research Projects Agency, "New DARPA Grand Challenge to Focus on Spectrum Collaboration," *DARPA*, March 23, 2016, https://www.darpa.mil/news-events/2016-03-23.
107. For technical studies on AI-ML for detection and classification of malware, see Daniel Gibert, Carles Mateu, and Jordi Planes, "The Rise of Machine Learning for Detection and Classification of Malware: Research Developments, Trends and Challenges," *Journal of Network and Computer Applications* 153 (March 2020); and Hemant Rathore et al., "Malware Detection Using Machine Learning and Deep Learning," *arXiv* (April 4, 2019), http://arxiv.org/abs/1904.02441.
108. Boulanin, ed., *Artificial Intelligence, Strategic Stability and Nuclear Risk*, p. 98.

109. Researchers have already demonstrated that GANs technology can develop offensive strategies that slip by existing intrusion detection systems, regardless of whether they are ML-based. See Weiwei Hu and Ying Tan, "Generating Adversarial Malware Examples for Black-Box Attacks Based on GAN," *arXiv* (February 20, 2017), http://arxiv.org/abs/1702.05983. GANs are not the only type of AI architecture used to generate adversarial attacks to evade malware detection or intrusion detection. See, for example, Hyrum S. Anderson et al., "Learning to Evade Static PE Machine Learning Malware Models via Reinforcement Learning," *arXiv* (January 30, 2018), https://arxiv.org/abs/1801.08917.
110. See US Department of Defense and Joint Chiefs of Staff, "Declaratory Policy, Concept of Operations, and Employment Guidelines for Left-of-Launch Capability," Report to the US Congress, May 10, 2017, p. 1; and Riki Ellison, "Left of Launch," *Missile Defence Advocacy Alliance*, March 16, 2015, https://missiledefenseadvocacy.org/alert/3132/.
111. Johnson, "'Catalytic Nuclear War' in the Age of Artificial Intelligence & Autonomy."
112. See, for example, Anne Johnson and Emily Grumbling, *Implications of Artificial Intelligence for Cybersecurity: Proceedings of a Workshop (2019)* (Washington DC: National Academies Press, 2019), pp. 25–26; and Mark Fitzpatrick, "Artificial Intelligence and Nuclear Command and Control," *Survival Editor's Blog*, April 26, 2019, https://www.iiss.org/publications/r/2019/survival-global-politics-and-strategy-junejuly-2019/613-05-fitzpatrick-cm; "Fake News: You Ain't Seen Nothing Yet," *The Economist*, July 1, 2017, https://www.economist.com/science-and-technology/2017/07/01/fake-news-you-aint-seen-nothing-yet; and Jennifer Langston, "Lip-Syncing Obama: New Tools Turn Audio Clips into Realistic Video," *UW News*, July 11, 2017, https://www.washington.edu/news/2017/07/11/lip-syncing-obama-new-tools-turn-audio-clips-into-realistic-video/.
113. Boulanin, ed., *Artificial Intelligence, Strategic Stability and Nuclear Risk*, p. 30.
114. Russia, for instance, already includes anti-saboteur robots in the formations accompanying the mobile Years ICBMs. These include both unarmed UAVs and combat land systems. See "Russia's Strategic Missile Force to Test Mobile Robot at Forthcoming Exercises," *TASS*, March 30, 2016, https://tass.com/defense/865997.
115. Barry R. Posen, *Inadvertent Escalation* (Ithaca, NY: Cornell University Press, 1991).

1

Strategic stability

A perfect storm of nuclear risk?

How might AI affect the risk of nuclear war? This chapter provides a systematic assessment of the potential stabilizing and, especially, destabilizing effects of AI-enabled capabilities employment against nuclear-armed adversaries. It considers the ways military AI might exacerbate tension between nuclear-armed powers (especially the US and China), thereby compromising stability. A significant gap exists between the expectations and fears of policymakers and global defense communities about it, and its actual military capabilities, particularly in the nuclear sphere.[1] The misconceptions that exist today are largely caused by the hyperbolic depictions of AI in popular culture and science fiction.[2] Misrepresentations of the potential opportunities and risks in the military sphere (or "military AI"[3]) can diminish transparency, mutual understanding, and trust, and obscure constructive and crucial debate on these topics.[4] A key challenge will be, therefore, balancing a trade-off between harnessing the qualitative operational and tactical benefits of AI, while managing the risks posed to "strategic stability" (including first-strike stability, crisis stability, and arms-race stability—see Figure 1.1) and nuclear security—or the "AI-nuclear dilemma." This chapter demystifies the hype surrounding AI in the context of nuclear weapons.

Much like previous transformative technologies (e.g., multiple independently targetable reentry vehicles (MIRVs) on intercontinental ballistic missiles (ICBMs), anti-satellite weapons (ASATs), ballistic missile technology, and cyber weapons), the uncertainty caused by the perceived utility—pace, proliferation, diffusion, and potential impact of deterrence and decision-making—of AI applications will likely have a significantly outsized impact on strategic stability.[5] AI's impact on strategic stability, and, in turn, the risk of nuclear war, will hinge on the fear (or risk acceptance) perceived by others and the manipulation of this fear in competitive strategic dyads—US vs. China, India vs. Pakistan, Russia vs. NATO, and Israel vs. Iran. A core finding

AI and the Bomb. James Johnson, Oxford University Press. © James Johnson (2023).
DOI: 10.1093/oso/9780192858184.003.0002

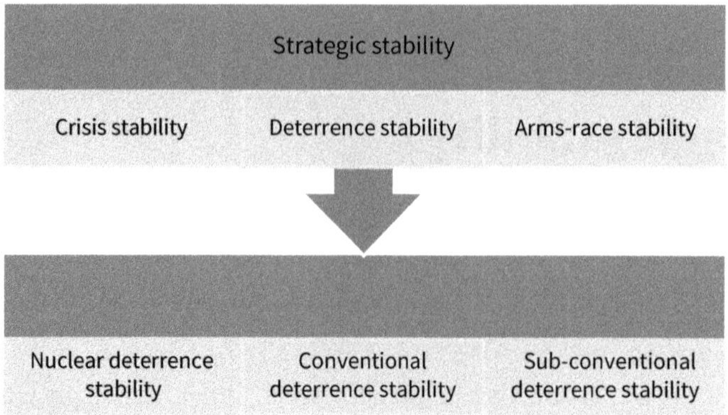

Figure 1.1 Components of "strategic stability"

of the research is that the extent to which military AI poses risks to strategic stability will largely depend on this technology's pace and scope to facilitate new ways to improve the delivery of and defense against nuclear weapons and strategic non-nuclear weapons.

The remainder of this chapter proceeds in three sections. The first section contextualizes the Cold War notion of "strategic stability"—the presence or absence of an incentive for an actor to engage in provocative behavior—with the use of advanced technology including AI in the nuclear domain, and the viability of this term in understanding power distribution among great and rising powers in the modern nuclear age. The second section examines the destabilizing characteristics of AI, such as machine-speed, bias, vulnerability, and non-human decision-making. Eschewing technological-deterministic schools of thought, the chapter conceptualizes AI technology as a natural manifestation—rather than the cause or origin—of an established trend in emerging technology, potentially leading states to adopt destabilizing launch postures. The third section describes how these characteristics can worsen some of the key conditions for nuclear instability (or strategic instability) in the digital age, particularly because of competitive and contested nuclear multipolar world order as well as differing escalation risk tolerance between great military powers—especially China and the US. It argues that coalescing of these conditions—in the presence of the inherently destabilizing military AI systems—will likely create conditions for miscalculations and misperceptions between nuclear-armed states—increasing escalation risk and the danger of deterrence failure. The chapter concludes with some early thoughts on what can be done to limit these risk factors in a future that will undoubtedly become more AI-oriented.

Rethinking "strategic stability"

The concept of "strategic stability" emerged in the latter half of the twentieth century and, despite being theoretically and politically contested to this day,[6] remains a useful intellectual tool for analyzing the potential of technically advanced weapons to undermine stability (i.e., evaluating nuclear force structures, deployment decisions, and a rationale for arms control).[7] Strategic stability is inextricably connected to the strategic thinking and debates that surrounded the "nuclear revolution," including:[8] how a nuclear war might be fought, the requirements and veracity of credible deterrence, the potential risks posed by pre-emptive and accidental strikes,[9] and how to ensure the survivability of retaliatory forces.[10] In short, strategic stability provides an over-arching theoretical framework for understanding the nature of security in the modern—and arguably "third"—nuclear age.[11]

During the twentieth century, the world eventually found a way to manage a paradigm shift in military technology—the atomic bomb. Nuclear warfare was avoided through a combination of deterrence, arms control, and safety and verification measures. In the "second nuclear age"—the spread of nuclear weapons for reasons that have nothing to do with the Cold War—states continue to face many of the same challenges and trade-offs in the use of advanced technology in the nuclear domain that the Cold War warriors wrestled with, *inter alia*:[12] (1) effective international regulation of atomic weapons; (2) safe and reliable military-civilian control of nuclear weapons and their attendant systems; (3) the prospect of nuclear blackmail of non-nuclear rivals; (4) the perceived risk of disarming pre-emptive strikes by adversaries in asymmetric situations; (5) a shifting strategic military balance between nuclear-armed adversaries; (6) the use of nuclear weapons in warfighting doctrine; (7) the use of nuclear weapons in the context of conventional conflict; and (8) striking an appropriate balance between signaling resolve and deterrence, and escalating a situation.[13]

In the broadest use of the term, strategic stability exists in the absence of a significant incentive for an adversary to engage in provocative behavior.[14] In other words, there is a lack of armed conflict, and a lack of perceived incentive to use nuclear weapons first, between nuclear-armed states.[15]. At its core, the concept is a phenomenon that focuses on finding a *modus vivendi* for the complex interactions and incentives of two (or more) actors.[16] The term is often associated with the relative power distribution among great and rising powers, particularly those in possession of nuclear weapons or the potential to acquire them. Therefore, strategic stability is a product of a complex interplay of political, economic, and military dynamics in which *technology* performs several functions—an equalizer, a counterweight, and a principal agent of change.[17] In short, there is far more to strategy than technology, though the appreciation of how much more varies widely.[18]

Technology has long been used to augment, automate, and enhance human behavior and decision-making in a military context; thus far, the "human factor" has trumped technologies' impact upon strategic stability. That is, the underlying forces behind fundamental shifts in strategic stability are generally less concerned with quantitative or qualitative assessments of military capabilities—and other measures of relative power—and instead are more focused on how nuanced institutional, cognitive, and strategic variables impact strategic decision-making and may cause misperceptions of others' intentions.[19] "Stability" is concerned with the relationship amongst these factors. Further, "strategic stability" and "strategic instability" are not necessarily mutually exclusive paradigms; they can both exist in a nuclear multipolar system. The role of technological change and strategic stability can be conceptualized, therefore, as part of a complex interaction of disruptive forces (or agents of change), which during periods of heightened geopolitical rivalry, great-power transitions, and strategic surprise may erode strategic stability and make conflict more likely.

Three broad forms (or subsets) distinguish strategic stability: first-strike stability, crisis stability, and arms-race stability. *First-strike stability* exists in situations when no one state can launch a surprise (or pre-emptive) attack against an opponent without the fear of devastating reprisals from survivable second-strike forces. That is, it is a lack of *both* incentives and pressures to use nuclear weapons first in a crisis.[20] Thus, fear that the advantage of first-strike capabilities could be eroded or neutralized by an adversary would destabilize and increase incentives to launch a pre-emptive strike.[21] Today, as throughout the Cold War, the vulnerability of Command and Control (C2) structures to counterforce attack remains high and is compounded by the increasing complexity of these systems and the propensity of states to use them to support nuclear and non-nuclear forces.

Crisis stability aims to prevent (or de-escalate) escalation during crises—such as those in Berlin and Cuba during the Cold War. Therefore, crisis stability depends on reciprocal fear between states; when crises arise, the system does not escalate a situation. Conversely, crisis instability refers to what Thomas Schelling called the "reciprocal fear of surprise attack"[22]—the belief that conflict is inevitable, and thus striking first and pre-emptively would create a strategic advantage.[23] The central concern of crisis stability is, therefore, not necessarily to deter conflict per se, but rather to avert the escalation of an ongoing (or emergent) crisis beyond the nuclear threshold. Thus, aside from technical characteristics of weapon systems (and like strategic stability), other psychological, political, and strategic factors—which are more challenging to quantify and model—are also relevant to understanding crisis stability. The research adopts a broader definition of the term "emotion, uncertainty, miscalculation, misperception, or the posture of forces" that do not incentivize states "to strike first, to avoid the worse consequences of incurring a first strike."[24] Finally, *arms-race stability* (during peacetime) can emerge when there are no exploitable inequalities (or asymmetries) separating adversaries' military forces—qualitatively or quantitatively.[25]

Crisis instability and arms-race instability can arise when states use strategic capabilities and instill fear into a situation.[26] This fear closely correlates with the incentives these decisions create. Thus, if both sides possess first-strike capabilities, either side's incentive to gain by choosing to strike first depends on the fear that hesitation might allow a rival to gain the upper hand.[27] In other words, *both* the incentive and fear created by strategic weapons can exacerbate escalation risk.[28] Crisis (in)stability is, therefore, fundamentally a psychological problem, which can raise doubts during a crisis about the assumptions that decision-makers share with an adversary a common notion of rationality—such as assessing risk, probability, and how others perceive their actions.[29] Similarly, in an arms race the incentive to gain the upper hand vis-à-vis a rival is correlated with the fear that not doing so will put an adversary in an advantageous position. Because of this connection, strategic stability is often characterized as crisis stability plus arms-race stability. Given the natural tension and trade-offs between these objectives, calibrating an effective nuclear strategy in bargaining situations—especially between deterrence and assurance—has eluded policymakers.[30]

Several psychological factors can compound these dangers and worsen crisis and arms-race instability, including: (1) psychological biases that impair the quality of decision-making during high-stress time-pressured situations; (2) a tendency to exaggerate the probability that an adversary is about to launch an attack and, in turn, overestimate the advantages of striking pre-emptively; and (3) the failure to see how their actions are in response to this bias and may cause an adversary to view conflict as inevitable.[31] Advanced AI-augmented autonomous weapons (e.g., cyber-offensive capabilities, hypersonic glide vehicles, and ASATs), which blur the distinction between nuclear and conventional warfare, can heighten strategic ambiguity during crises, creating first-mover advantage incentives, which may lead states to overestimate an adversary's capabilities and thus incentivize pre-emption.[32]

When only one side of a competitive dyad possesses AI-augmented autonomous weapons (e.g., unmanned aerial vehicles (UAVs), smart munitions, and loitering weapons), they might increase the credibility of the state's deterrent threats.[33] Moreover, even when the perception exists that AI is a force multiplier, if an adversary's AI systems are not, amongst other things, transparent, reliable, or easily verifiable, it would be difficult to determine an adversary's capabilities and objectively assess its credibility as a deterrent threat. This uncertainty could increase these systems' deterrent values if an adversary *overestimates* their capabilities or decreases their deterrent utility if actors *underestimate* them.[34]

During peacetime, the proliferation of advanced weapon systems could generate the search for counter-measures that amplify states' fear and uncertainties, leaving them feeling more vulnerable—known as a "security dilemma" (see Chapter 4) associated with arms racing and first-strike instability.[35] Conversely, where *both* sides possess these capabilities, they could be viewed as a relatively low-risk tactic to launch probing (or "salami-slicing") attacks against an adversary, creating dynamics conducive to crisis instability and unintentional escalation.[36] Autonomous drone systems could, for example, be deployed in low-intensity salami-slicing tactics to chip away at an adversary's will (or resolve), but without crossing a threshold (or psychological red line) that would provoke escalation.[37]

Scholars Dale Walton and Colin Gray have argued that "true strategic stability is a Platonic ideal, useful as a yardstick for judging real-world conditions, but inherently unattainable as a policy goal"—especially in a multipolar nuclear world order.[38] Though a stable strategic environment is not necessarily a catalyst for positive geopolitical change, a conflict between great powers is more likely when the current strategic environment is unstable and uncertain.[39] Political thought scholarship has demonstrated that security competition—motivated by the desire to control warfare—tends to be ratcheted up because of the complexity and uncertainty of military technology and operations overtime; akin to the Clausewitzian conditions of "fog and friction" (see Chapter 3).[40]

Furthermore, we are unable to assume *ex-ante* that a specific policy (e.g., arms control or confidence-building measures), which might improve (or worsen) strategic stability, will necessarily have equally positive (or negative) outcomes for the other variables involved in nuclear policy (e.g., credible deterrence, bureaucratic feasibility, alliance politics, and domestic political motives).[41] A strategy of offshore balancing could be used, for example, to restore the balance of power (i.e., stabilizing) within a regional alliance structure vis-à-vis a rising revisionist power. Conversely, band-wagoning pressures within the system may compound disruptive changes in the balance of power caused by this behavior and destabilization.

According to scholar Robert Jervis, the deterrence implications of invulnerable nuclear arsenals are "many and far-reaching."[42] That is, the existence of survivable nuclear arsenals (or second-strike capabilities) should ensure that war does not occur, crises will be rare, and the status quo will be relatively easy to maintain. Scholars have reconciled the persistently high levels of strategic competition and seemingly anomalous states' behavior during the nuclear era (i.e., arms racing, territorial struggles, and alliance building), with misguided (or mistaken) decisions by leaders, bureaucratic pathologies, and domestic political factors.[43] More recently, scholars have argued that the idea of a "nuclear revolution theory" is itself a flawed intellectual construct.[44]

In the seven decades since the introduction of nuclear weapons, there have been several technological threats to stability.[45] Scholars Keir Lieber and Daryl Press argue that global nuclear forces have become increasingly vulnerable due to rapid improvements in conventional counterforce technologies, explaining why strategic tensions have persisted.[46] Taken together, advances in technologies such as imagery and sensor technology, data processing, communications, unmanned vehicles, cyberspace, and AI have increased the uncertainty about states' future capabilities, *both* real and perceived. Strategic instability, therefore, involves a quantitative (i.e., capabilities) as well as a qualitative (i.e., intentions) assessment. Thus, nuclear powers must maintain a diverse field of cutting-edge nuclear and conventional weapon systems (both kinetic and non-kinetic) to hedge against the *possibility* of future technological breakthroughs.[47]

Research on bargaining and war also suggests that uncertainty about states' capabilities makes it more challenging to reach diplomatic solutions once a crisis or conflict begins.[48] Further, the increasing co-mingling (or blurring) of the nuclear and non-nuclear domains since the end of the Cold War (e.g., nuclear launch and delivery structures, and their attendant command, control, communications, and intelligence networks) will continue to have destabilizing implications for nuclear escalation.[49] Thomas Schelling's and Morton Halperin's Cold War-era arguments about emerging technology and new challenges to strategic stability and arms control resonate today:

> The present race seems unstable *because of the uncertainty in technology and the danger of a decisive breakthrough*. Uncertainty means that each side must be prepared to spend a great deal of money; it also means a constant fear on either side

> that the other has developed a dominant position, or will do so, or will fear the first to do so, with the *resulting danger of premeditated or pre-emptive attack* (emphasis added).[50]

The goal of preserving stability to prevent catastrophic nuclear war and the instability that emerged during the Cold War has important implications for the (albeit lightly researched) discourse on AI and strategic stability.[51] Four interrelated lessons for stability and advanced technology emerged from the Cold War experience. First, the destabilizing impact of weapons proliferation (both vertical and horizontal) becomes more acute in situations where technologies have the potential to create asymmetric military advantages between rival states such as: improved ICBM accuracy; ballistic missiles armed with MIRVs; ballistic anti-ship missiles; and missile defense technology. Technologies like these can incentivize states to diversify, secure, and harden their strategic forces.[52] Ballistic missiles, for example, caused a trade-off between the perceived utility of independently targeted reentry vehicles to maximize the success (or perceived success) of a disabling first strike, and the detrimental impact of this asymmetry for strategic stability.[53]

Second, where the risk exists—or is perceived—that a conventional conflict might cross the nuclear threshold, the implications for crisis instability and military escalation can be severe.[54] The US Office of Technology Assessment described this risk's nature in the following way:

> The degree to which strategic force characteristics might, in a crisis situation, reduce incentives to initiate the use of nuclear weapons. Weapon systems are *considered destabilizing* if, in a crisis, they would *add significant incentives to initiate a nuclear attack*, and particularly to attack quickly before there is much time to collect reliable information and carefully weigh all available options and their consequences (emphasis added).[55]

Third, any meaningful definition of strategic stability must include advanced strategic non-nuclear weapons—notably conventional counterforce capabilities—that one side could use to destroy an adversary's nuclear capabilities (e.g., ballistic missile defense, cyber, AI, ASATs, anti-submarine weapons, and precision strike munitions).[56] That is, technologically advanced offensive and defensive non-nuclear weapons can reduce the vulnerability of states to a nuclear attack.[57]

Finally, and related, nuclear modernization programs that are primarily motivated by the technological advancement (or technological determinism) associated with increased levels of automation often trigger arms racing dynamics that worsen crisis stability.[58] The intense bipolar military-technical strategic rivalry (both offensive and defensive) during the Cold War can give us a better sense of how the recent developments in military AI could affect future strategic stability.[59]

The recent history of emerging technology and revolution in military affairs counsels caution in extrapolating from current (or past) technological trends.[60] Technology rarely evolves in the way futurists predict, and many defense innovations have had countervailing or conditional effects that failed to live up to exaggerated predictions or proved wrong.[61] With aircraft came radar and anti-aircraft guns; with submarines came anti-submarine warfare tactics such as depth charges, and gas masks followed the development of poisonous chemical gas. Many predicted, for instance, that chemical weapons would instantly and dramatically change the nature of warfare and deterrence after the British used poison gas during World War I. However, chemical weapons proved far less practical, impactful, disruptive, and relatively easier to defend against than conventional explosives.[62] Similarly, inter-war strategists believed that the tank would revolutionize warfare. In the event, the German anti-tank weapon and modern warfare tactics were able to effectively countervail British tanks during *Operation Goodwood*. More recently, US "network-centric warfare" did not prove to be the strategic game-changing innovation predicted by many military thinkers.[63]

As a corollary, AI-powered drone swarms will likely be countervailed by parallel advances in defensive uses of image and remote sensor technology to detect and track swarms and digital jammers to interfere with their networks' high-energy lasers to destroy them. Further, offensive cyber could also be used in "left of launch" operations to prevent these swarms from mounting an attack in the first place.[64] Technological-deterministic schools of thought that conceptualize technology as the *sole* driver of strategic stability and the conduct of warfare oversimplify this complicated human endeavor.[65] As we discovered in the introduction, even if AI ultimately has a transformative strategic effect, it will likely be at least far more non-linear, iterative, and prosaic than many expect.[66] As Bernard Brodie cautioned during the Cold War-era:

> What may look like extraordinarily important changes in the tools of war or related technologies may appear to lack a significant impact on strategy and politics ... *even though those improvements may look quite significant* to a scientist or an engineer (emphasis added).[67]

From this perspective, the increasing speed of warfare, shortening of the decision-making timeframe, and the co-mingling of military capabilities that have occurred within the broader context of emerging technology would likely have occurred whether or not AI was involved.[68] Put another way, AI technology can be viewed as a natural manifestation—rather than the cause or origin—of an established trend in emerging technology, potentially leading states to adopt destabilizing launch postures.[69] These trends also demonstrate the difficulty of applying traditional definitions of strategic stability to advanced technology, such as AI, which also have dual-use and cross-domain uses. In sum, the extent military AI poses risks to strategic stability will largely depend on the pace and scope of this technology to facilitate new ways of improving the delivery of and defense against nuclear weapons and strategic non-nuclear weapons.

While the potential strategic effects of AI are not unique or exclusive to this technology, the confluence of five interrelated trends, examined in the remainder of this book, weighs heavily on the pessimistic side of the instability-stability ledger: the rapid technological advancements and diffusion of military AI;[70] the inherently destabilizing characteristics of AI technology; the multifaceted possible intersections of AI with nuclear weapons; and related, the interplay of these intersections with strategic non-nuclear capabilities; and, finally, the backdrop of a competitive multipolar nuclear world order, which may entice states to prematurely deploy unverified, unreliable, and unsafe AI-augmented weapons into combat situations.

AI's destabilizing characteristics

AI does not exist in a vacuum. In isolation, therefore, AI is unlikely to be a strategic game changer. Instead, it will likely reinforce the destabilizing effects of existing advanced conventional (especially counterforce) capabilities, thereby increasing the speed of warfare and compressing the decision-making timeframe.[71] Unlike the tactical and operational levels of warfare, AI's effect at a strategic level remains uncertain. On the one hand, AI-augmented C2 systems could mitigate many of the shortcomings inherent to human strategic decision-making, such as the susceptibility to invest in sunk costs, skewed risk judgment, cognitive heuristics, and group-think—with stabilizing effects. On the other hand, AI systems that enable commanders to predict the potential production, commissioning, deployment, and use of nuclear weapons by adversaries will likely lead to unpredictable system behavior and subsequent outcomes, which, *in extremis*, could undermine first-strike stability—the premise of mutually assured destruction—and make future nuclear war winnable.[72]

Machine-speed AI warfare

AI introduces a unique means to operate and respond at machine speed in the use of military force—across all military domains—during conflict, crisis, and war, and within the related broader national security arenas such as cybersecurity, counterterrorism, crime, and counterespionage. In military arenas where a premium on autonomy and speed exists (e.g., missile defense, autonomous weapon systems, and cyberspace), faster reaction times will likely have outsized tactical and strategic effects. Given the speed at which AI-enabled digital operations occur, human involvement may become impossible and, increasingly, levels of automation a tactical prerequisite. Put differently, while defense planners generally want to keep open as many options as possible for as long as they can and to preserve their decision space, the tactical possibilities of AI-powered capabilities will likely increase the geopolitical pressures (see Chapter 4) to choose speed over keeping options open and buying time. By their very nature, AI algorithms discover patterns and make predictions through undecipherable calculations derived from complex data at a speed and scale that humans cannot hope to replicate or even fully comprehend.[73] Two states both employing algorithms that act in automated "self-defense"[74] or countermeasure might conceivability find themselves rapidly and inadvertently in an armed conflict—caused by algorithms behaving unpredictably when interacting with other algorithms.[75] Moreover, developers may also be able to embed algorithms with an automated anticipatory self-defense mechanism to respond instantly to imminent threats, enabling pre-emptive counter-strike operations.[76]

Theoretically speaking, much like the algorithms that are performing in DeepMind's StarCraft II strategic game,[77] it would be possible for algorithms to "game out" (or manipulate) various responses—using early-warning systems, intelligence surveillance reconnaissance, and other alert systems—and select the optimal one to attack an adversary state, and to identify potential consequences. While the current generation of automated missile defense systems (e.g., terminal high-altitude area defense and Patriot systems) are capable of tracking and prosecuting hostile targets without human intervention, they are unable to autonomously monitor and improve their performance, independent of human intervention. AI would give the next generation that ability.

AI technology fused with advanced weapons—potentially simultaneously across multiple combat zones—could allow these systems to react at machine speed and boost the overall pace of combat. For example, a recent wargaming exercise hosted by the US Army to test the impact of AI-enabled tools on operational tempo—"Project Convergence"—demonstrated that better digital connections between weapons augmented with AI technology could reduce, from minutes to seconds, the time it took to identify, target, and destroy a target.[78] Despite the tactical advantages of being able to react in real-time, especially in asymmetric contested environments, experts warn that massive increases in the speed of combat could result in machines reacting to combat situations at a pace that surpasses human comprehension—so much so that commanders might be unable to control, contain (i.e., de-escalate or not escalate further), or terminate a situation.[79] In other words, a fundamental tension exists between the desire to prepare to escalate a situation if required, and the concomitant goal of minimizing the risk of allowing a situation to spiral out of control—a tension that is especially acute when there are perceived battlefield advantages in acting quickly and decisively.[80]

Because human commanders would be unable to react quickly enough, the decision to delegate control to an autonomous system would be a challenging ethical and tactical dilemma. Put another way, while AI-enabled autonomous early-warning systems would theoretically allow defense planners to identify and monitor threats faster and more reliably than before, the lack of human judgment and supervision, coupled with the inherent "brittleness" (i.e., a lack of real-world common sense to deal with novel situations) and "black box" (or opaque and unexplainable) characteristics of AI-ML algorithms, mean that the risk of destabilizing accidents and false alarms will likely rise.[81] Given these risks, ensuring the quality and reliability, and transparency and "explainability," of AI algorithms is a critical task for policymakers.

As strategist Thomas Schelling argued, "when speed is critical, the victim [in this case a nuclear-armed state] of an accident or a false alarm is under terrible pressure."[82] This point is especially true of the time pressures associated with AI. Fear created by the unpredictability and uncertainty of an enemy conducting warfare at machine speed may, therefore, tempt nuclear states to automate their nuclear retaliatory capability. That is, all else being equal, a nuclear-armed state less confident in its second-strike capabilities (i.e., China, North Korea, Pakistan, and perhaps Russia) will be more inclined to use automation.

The high speed associated with AI and autonomy can have unexpected and escalatory outcomes. For example, according to the US Securities and Exchange Commission the 2010 stock market "flash crash"—which reduced the stock market value by one trillion dollars within a matter of minutes—was enabled and amplified by the use (or misuse) of autonomous financial trading algorithms.[83] Unlike the financial markets, however, there is no overarching authority in global security to enforce pre-programed fail-safe mechanisms based on a shared set of rules. Thus, the prospect of a flash crash-like event occurring in adversarial, loosely regulated, offense-dominant, and strategically competitive domains, such as cyberspace, hypersonic weapons, missile defense, or ASATs, is especially alarming. Russia's embrace of drones in Ukraine since 2014, for example, has significantly shortened the time between when their reconnaissance drones first detect Ukrainian forces and when they cue their precision rockets to destroy those forces.[84]

Military AI systems functioning at machine speed could push the pace of combat to a point where the actions of machines eclipse the ability of human decision-makers to control (or even comprehend) events. *In extremis*, human commanders might lose control of the outbreak, course, and termination of warfare. Were humans to effectively lose (or predelegate) control of warfare to machines, inadvertent escalation pathways and crisis instability would increase, potentially with catastrophic results (see Chapter 3). Compelled by the speed and precision of AI to make decisions in a compressed timeframe, a state might accept higher risks and escalate a conflict with the belief it was in a "use-it-or-lose-it" situation, or with a lack of confidence in its ability to guarantee the safety and control of its nuclear arsenals.[85]

Given the risk of escalation in fusing AI with advanced capabilities, and greater degrees of autonomy, it is critical that safeguards are put in place to limit the parameters with which algorithms respond to threats. For example, triggering an alert when certain conditions are present or absent to mitigate false negatives and positives—for example, calibrating a counter-defense system to have the lowest possible effect required to respond to an attack against the system until a human can evaluate the situation.[86] Although it may be technically possible to keep a "human in/on the loop" in the context of autonomous weapon systems such as air-defense systems, unmanned vehicles, and automated cyber, the intrinsic tactical value of AI augmentations lies in their ability to respond to threats in real-time. The propensity for defense planners to rely on the algorithmic prescriptions to guide decision-making will, therefore, be strong—also known as "automation bias" discussed below.[87] In short, because of the transparency and "explainability" (or "black box") issues described in the introduction, policymakers might focus on the reliability of AI (the process) to instill greater confidence—such as calibrating parameters and limits within applicable legal frameworks—rather than their predictability (the outcome).

Bias in AI-ML

Despite the speed, diverse data pools, and processing power of algorithms compared to humans, ML systems will still depend on the assumptions, institutional values, and preferences encoded into them by human engineers—that is, inadvertently baking the biases of states and designers into the algorithmic architecture of weapon systems. For instance, whereas one state trains its algorithms on historical examples that reflect their strategic preferences for force employment philosophy and world views, another's algorithm interprets military and customary law differently (i.e., on the continuum of permissibility), analyzing situations and events through a different political, cultural, cognitive, and military doctrinal lens. These profiling dynamics increase the risk of bias in how AIs interpret others' signals and actions—particularly if states incorporate (possibly erroneous) preconceptions and assumptions about others into the machine learning (ML) training process.

Implicit or explicit biases incorporated into military support systems may cause errors from feedback loops (i.e., action and counter-reaction cycles), especially in cluttered and complex battlefield environments. For example, during the Cold War the Soviets developed a computer program known as "VRYAN," designed to notify Soviet leaders of a pre-emptive US nuclear strike. However, the data used to feed the system was often biased, and thus propelled a feedback loop which heightened the Kremlin's fear that the US was pursuing first-strike superiority.[88] These feedback loops generated by AI systems could potentially trap human operators into machines' bias and flawed assumptions.[89]

As we have seen, ML systems function much like "black boxes," making them technically highly unpredictable and potentially unreliable. As machines become increasingly embedded into the decision-making kill-chain (i.e., the process of detecting, targeting, and striking a threat), the risk that algorithmic biases fed into recommendation and decision-making support systems triggers a cascading error effect, whereby a skewed or erroneous input fed into one system is used by another system to inform a decision further up in the interconnected and interdependent chain—that is, compounding the effect of a small error caused by bias.[90] The degree of trust and reliance on AI technology will, therefore, also be a key factor in determining the impact of the problem of bias in AI systems, which is considered below.

Vulnerability to cyber-attacks

AI has the potential to both reduce and increase a military's vulnerability to cyber-attacks. On the one hand, AI cyber-defense tools (or "counter-AI"), which are designed to recognize changes to patterns of behavior in a network and detect anomalies, can automatically recognize software code vulnerabilities, potentially forming a more robust defense against cyber subversions. The US DoD's Defense Innovation Unit is prototyping an application that leverages AI to decipher high-level strategic questions, map probabilistic chains of events, and develop alternative strategies to make DoD systems more resilient to AI-augmented cyber-attacks and allow them to fix errors faster than humans.

On the other hand, an adversary could, for example, use malware to take control, manipulate, or fool the behavior and pattern recognition systems of autonomous systems. In one case, analysts found it extremely challenging to detect malware that infected systems controlling the cockpits of American UAVs (Predator and Reaper drones) flying missions over Middle Eastern warzones at the Creech US Air Force Base in Nevada in 2011.[91] Offensive attacks such as this would be relatively easy to execute but difficult to detect, attribute, or effectively counter. It is now thought to be possible that a cyber-attack could infiltrate a nuclear weapon system, threaten the integrity of its communications, and ultimately (and possibly unbeknownst to its target) gain control of its nuclear as well as non-nuclear C2 systems. Advances in AI might, therefore, exacerbate this challenge by enabling improvements to the cyber offense, thereby conferring further advantages to first-movers in this domain.[92] For example, ML and AI could dramatically reduce the high levels of labor intensity and technical skill required to execute advanced persistent threat (APT) operations—or "hunting for weaknesses."[93] Future AI APT tools could offer a would-be attacker a cheap and easy way to replicate powerful cyber weapons, which would require minimal technical knowledge to employ.

While manipulations and system subversions are possible with existing cyber offense tools, AI and increasing degrees of military autonomy could amplify the potential speed, power, and scale of future attacks in cyberspace. In response to these anticipated vulnerabilities and to achieve the first-mover advantage, China, Russia, and the US have continued to harden their AI cyber defenses.[94] For example, open sources indicate that Chinese analysts view the vulnerability of China's Nuclear Command, Control, and Communication

(NC3) systems to cyber infiltrations—even if an attacker's objective was limited to cyber espionage—as highly escalatory.[95] China's fear that its early-warning missile systems are inadequate to respond to a disarming first strike by the US has prompted Beijing to prioritize mitigating false negatives, when early-warning systems fail to warn of an imminent attack—as opposed to a false positive that warns of an attack that does not exist.[96] AI-enhanced cyber capabilities simultaneously bolstering deterrence and incentivizing others to attack could intensify a paradox of enhanced capabilities and increased vulnerabilities in the cyber domain.

Non-human decision-making

As AI systems become further integrated into the strategic decision-making process, the potential vulnerabilities and risks associated with military autonomy will likely increase. According to the former director of the Defense Advanced Research Projects Agency, Arati Prabhakar, despite AI's potential tactical and operational impact, the technology remains "fundamentally limited" due to its propensity to misidentify objects and be easily spoofed.[97] Prabhakar adds that decision-making errors committed by AI systems are often inexplicable and unlike errors that humans would make. In one, now infamous, example, Prabhakar demonstrated how a picture of a baby holding a toothbrush was misidentified by an ML algorithm as a baby with a baseball bat.[98] Unpredictable errors and decisions made by AI systems deployed at scale and across multiple domains and combat theaters would be inherently destabilizing.[99]

The inability of AI to understand context (i.e., the rationale and consequences of actions) or empathize (i.e., determine intent) would likely become a liability during wartime, when a degree of flexibility down the chain of command is generally considered positive. For example, the near catastrophic ICBM test at the US Vandenberg Air Force Base during the 1962 Cuban Missile Crisis was attributed to officers following predefined protocols without questioning these guidelines in the context of new information.[100] Human errors typically occur at an individual level and seldom repeat in the same way; by contrast, AI systems may conceivably fail simultaneously and repeat this failure indefinitely. In his vision of an AI future, for example, philosopher Nick Bostrom argues that "general" AI applications could overrun the universe in their single-minded pursuit of the efficient manufacturing of paper clips.[101] That is, AI-enhanced systems could be making decisions in pursuit of predetermined human goals in fundamentally non-human ways, with uncertain and potentially unintended outcomes.

The notion (albeit speculative) of "superintelligent" AI systems (i.e., machines that possess intelligence which surpasses human cognitive abilities) pitted against humans in the context of nuclear weapons would raise similar questions to those faced by human commanders in modern warfare:[102] What are AI's intentions (real or perceived) and motives? How might intelligent machines be deterred, coerced, or manipulated? And how could they be coopted (or manipulated) to de-escalate a situation? Artificial general intelligence systems functioning at machine speed and armed with oracle-like predictive foresight could dominate the strategic decision-making process, outsourcing the role of human cognition and agency in the mechanisms designed to manage and control warfare.[103]

A broader point here, which is addressed further in Chapter 2, is that the higher the uncertainty generated by AI-powered weapons, the more using them during a crisis may bolster deterrence and thus encourage the other side to eschew assertiveness and aggression for caution and prudence, with stabilizing effects. The uncertainty caused by the introduction of AI into a situation might, therefore, incentivize states—especially those facing a superior adversary—to delegate decisions to machines for the perceived deterrence utility.[104] Because of the difficulty in demonstrating a posture like this *before* a crisis or conflict ensues, however, deterrence signaling with AI technology might equally destabilize a situation.[105]

If commanders decide to delegate greater authority to inherently inflexible AI systems, the dehumanization of future defense planning will undermine stability by significantly inhibiting induction. Human induction (i.e., the ability to form general rules from specific pieces of information) is a crucial aspect of defense planning, primarily to manage situations that require high levels of visual and moral judgment and reasoning.[106] Several analysts have warned that if human commanders place too much confidence in AI analysis without fully comprehending how machines reach a particular outcome, then machine-generated data could be trusted implicitly and without scrutiny by decision-makers.[107] For example, the Tesla Model 3 crash in 2018—where a driver in autopilot mode plowed into a fire truck on a freeway—clearly demonstrated the risk of placing too much trust in autonomous technology, though not necessarily AI.[108] In this scenario, like other incidents involving vehicles in autopilot mode, the skill-based reasoning automated systems relied on for "bottom-up" processing (or "skills-based reasoning") failed, and fatalities occurred because inattentive drivers did not realize that an automated driving assist feature still needs "top-down reasoning"—or human intuition, common sense, experience, and judgment.[109]

The tendency of humans to anthropomorphize machines and use automation as a heuristic replacement for vigilant information-seeking, cross-checking, and adequate processing supervision is known as "automation bias."[110] People's deference to machines can result from the presumption that: (a) decisions result from hard empirically based science; (b) AI algorithms function at speeds and complexities beyond human capacity; or (c) because people fear being overruled or outsmarted by machines.[111] This phenomenon in a military context could mean that planners become more predisposed to view the judgments of AIs as analogous (or even superior) to those of humans. As the eighteenth-century Scottish philosopher David Hume opined, "there is a universal tendency among mankind to conceive all beings like themselves. We find human faces in the moon, armies in the clouds."[112] While anthropomorphism might, on the one hand, help reduce over-trust caused by faith in the objectivity and infallibility of machines, it may on the other hand lead to other problems related to the perceived capabilities imbued in machines, such as intentions, restraint, and signaling costing signals (see Chapter 2).[113] Therefore, defense planners need to be trained to recognize, resist, and pre-empt unwarranted levels of automation bias and anthropomorphizing of AIs at each stage of the decision-making process.[114]

What level of "explainability" is required to make an AI "useable"? And must an AI produce more accurate and reliable predictions than humans *before* a state employs them? Absent human judgment, intuition, and oversight (or "meaningful human control"), an over-reliance on the decisions of machines, coupled with AIs that use overly simple value-maximizing decision processes in *a priori* situations of complexity and uncertainty—weapon systems may travel or behave erratically, enemy forces may dissemble or reposition, and battlefield obstacles can move rapidly and unpredictably—and when priorities may change or commands require recalibration, will likely create conditions ripe for strategic instability.[115]

Conditions for nuclear instability

Two existing conditions for nuclear instability, likely to be exacerbated by AI, are nuclear multipolarity and differing escalation risk tolerance between military powers. The coalescing of these conditions, in the presence of the inherently destabilizing military AI systems described earlier, will lead to an interplay of forces ripe for miscalculations and misperceptions between nuclear-armed states—increasing the dangers of escalation and deterrence failure under the nuclear shadow.

Nuclear multipolarity

The emergence of nuclear multipolarity in the "second nuclear age" has created multifaceted escalation pathways to a nuclear confrontation involving an expanding number of nuclear-armed poles, compared with bipolarity during the Cold War era.[116] This multipolarity is important precisely because each state will choose a different response to the new choices emerging in the digital age. Motivated states could eschew the limitations of AI, compromising safety and verification standards to protect or capture the upper hand (the first-mover advantage) on the future digitized battlefield (see Chapter 4).[117] The historical record attests that strategic competition—motivation to control warfare—tends to be ratcheted up because of the complexity of military technology and operations over time.[118] Thus, the pursuit of AI technology by great powers—especially China, the US, and Russia—will likely compound the destabilizing effects of AI in increasing great-power competition.

Against this inopportune geopolitical backdrop, perceived strategic benefits of AI-enhanced weapons generate risks to nuclear security and stability—the greatest of which today is the premature adoption of unsafe, unverified, and unreliable AI technology in the context of decisions to use nuclear weapons, which could be catastrophic. For example, the proliferation of low-risk and low-cost AI-augmented autonomous weapons such as drone swarms—with ambiguous rules of engagement—will become an increasingly enticing asymmetric option to undermine an adversary's military readiness, deterrence, and resolve.[119] For example, in the private sector, competitive first-mover advantage pressures have created inflated and unrealistic expectations that impel companies (e.g., driverless cars) to rapidly transition AI-infused applications to production before maturity; this has been a common trend for many Silicon Valley-backed innovations—known as the "fake-it-'til-you-make-it" culture.[120] History is replete with examples of promises of operational readiness ending in costly system failure and these cases should serve as a warning to those in the defense and nuclear communities who overlook the limitations of AI technology described in the introduction, for ephemeral (and possibly illusory) tactical and operational gains.[121]

Competing states making decisions under the nuclear shadow will be more inclined to assume the worst of others' intentions, especially in situations where the legitimacy of the status quo is contested (e.g., maritime Asia).

According to scholar John Mearsheimer, "as long as the system remains anarchic, states will be tempted to use force to alter an unacceptable status quo."[122] Thus, efforts by one state to enhance the survivability of its strategic forces with state-of-the-art dual-use technology like AI could easily be perceived by the other side as a potential threat to its ability to survive, and respond to, a nuclear first strike—or second-strike capability.[123]

In a world of revisionist and dissatisfied nuclear-armed states, it seems improbable that improvements in intelligence collection and analysis derived from advances in AI would have a stabilizing impact.[124] For this to happen, equal access to intelligence and shared confidence in the accuracy and credibility of these systems would be required. Furthermore, the intentions of all parties would need to be benign for any reassurances or confidence-building efforts to succeed. Because nuclear interactions increasingly involve the complex interplay of nuclear and non-nuclear (and state and non-state) actors, the leveraging of AI in this multipolar context will increasingly place destabilizing pressures on nuclear states. Taken together, these interactions will likely complicate escalation management efforts during future crises or conflict—especially involving China and the US.

Differing escalation risk tolerance in the US-China dyad

While US defense analysts and their Chinese counterparts are aware of the potential escalation risks between nuclear-armed great powers, their respective doctrines do not address how an adversary might respond to escalatory behavior. Instead, these rival strategic communities generally assume that escalation in future conflict can be effectively countered and contained by establishing and sustaining escalation dominance. Chinese doctrinal emphasis on seizing the initiative early and pre-emptively in conventional warfare to achieve escalation dominance may result in the opposite outcome: triggering rapid and possibly uncontrollable escalation to a nuclear level of conflict.[125] However, the evidence does not suggest that China intends to use nuclear missiles to achieve escalation dominance, or for deliberate escalation.

A central challenge in defining strategic stability between the US and China is that both parties lack a common definition of "strategic stability"—unlike Russia the US does not maintain an official policy of mutual vulnerability with China—and have mutually exclusive concerns about this strategic dyad. Beijing's overarching concern relates to perceptions that the US seeks superior conventional (especially counterforce and missile defenses) and nuclear capabilities that could be used to conduct first strikes against China's comparatively smaller and less diversified nuclear deterrent, thus undermining Chinese retaliatory capability.[126] Washington's principal concern is that Beijing will not eschew the use of its expansive conventional forces and power projection capacity to enforce its claims in regional territorial disputes with its neighbors (i.e., the East and South China Seas), several of whom are US treaty allies (South Korea and Japan), or strategic partners (Taiwan, Singapore, and the Philippines)—that is, China's nuclear and conventional capabilities could be used to neutralize US operational conventional capacity to defend its allies and unilaterally alter the status quo in a future regional contingency.[127] Further, these divergent threat perceptions and views on strategic stability are complicated by the significant disparities between the two states' nuclear arsenals—the US and Russia still possess more than 90 percent of the world's total stockpile of nuclear weapons.[128]

The broadening of US-China great-power strategic competition (i.e., trade relations, regional disputes, cyber espionage, and the coronavirus pandemic and general political confrontation) evinces classical "security dilemma" dynamics (see Chapter 4)—that is, both sides perceive the other's behavior as threatening and respond to secure themselves in ways that can lead to action-reaction dynamics, spirals of mistrust and arms racing, and other destabilizing escalatory behavior.[129] Consequently, both sides lack confidence in the other's willingness to stabilize strategic relations; nor are they incentivized to pursue measures (arms control, build mutual trust, transparency, or display unilateral restraint) to enhance strategic stability.[130] Absent commonly held escalation thresholds and a mutual framework to deter either side from violating them, a US-China crisis operating under the assumption that they can effectively control escalation (e.g., in the South China Seas, the Korean Peninsula, or Taiwan Straits) would likely increase the risks of inadvertent escalation—because both sides underestimate their ability to prevent a situation spiraling out of control.[131]

According to Avery Goldstein, Professor of Political Science at the University of Pennsylvania—who uses deterrence theory to analyze factors leading to first-strike instability in the US-China strategic dyad—Chinese overconfidence in its ability to prevent conventional military confrontation escalating to nuclear war might increase the risk of a conflict or crisis inadvertently or accidentally crossing the nuclear Rubicon.[132] Moreover, divergent views of US-China attitudes about controlling escalation below and above the nuclear threshold could also be detrimental to crisis stability. The US defense community is generally more concerned that a low-level conventional conflict might escalate to a nuclear level but is, apparently, less concerned about the ability of the US to control escalation above the nuclear threshold.

Paradoxically, during a US-China crisis Washington may overstate the possibility that Beijing would use nuclear weapons, and simultaneously understate the scale of a Chinese retaliatory nuclear response.[133]

Furthermore, in a future US-China conflict the US would have a strong incentive to pre-emptively attack China's mobile missiles and attendant (and likely dual-use) command, control, communications, and intelligence systems to achieve escalation dominance, which Beijing could misperceive as a conventional counterforce attack or, worse, as a precursor to a first nuclear strike.[134] Chinese analysts generally assume that the US intends to undermine China's relatively small nuclear deterrent and attendant support systems with advanced conventional weapons—especially US conventional prompt global strike and missile defenses.[135] In short, divergent US-China attitudes about the ability to de-escalate a low-intensity conventional or nuclear conflict makes a conventional conflict more likely to escalate to a high level of intensity.[136]

Managing military (especially inadvertent) escalation risk has not been a traditional feature of Chinese strategic thinking.[137] China's strategic community is believed to share a high level of confidence in the ability of China's long-standing no-first-use nuclear pledge to control escalation.[138] Because Chinese analysts view China's no-first-use commitment as a *de facto* firebreak between the use of its conventional and nuclear capabilities to de-escalate a situation, the resultant overconfidence might increase inadvertent escalation risks.[139] This overconfidence could make it less likely that Chinese leaders recognize escalation risks caused by miscalculation or misperception of US intentions. China's sanguine attitude to managing escalation can in part be attributed to the belief that once the nuclear Rubicon is crossed, the use of nuclear weapons cannot be easily controlled by either side. Because of this belief, Chinese analysts do not believe that a limited nuclear war would stay limited. Moreover, China's operational doctrine does not contain plans to wage a limited nuclear war, which China might pursue if it believed nuclear escalation could be controlled.[140]

Another complicating factor is that new escalation thresholds and operating norms for AI-augmented weapons have yet to emerge. Today's thresholds in the context of autonomous weapon systems are considered inappropriate and ambiguous.[141] Without commonly held operational norms and an adversary's strategic priorities and political objectives, militaries deploying military AI could inadvertently cross already vague escalation thresholds.[142] In 2016, for example, China captured a US underwater drone, asserting that it posed a hazard to Chinese maritime navigation. Washington, in response, called China's behavior "unlawful," claiming the drone was a "sovereign immune vessel."[143] This episode ended with China returning the drone after days of diplomatic contention, and demonstrated the potential risk of inadvertent escalation caused by the ambiguities surrounding the deployment of new (and especially dual-use) technology in contested territory between strategic rivals that is considered in Chapter 4. In sum, the combination of first-strike vulnerability and opportunity enabled by a growing portfolio of technologies such as AI for warfighting, coercion, and influence will have significant implications for escalation dynamics in future warfare.

Conclusion

This chapter conceptualized AI and modern technological change in the context of nuclear weapons and strategic stability. It argued that strategic stability is ultimately a product of a complex interplay of myriad forces, and technology is best viewed as a variable that can have both stabilizing *and* destabilizing effects. Technology has long been used to augment, automate, and enhance human behavior and decision-making in a military context; thus far, the "human factor" has trumped technology in its impact on strategic stability. The chapter also found that the role of technological change and strategic stability can best be understood as part of a complex interaction of disruptive forces (or agents of change), which, during periods of heightened geopolitical rivalry, great-power transitions, and strategic surprise, can erode strategic stability and make conflict more likely.

In addition, the research demonstrated that the instability that emerged during the Cold War has important implications for the contemporary AI-strategic nexus. First, the destabilizing impact of weapons proliferation becomes more acute in situations where technology has the potential to create asymmetric military advantages between rivals. Second, where the risk exists (or is perceived to exist) that a conventional conflict might cross the nuclear threshold, the implications for crisis instability and military escalation can be severe. Third, any meaningful definition of strategic stability must include advanced strategic non-nuclear weapons that adversaries would likely use to destroy an enemy's nuclear capabilities. Finally, nuclear modernization programs which are primarily motivated by technologically advanced capabilities can trigger arms racing dynamics that worsen crisis stability.

The chapter revisited the existing literature on the intersection of emerging technology and nuclear stability to conceptualize the AI-strategic nexus. The research demonstrated that military AI can be conceptualized as a natural manifestation of an established trend in emerging technologies (i.e., co-mingling of nuclear and conventional capabilities, conventional weapons increasingly can threaten nuclear arsenals, compression decision-making, and the speed of war), which might persuade states to adopt destabilizing nuclear postures and more coercive nuclear strategies.[144] These shifts could, in turn, trigger "security dilemma" dynamics and competitive arms racing—both in the development of damage limitation conventional counterforce capabilities and more survivable nuclear forces and NC3 systems.[145] In short, while the potential strategic effects of AI are not unique to this technology, the confluence of these trends suggests that its trajectory will likely be fundamentally destabilizing.

One overarching argument is that military AI is likely to exacerbate the destabilizing and escalatory effects of an increasingly complex interplay of advanced military technology in a multipolar nuclear world order. Nuclear-armed states leveraging AI to achieve or sustain first-mover advantages in this multipolar context will likely destabilize this fragile order with uncertain outcomes. Given the multifaceted interplay AI-augmented enabling capabilities might have with strategic weapons (both nuclear and conventional weapons with strategic effects), it will be particularly important for analysts, academics, and decision-makers alike to (a) develop a robust understanding of the confluence between these diverse capabilities, and (b) understand how competing strategic communities view these dynamics, as well as the implications of these trends for nuclear and conventional strategy and posture, arms races, arms control, escalation management, and cross-domain as well as extended deterrence.

To pre-empt and mitigate, or at least manage, the destabilizing and escalatory risks posed by military AI, great military powers (especially the US, Russia, and China) must closely coordinate their confidence-building measures to pre-empt some of the risks to stability described in this chapter. Specifically, great powers should establish an international framework for governance, norms, regulation, and transparency in the development and deployment of AI-augmented military capabilities. These frameworks must encompass not only the present but also potential future developments, particularly what is and is not being baked into AI algorithms and the best method to temper public debate from becoming too fixated on killer robots and machine overlords. Besides, if an adversary deploys an AI-powered system, the opacity inherent in its underlying algorithms—and even if the other side was able to gain access to these networks—might inhibit the target's ability to determine whether an attack was intentional or accidental, thus frustrating accountability efforts.[146] In short, AI technology provides new opportunities for obfuscation and misperception across the escalation management spectrum. Chapter 3 considers the how AI technology could exacerbate unintentional escalation risk.

Ultimately, success in these efforts will require all stakeholders to be convinced of the need and the potential mutual benefits of taking steps toward the establishment of a coherent governance architecture to institutionalize, internalize new norms, and ensure compliance with the design and deployment of AI and autonomy in the military sphere: as John Maynard Keynes opined, the ability to make things happen that otherwise would not needs a combination of technological know-how, policy, and bureaucratic skills matched by political capital and investment—the concluding chapter will come back to these issues.[147] The next chapter considers how AI technology might affect nuclear deterrence. How might the introduction of non-human agents into a crisis or conflict between nuclear powers impact deterrence dynamics?

Notes

1. Sections of this chapter are derived in part from James Johnson, "Artificial Intelligence in Nuclear Warfare: A Perfect Storm of Instability?" *The Washington Quarterly* 43, no. 2 (2020): pp. 197–211. Copyright Taylor & Francis, doi: 10.1080/0163660X.2020.1770968.
2. See, for example, George Zarkadakis, *In Our Image* (New York: Pegasus Books, 2015); Christianna Ready, "Kurzweil Claims That the Singularity Will Happen by 2045," *Futurism*, October 5, 2017, https://futurism.com/kurzweil-claims-that-the-singularity-will-happen-by-2045.
3. James Johnson, *Artificial Intelligence & the Future of Warfare: USA, China, and Strategic Stability* (Manchester: Manchester University Press, 2021), Ch. 1.
4. Izumi Nakamitsu, "Keynote Speech at the Virtual UK Project on Nuclear Issues 2020 Annual Conference Royal United Services Institute for Defence and Security Studies," *Royal United Services Institute*, June 10, 2020, https://front. unarm.org/wp-content/uploads/2020/06/10-June-High-Representative- Keynote-at-RUSI-UK-PONI-Annual-Conference-2020.pdf.
5. See Michael Horowitz, Paul Scharre, and Alex Velez-Green, *A Stable Nuclear Future? The Impact of Automation, Autonomy, and Artificial Intelligence* (Philadelphia, PA: University of Pennsylvania, 2017); Mary L. Cummings, *Artificial Intelligence and the Future of Warfare* (London: Chatham House, 2017); Lawrence Freedman, *The Future of War* (London: Penguin Random House, 2017); Greg Allen and Taniel Chan, *Artificial Intelligence and National Security* (Cambridge, MA: Belfer Center for Science and International Affairs, 2017); Kareem Ayoub and Kenneth Payne, "Strategy in the Age of Artificial Intelligence," *Journal of Strategic Studies* 39, no. 5–6 (2016): pp. 793–819, doi: 10.1080/01402390.2015.1088838; Edward Geist and Andrew J. Lohn, *How Might Artificial Intelligence Affect the Risk of Nuclear War?* (Santa Monica, CA: RAND Corporation, 2018), doi: 10.7249/PE296; Frank G. Hoffman, "Will War's Nature Change

in the Seventh Military Revolution?" *Parameters* 47, no. 4 (Winter 2017/2018), https://publications.armywarcollege.edu/pubs/3529.pdf.

6. There is no single, universally accepted definition of "strategic stability," the factors that contribute to stability (or instability), or agreed-upon metrics to measure it. For example, see Henry A. Kissinger, "Arms Control, Inspection and Surprise Attack," *Foreign Affairs* 38, no. 4 (July 1960): pp. 557–575; Glenn H. Snyder, *Deterrence and Defense: Toward a Theory of National Security* (Princeton, NJ: Princeton University Press, 1961), pp. 107–108; and Herman Kahn, *On Escalation: Metaphors and Scenarios* (New York: Praeger Publishers, 1965).
7. For an influential study on the promotion and analysis of nuclear arms control and strategic stability, see Thomas C. Schelling and Morton Halperin, *Strategy and Arms Control* (New York: Twentieth Century Fund, 1961).
8. Robert Jervis, *The Meaning of the Nuclear Revolution: Statecraft and the Prospect of Armageddon* (Ithaca, NY: Cornell University Press, 1989).
9. See Eric Schlosser, *Command and Control: Nuclear Weapons, the Damascus Accident, and the Illusion of Safety* (New York: Penguin, 2013); and Scott D. Sagan, *The Limits of Safety: Organizations, Accidents, and Nuclear Weapons* (Princeton, NJ: Princeton University Press, 1993). On nuclear terrorism, see Graham Allison, *Nuclear Terrorism: The Ultimate Preventable Catastrophe* (New York: Owl, 2004); John Mueller, *Atomic Obsession: Nuclear Alarmism from Hiroshima to Al-Qaeda* (New York: Oxford University Press, 2010); and Keir A. Lieber and Daryl G. Press, "Why States Won't Give Nuclear Weapons to Terrorists," *International Security* 38, no. 1 (Summer 2013): pp. 80–104.
10. Michael Gerson, "The Origins of Strategic Stability," in *Strategic Stability: Contending Interpretations*, edited by Colby Elbridge and Michael Gerson (Carlisle, PA: Army War College, 2013), pp. 1–46.
11. Johnson, *Artificial Intelligence & the Future of Warfare*, Ch. 2. On the "third nuclear age," see Andrew Futter and Benjamin Zala, "Strategic Non-Nuclear Weapons and the Onset of a Third Nuclear Age," *European Journal of International Security* 6, no. 3 (2021): pp. 1–21.
12. For literature on the "second nuclear age," see Paul Bracken, *The Second Nuclear Age: Strategy, Danger, and the New Power Politics* (New York: Times Books, 2012); Colin S. Gray, *The Second Nuclear Age* (Boulder, CO: Lynne Rienner, 1999); and Keith Payne, *Deterrence in the Second Nuclear Age* (Washington, DC: Georgetown University Press, 1996).
13. See Lawrence Freedman, *The Evolution of Nuclear Strategy* (Basingstoke: Palgrave Macmillan 2003); Michael Quinlan, *Thinking about Nuclear Weapons: Principles, Problems, Prospects* (Oxford: Oxford, University Press, 2009).
14. The 2010 Nuclear Posture Review states that the goal of US nuclear strategy is to "strengthen deterrence of regional adversaries," such as North Korea, while "reinforcing strategic stability" with Russia and China.
15. Thomas C. Schelling, *Arms and Influence* (New Haven, CT: Yale University Press, 1966), p. 234.
16. Gerson, "The Origins of Strategic Stability," p. 26.
17. Lehman F. Ronald, "Future Technology and Strategic Stability," in *Strategic Stability: Contending Interpretations*, edited by Elbridge and Gerson, pp. 147–199; and Lawrence Freedman, "Has Strategy Reached a Dead-End?" *Futures* 11, no. 2 (1979): pp. 122–131.

18. For example, Michael Howard identified three additional dimensions of strategy, while Colin Gray proposed seventeen. See Michael Howard, "The Forgotten Dimensions of Strategy," *Foreign Affairs* (1979), https://www.foreignaffairs.com/articles/1979-06-01/forgotten-dimensions-strategy; and Colin S. Gray, *Theory of Strategy* (Oxford: Oxford University Press, 2018).
19. Charles Duelfer and Stephen Dyson, "Chronic Misperception and International Conflict: The U.S.-Iraq Experience," *International Security* 36, no. 1 (Summer 2011): pp. 75–78.
20. Stephen J. Cimbala, *The Dead Volcano: The Background and Effects of Nuclear War Complacency* (Westport, CT: Praegar, 2002), p. 66.
21. Robert Jervis, *How Statesmen Think: The Psychology of International Politics* (New Jersey, Princeton, NJ: Princeton University Press, 2017), Ch. 4.
22. Thomas C. Schelling, *The Strategy of Conflict* (Cambridge, MA: Harvard University Press, 1960).
23. Crisis stability does not imply that crises are impossible or will necessarily be infrequent; rather it means that when crises do arise, the mechanisms do not make them worse. Aaron R. Miles, "The Dynamics of Strategic Stability and Instability," *Comparative Strategy* 35, no. 5 (2016): pp. 423–437, 425.
24. Glenn A. Kent and David E. Thaler, "First-Strike Stability: A Methodology for Evaluating Strategic Forces," *RAND Project Air Force*, August 1989, https://www.rand.org/pubs/reports/R3765.html.
25. Like "strategic stability," the concept of "arms race" is also contested. Barry Buzan and Eric Herring, *The Arms Dynamic in World Politics* (London, Boulder, CO: Lynne Reinner, 1998), p. 77.
26. Schelling and Halperin, *Strategy and Arms Control*.
27. Jervis, *How Statesmen Think*, p. 95.
28. Miles, "The Dynamics of Strategic Stability and Instability."
29. These psychological factors raise questions about the assumption, during a crisis, that a commander can depend on a common sense of rationality with the other side. Bradley A. Thayer, "Thinking about Nuclear Deterrence Theory: Why Evolutionary Psychology Undermines its Rational Actor Assumptions," *Comparative Strategy* 26, no. 4 (2007): pp. 311–323.
30. Bargaining situations in the context of nuclear weapons involve frequently competing for incentives, including: (1) deterrence: dissuading an adversary from doing something they want to do; (2) compellence: forcing an adversary to do something they do not wish to do; (3) assurance: convincing allies that security guarantees are credible, and (4) reassurance: convincing an adversary they will not be attacked so long as they refrain from provocative behavior. See Robert Jervis, *Perception and Misperception in International Politics* (Princeton, NJ: Princeton University Press, 1976), Ch. 3; and Andrew H. Kydd, and Roseanne W. McManus, "Threats and Assurances in Crisis Bargaining," *Journal of Conflict Resolution* 61, no. 2 (2017): pp. 68–90.
31. Jervis *How Statesmen Think*, p. 219.
32. Joseph Johnson, "MAD in an AI Future?" Center for Global Security Research, Lawrence Livermore National Laboratory, June 3, 2019.
33. For example, see Michael J. Boyle, "The Race for Drones," Orbis, November 24, 2014, pp. 76–94.

34. Yuna Huh Wong et al., *Deterrence in the Age of Thinking Machines* (Santa Monica, CA: RAND Corporation, 2020), doi: 10.7249/RR2797, p. 82.
35. Jervis Robert, "Cooperation under the Security Dilemma," *World Politics* 30 (1978): pp. 167–214.
36. There is a distinction between the credibility of *capability* (in this case) and the credibility of *resolve* discussed below.
37. For a discussion of the impact of the arms race and effect on arms racing dynamics and military drones, see Boyle, "The Race for Drones."
38. Benjamin Zala, "How the Next Nuclear Arms Race Will be Different from the Last One," *Bulletin of the Atomic Scientists* 75, no. 1 (2019): pp. 36–43; and Colin S. Gray and Dale C. Walton, "The Geopolitics of Strategic Stability," in *Strategic Stability: Contending Interpretations*, edited by Elbridge and Gerson, p. 93.
39. Ibid., p. 110.
40. Winner Langdon, *Autonomous Technology: Technics-Out-of-Control as a Theme in Political Thought* (Cambridge, MA: MIT Press, 1977).
41. James M. Acton, "Reclaiming Strategic Stability," in *Strategic Stability: Contending Interpretations*, edited by Elbridge and Gerson, pp. 138–139.
42. Jervis, *The Meaning of the Nuclear Revolution*, p. 45. On the theory of the nuclear revolution, see Kenneth N. Waltz, "Nuclear Myths and Political Realities," *American Political Science Review* 84, no. 3 (September 1990): pp. 731–745.
43. Robert Jervis, *The Illogic of American Nuclear Strategy* (Ithaca, NY: Cornell University Press, 1984); Waltz, "Nuclear Myths and Political Realities"; Charles L. Glaser, *Analyzing Strategic Nuclear Policy* (Princeton, NJ: Princeton University Press, 1990); and Stephen Van Evera, *Causes of War: Power and the Roots of Conflict* (Ithaca, NY: Cornell University Press, 1999).
44. Lieber A. Keir and Daryl G. Press, "The New Era of Counterforce: Technological Change and the Future of Nuclear Deterrence," *International Security* 41, no. 4 (2017): pp. 9–49.
45. Ronald, "Future Technology and Strategic Stability," pp. 147–199.
46. In this view, a combination of pre-emptive counterforce technology and missile defense capabilities that blunted an adversary's nuclear deterrent (or second-strike capacity) would be highly destabilizing. Keir and Press, "The New Era of Counterforce," pp. 9–13.
47. Thomas Schelling argued that risk manipulation and *uncertainty* play critical roles in deterrence. Schelling, *Arms and Influence*, pp. 92–125.
48. Bargaining situations are those where actors gain from cooperating, as opposed to zero-sum scenarios where various possible outcomes favor one side more than the other. See Fearon D. James, "Rationalist Explanations for War," *International Organization* 49, no. 3 (1995): pp. 379–414; and Erik Gartzke, "War Is In the Error Term," *International Organization* 53, no. 3 (1999): pp. 367–587.
49. Four trends in military technology and doctrine will likely exacerbate the future threats posed by "entanglement": (1) technological advances in conventional precision missiles and cyber-weapons; (2) increasing vulnerability of nuclear and conventional command, control, and intelligence systems (C3I) to cyber-attacks; (3) growing military reliance on dual-use C3I capabilities; and (4) the recent emergence of conventional warfighting military doctrines (China, Russia, and the US), which explicitly envisage (early and pre-emptive) attacks on an adversary's C3I dual-use assets. James M. Acton, "Escalation

through Entanglement: How the Vulnerability of Command-and-Control Systems Raises the Risks of an Inadvertent Nuclear War," *International Security* 43, no. 1 (Summer 2018): pp. 56–99.

50. Schelling and Halperin, *Strategy and Arms Control*, p. 37.
51. Notable exceptions include: Vincent Boulanin, ed., *The Impact of Artificial Intelligence on Strategic Stability and Nuclear Risk Vol. I Euro-Atlantic Perspectives* (Stockholm: SIPRI Publications, May 2019); Geist and Lohn, *How Might Artificial Intelligence Affect the Risk of Nuclear War?*; and James Johnson, "Artificial Intelligence: A Threat to Strategic Stability," *Strategic Studies Quarterly* 14, no. 1 (2020): pp. 16–39.
52. Austin Long and Brendan Rittenhouse Green, "Stalking the Secure Second Strike: Intelligence, Counterforce, and Nuclear Strategy," *Journal of Strategic Studies* 38, no. 1–2 (2015): pp. 38–73; Owen R. Coté Jr., *The Third Battle: Innovation in the US Navy's Silent Cold War Struggle with Soviet Submarines* (Newport, RI: Naval War College, 2003); Peter Sasgen, *Stalking the Red Bear: The True Story of a US Cold War Submarine's Covert Operations against the Soviet Union* (New York: St. Martin's, 2009).
53. Altmann Juürgen and Frank Sauer, "Autonomous Weapon Systems and Strategic Stability," *Survival* 59, no. 5 (2017): pp. 117–142.
54. For definitions of the concepts of escalation threshold, escalation ladder, and escalation path, see Kahn, *On Escalation: Metaphors and Scenarios*, p. 37.
55. US Congress, Office of Technology Assessment, Ballistic Missile Defense Technologies, OTA-ISC-254 Washington DC: US Government Printing Office, September 1985, pp. 119 and 128.
56. James M. Acton, *Silver Bullet? Asking the Right Questions About Conventional Prompt Global Strike* (Washington, DC: Carnegie Endowment for International Peace, 2013).
57. Examples of "strategic capabilities" include: long-range nuclear and conventional munitions (e.g., ICBMs), long-range penetrating bombers, shorter-range tactical (or theater) weapons that are or can be forward deployed. Offensive cyber and counter-space (i.e., ASATs) have also emerged as strategic capabilities. Finally, defense systems (e.g., ballistic missile defense systems) can also be viewed as strategic in as much they are intended (or able) to impair a state's ability to respond at a strategic level.
58. The world's official nuclear states—the US, Russia, China, India, Britain, and France—all eschewed recent calls by Washington for a "world without nuclear weapons" and instead have embarked on nuclear modernization programs. To date, the US has lagged in this effort, and the detail of its modernization program remains inconclusive. Zala, "How the Next Nuclear Arms Race Will be Different from the Last One."
59. Ronald, "Future Technology and Strategic Stability."
60. Stephen D. Biddle, "The PAST as Prologue: Assessing Theories of Future Warfare," *Security Studies* 8, no. 1 (1998): pp. 1–74.
61. See Steven Metz, *Armed Conflict in the 21st Century: The Information Revolution and Post-Modern Warfare* (Carlisle: Strategic Studies Institute, 2000); Steven Metz and James Kievit, *Strategy and the Revolution in Military Affairs: From Theory to Policy* (Carlisle: Strategic Studies Institute, 1995); and Stephen Biddle, *Military Power: Explaining Victory & Defeat in Modern Battle* (Princeton, NJ: Princeton University Press, 2004).
62. Bernard Brodie and Fawn M. Brodie, *From Crossbow to H-Bomb* (Bloomington: Indiana University Press, 1973), Ch. 23.

63. Stephen P. Rosen, "The Impact of the Office of Net Assessment on the American Military in the Matter of the Revolution in Military Affairs," *Journal of Strategic Studies* 33, no. 4 (2010): pp. 469–482.
64. Andrew Liptak, "The US Air Force Has a New Weapon Called THOR That Can Take Out Swarms of Drones," *the verge*, June 21, 2019, https://www.theverge.com/2019/6/21/18701267/us-air-force-thor-new-weapon-drone-swarms.
65. Since the 1300s, only five RMAs have occurred, and technological changes primarily drove only one to the tools of war (i.e., nuclear weapons). See Knox Macgregor, and Williamson Murray, eds, *The Dynamics of Military Revolution, 1300–2050* (Cambridge: Cambridge University Press, 2001).
66. Fu Wanjuan, Yang Wenzhe, and Xu Chunlei, "Intelligent Warfare, Where Does It Not Change?" *PLA Daily*, January 14, 2020, http://www.81.cn/jfjbmap/content/2020-01/14/content_252163.htm.
67. Bernard Brodie, "Technological Change, Strategic Doctrine, and Political Outcomes," in *Historical Dimensions of National Security Problems*, edited by Klaus Knorr (Lawrence, KS: University Press of Kansas, 1976), p. 263.
68. Lieber A. Keir and Daryl G. Press, "The New Era of Counterforce: Technological Change and the Future of Nuclear Deterrence," *International Security* 41, no. 4 (2017): pp. 9–49.
69. H. M. Kristensen, M. McKinzie, and T. A. Postol, "How US Nuclear Force Modernization Is Undermining Strategic Stability: The Burst-Height Compensating Super-Fuze," *Bulletin of the Atomic Scientists* (March 1, 2017).
70. Where primarily commercial forces drive military innovation, technology tends to mature and diffuse at a faster pace compared to technologies with only a military utility, such as stealth technology. Michael C. Horowitz, *The Diffusion of Military Power: Causes and Consequences for International Politics* (Princeton, NJ: Princeton University Press, 2010).
71. Forrest E. Morgan et al., *Dangerous Thresholds: Managing Escalation in the 21st Century* (Santa Monica, CA: RAND Corporation, 2008); Barry R. Posen, *Inadvertent Escalation: Conventional War and Nuclear Risks* (Ithaca, NY: Cornell University Press, 1991); and Lieber and Press, "The New Era of Counterforce."
72. Johnson, *Artificial Intelligence & the Future of Warfare*, p. 28.
73. The inability of humans to comprehend the so-called "black box" outputs of AI algorithms will likely be exacerbated as the militaries continue to use ML systems that can generate new algorithms independent of human input—or "unsupervised" ML. See Memoona Khanum and Tahira Mahboob, "A Survey on Unsupervised Machine Learning Algorithms for Automation, Classification and Maintenance," *International Journal of Computer Applications* (0975–8887) 119, no. 13 (2015): pp. 34–39.
74. To meet the legal requirements of self-defense, an algorithm would need to be programed to respond in a manner consistent with the laws of armed conflict, including distinction and proportionality. Specifically, an AI software engineer would need to design an algorithm that effectively incorporates the lawful scope of a forcible defensive response needed to "convert contested and malleable concepts such as necessity and proportionality into discrete, binary decision trees." Ashley Deeks, Noam Lubell, and Daragh Murray, "Machine Learning, Artificial Intelligence, and the Use of Force by States," *Journal of National Security Law & Policy*, Virginia Public Law and Legal Theory Research Paper no. 10, 2019, p. 14.

75. See Mark Buchanan, "Physics in Finance: Trading at the Speed of Light," *Nature*, February 11, 2015, https://www.nature.com/news/physics-in-finance-trading-at-the-speed-of-light-1.16872.
76. See Micah Musser and Ashton Garriott, "Machine Learning and Cybersecurity: Hype and Reality," *Center for Security & Emerging Technology*, June 2021, https://cset.georgetown.edu/publication/machine-learning-and-cybersecurity/.
77. AlphaStar Team, "Alphastar: Mastering the Real-Time Strategy Game Starcraft II," *DeepMind Blog*, January 24, 2019, https://deepmind.com/blog/article/alphastar-mastering-real-time-strategy-game-starcraft-ii.
78. Patrick Tucker, "It's Time to Wargame Against an AI-Enabled China," *Defense One*, June 23, 2021, https://www.defenseone.com/technology/2021/06/its-time-wargame-against-ai-enabled-china/174922/.
79. De-escalation and war termination present separate but linked challenges; an adversary may choose to de-escalate or not escalate further but not abandon war aims or ongoing operations. That is, a separate choice is required to terminate a conflict, which will likely depend on whether political conditions acceptable to the conceding state are available.
80. Robert Jervis, "Some Thoughts on Deterrence in the Cyber Era," *Journal of Information Warfare* 15, no. 2 (2016): pp. 63–73.
81. In the medical domain, for example, a new study on the use of ML to identify cases of sepsis in patients found that the system missed two-thirds of positives cases, rarely found cases that medical staff did not notice, and frequently issued false alarms. Tom Simonite, "An Algorithm That Predicts Deadly Infections is Often Flawed," *Wired*, June 21, 2021, https://www.wired.com/story/algorithm-predicts-deadly-infections-often-flawed/?.
82. Schelling, *Arms and Influence*, p. 234.
83. Bob Pisani, "What Caused the Flash Crash? CFTC, DOJ Weigh In," *CNBC*, April 21, 2015, https://www.cnbc.com/2015/04/21/what-caused-the-flash-crash-cftc-doj-weigh-in.html.
84. John Wendle, "The Fighting Drones of Ukraine: In Garages and Warehouses around Kiev, an Army of Gadgeteers Takes on the Russian War Machine," *Air & Space Magazine*, February 2018, https://www.airspacemag.com/flight-today/ukraines-drones–180967708/.
85. Forrest E. Morgan et al., *Dangerous Thresholds: Managing Escalation in the 21st Century* (Santa Monica, CA: RAND Corporation, 2008), https://www.rand.org/pubs/monographs/MG614.html; Barry R. Posen, *Inadvertent Escalation: Conventional War and Nuclear Risks* (Ithaca, NY: Cornell University Press, 1991); James Johnson, "The AI-Cyber Nexus: Implications for Military Escalation, Deterrence and Strategic Stability," *Journal of Cyber Policy* 4, no. 3 (2019): pp. 442–460, doi: 10.1080/23738871.2019.1701693.
86. A wide range of specific tasks could be fully or partially automated with the use of AI-ML, including: vulnerability discovery, deception, and attack disruption. Musser and Garriott, "Machine Learning and Cybersecurity: Hype and Reality," p. iv.
87. Cortney Weinbaum, *The Ethics of Artificial Intelligence in Intelligence Agencies, The National Interest*, July 18, 2016, https://nationalinterest.org/blog/the-buzz/the-ethics-artificial-intelligence-intelligence-agencies-17021.
88. President's Foreign Intelligence Advisory Board, "The Soviet 'War Scare,'" February 15, 1990, vi, 24 et seq, https://nsarchive2.gwu.edu/nukevault/ebb533-The-Able-Archer-War-Scare-Declassified-PFIAB-Report-Released/.

89. Mark Fitzpatrick, "Artificial Intelligence and Nuclear Command and Control," *Survival* 61, no. 3 (June/July 2019): pp. 81–92, doi: 10.1080/00396338.2019.1614782.
90. Deeks, Lubell, and Murray, "Machine Learning, Artificial Intelligence, and the Use of Force by States," pp. 17–18.
91. Noah Shachtman, "Exclusive: Computer Virus Hits U.S. Drone Fleet," *Wired*, October 7, 2011, https://www.wired.com/2011/10/virus-hits-drone-fleet/.
92. Johnson, "The AI-Cyber Nexus," pp. 447–448.
93. Allen and Chan, *Artificial Intelligence and National Security*, p. 19.
94. Joshua Berlinger and Juliet Perry, "China Tried to Hack Group Linked to Controversial Missile Defense System, US Cybersecurity Firm Says," *CNN*, April 27, 2017, https://www.cnn.com/2017/04/27/asia/china-south-korea-thaad-hack/index.html.
95. James M. Acton et al., eds, *Entanglement: Russian and Chinese Perspectives on Non-Nuclear Weapons and Nuclear Risks* (Washington, DC: Carnegie Endowment for International Peace), p. 81.
96. Lora Saalman, "Fear of False Negatives: AI and China's Nuclear Posture," *Bulletin of the Atomic Scientists*, April 24, 2018, https://thebulletin.org/2018/04/fear-of-false-negatives-ai-and-chinas-nuclear-posture/.
97. Mark Pomerlau, "DARPA Director Clear-Eyed and Cautious on AI," *Government Computer News*, May 10, 2016, https://gcn.com/articles/2016/05/10/darpa-ai.aspx.
98. Andrew Ilachinski, *AI, Robots, and Swarms: Issues, Questions, and Recommended Studies* (Washington, DC: CNA, January 2017), p. 61, https://www.cna.org/cna_files/pdf/DRM-2017-U-014796-Final.pdf.
99. David Gunning, "Explainable AI Program Description," *DARPA*, November 4, 2017, https://www.darpa.mil/attachments/XAIIndustryDay_Final.pptx.
100. Cimbala, *The Dead Volcano: The Background and Effects of Nuclear War Complacency*, p. 66.
101. Nick Bostrom, "Ethical Issues in Advanced Artificial Intelligence," in *Science Fiction and Philosophy: From Time Travel to Superintelligence*, edited by Susan Schneider (Oxford: John Wiley & Sons, 2009), pp. 277–286.
102. Nick Bostrom, *Superintelligence: Paths, Dangers, Strategies* (Oxford: Oxford University Press, 2014).
103. Kenneth Payne, *Strategy, Evolution, and War: From Apes to Artificial Intelligence* (Washington, DC: Georgetown University Press, 2018), pp. 209–211.
104. Wong et al., *Deterrence in the Age of Thinking Machines*.
105. Ibid., pp. 60–61.
106. Cummings, *Artificial Intelligence and the Future of Warfare*, p. 7.
107. Raja Parasuraman and Victor Riley, "Humans and Automation: Use, Misuse, Disuse, Abuse," *Human Factors* 39, no. 2 (June 1997): pp. 230–253, doi:10.1518/001872097778543886.
108. Andrew J. Hawkins, "Tesla Model S Plows into a Fire Truck while Using Autopilot," *The Verge*, January 23, 2018, https://www.theverge.com/2018/1/23/16923800/tesla-firetruck-crash-autopilot-investigation.
109. Mary L. Cummings, "Rethinking the Maturity of Artificial Intelligence in Safety-Critical Settings," *AI Magazine* 42, no. 1 (2021), pp. 6–15.
110. See Linda J. Skitka, Kathleen Mosier, and Mark Burdick, "Automation Bias: Decision Making and Performance in High-Tech Cockpits," *International Journal of Aviation Psychology* 8, no. 1 (1998): pp. 47–63, doi: 10.1207/s15327108ijap0801_3; and David Watson,

"The Rhetoric and Reality of Anthropomorphism in Artificial Intelligence," *Minds and Machines* 29 (2019): pp. 417–440.

111. Lonnie Shekhtman, "Why Do People Trust Robot Rescuers More Than Humans?" *Christian Science Monitor*, March 1, 2016, https://www.csmonitor.com/Science/2016/0301/Why-do-people-trust-robot-rescuers-more-than-humans.
112. David Hume, *Stanford Encyclopaedia of Philosophy*, April 17, 2019, https://plato.stanford.edu/entries/hume/.
113. Ewart J. de Visser et al., "Almost Human: Anthropomorphism Increases Trust Resilience in Cognitive Agents," *Journal of Experiential Psychology Applied* 22, no. 3 (2016): pp. 331–349.
114. The source of over-trust and automation bias might also be a machine's (real or perceived) capability, regardless of anthropomorphism.
115. A fully autonomous unsupervised weapon system would not be certain it has found the most optimal course actions until it had tested all possible scenarios with all likely adversaries—which would appear an implausible task and thus impose outsized risk on commanders fielding machines that, during fast-paced combat situations, freeze or behave irrationally. See Michael C. Horowitz, "The Promise and Peril of Military Applications of Artificial Intelligence," *Bulletin of the Atomic Scientists*, April 23, 2018, https://thebulletin.org/2018/04/the-promise-and-peril-of-military-applications-of-artificial-intelligence/; and Bostrom, *Superintelligence: Paths, Dangers, Strategies*, p. 17.
116. Paul Bracken, *The Second Nuclear Age: Strategy, Danger, and the New Power Politics* (New York: Times Books, 2012); Michael Krepon, "Can Deterrence Ever Be Stable?" *Survival* 57, no. 3 (2015): pp. 111–132, doi: 10.1080/00396338.2015.1046228.
117. Geist and Lohn, *How Might Artificial Intelligence Affect the Risk of Nuclear War?* https://www.rand.org/pubs/perspectives/PE296.html; Ayoub and Payne, "Strategy in the Age of Artificial Intelligence," pp. 799–819.
118. Winner, *Autonomous Technology*.
119. Johnson, "Artificial Intelligence," pp. 16–39, https://www.jstor.org/stable/26891882.
120. See Olivia Solon, "The Rise of 'pseudo-A': How Tech Firms Quietly use Humans to do Bots' Work," *Guardian*, July 6, 2018, https://www.theguardian.com/technology/2018/jul/06/artificial-intelligence-ai-humans-bots-tech-companies; and Paul Eisenstein, "Not Everyone is Ready to Ride as Autonomous Vehicles Take to the Road in Ever-Increasing Numbers," *CNBC*, October 14, 2018, https://www.cnbc.com/2018/10/14/self-driving-cars-take-to-the-road-but-not-everyone-is-ready-to-ride.html.
121. See Brodie, "Technological Change, Strategic Doctrine, and Political Outcomes"; Biddle, *Military Power*; and Macgregor and Murray, eds, *The Dynamics of Military Revolution, 1300–2050*.
122. John J. Mearsheimer, *Conventional Deterrence* (Ithaca, NY: Cornell University Press, 1984), p. 210.
123. Thomas J. Christensen, "The Meaning of the Nuclear Evolution: China's Strategic Modernization and U.S.-China Security Relations," *Journal of Strategic Studies* 35, no. 4 (August 2012): pp. 467–471; Acton, *Silver Bullet?*; Caitlin Talmadge, "Would China Go Nuclear? Assessing the Risk of Chinese Nuclear Escalation in a Conventional War with the United States," *International Security* 41, no. 4 (Spring 2017): pp. 50–92, doi: 10.1162/ISEC_a_00274.

124. Donald J. Trump, *National Security Strategy of the United States of America* (Washington, DC: The White House, December 2017), https://www.whitehouse.gov/wp-content/uploads/2017/12/NSS-Final-12-18-2017-0905-2.pdf.
125. Christensen, "The Meaning of the Nuclear Evolution"; Fiona S. Cunningham and M. Taylor Fravel, "Assuring Assured Retaliation: China's Nuclear Posture and U.S.-China Strategic Stability," *International Security* 40, no. 2 (Fall 2015): pp. 40–45, doi: 10.1162/ISEC_a_00215; Morgan et al., *Dangerous Thresholds*, Ch. 3; Avery Goldstein, "First Things First: The Pressing Danger of Crisis Instability in U.S.-China Relations," *International Security* 37, no. 4 (Spring 2013): pp. 49–89, https://muse.jhu.edu/article/503798/pdf.
126. Caitlin Talmadge, "The US-China Nuclear Relationship: Why Competition is Likely to Intensify," Brookings Institution, September 2019, https://www.brookings.edu/wp content/uploads/2019/09/FP_20190930_china_nuclear_weapons_talmadge-2.pdf, p. 3.
127. See Cunningham and Fravel, "Assuring Assured Retaliation," pp. 15–23; Talmadge, "Would China Go Nuclear? Assessing the Risk of Chinese Nuclear Escalation in a Conventional War with the United States," *International Security* 41, no. 4 (2017): pp. 50–92; and Adam P. Liff and G. John Ikenberry, "Racing Toward Tragedy? China's Rise, Military Competition in the Asia Pacific, and the Security Dilemma," *International Security* 39, no. 2 (2014): pp. 52–91.
128. Robert Jervis, "Cooperation under the Security Dilemma," pp. 169–214; Charles L. Glaser, "The Security Dilemma Revisited," *World Politics* 50, no. 1 (1997): pp. 171–201; Ken Booth and Nicholas Wheeler, *The Security Dilemma: Fear, Cooperation, and Trust in World Politics* (New York: Palgrave Macmillan, 2008); and Shiping Tang, "The Security Dilemma: A Conceptual Analysis," *Security Studies* 18, no. 3 (2009): pp. 587–623.
129. Charles L. Glaser, "The Security Dilemma Revisited," *World Politics* 50, no. 1 (1997): pp. 171–201.
130. See Thomas J. Christensen, "China, the US-Japan Alliance, and the Security Dilemma in East Asia," *International Security* 23, no. 4 (1999): pp. 49–80; Thomas J. Christensen, "The Contemporary Security Dilemma: Deterring a Taiwan Conflict," *Washington Quarterly* 25, no. 4 (2002): pp. 7–21; James Johnson, *The US-China Military and Defense Relationship during the Obama Presidency* (New York: Palgrave Macmillan, 2018); and Andrew Scobell, "Perceptions and Misperception in US-China Relations," *Political Science Quarterly* 135, no. 4 (2020): pp. 637–664. Adam P. Liff and G. John Ikenberry, "Racing Toward Tragedy? China's Rise, Military Competition in the Asia Pacific, and the Security Dilemma," *International Security* 39, no. 2 (2014): pp. 52–91.
131. Morgan et al., *Dangerous Thresholds*, pp. 169–170.
132. Goldstein, "First Things First."
133. Jeffrey A. Larsen and Kerry M. Kartchner, eds, *On Limited Nuclear War in the 21st Century* (Palo Alto, CA: Stanford University Press, 2014).
134. Johnson, *Artificial Intelligence & the Future of Warfare*, Ch. 5.
135. Lora Saalman, "The China Factor," in *Missile Defense: Confrontation and Cooperation*, edited by Alexei Arbatov, Vladimir Dvorkin, and Natalia Bubnova (Moscow: Carnegie Moscow Center, 2013), pp. 226–252; Acton, *Silver Bullet?*
136. Fiona S. Cunningham and M. Taylor Fravel, "Dangerous Confidence? Chinese Views on Nuclear Escalation," *International Security* 44, no. 2 (2019): pp. 61–109, doi.org/10.1162/ISEC_a_00359.

137. M. Taylor Fravel, *Active Defense: China's Military Strategy since 1949* (Princeton, NJ: Princeton University Press, 2019).
138. James Johnson, "Chinese Nuclear 'War-fighting': An Emerging Intense U.S.-China Security Dilemma and Threats to Crisis Stability in the Asia Pacific," *Asian Security* 15, no. 3 (2019): pp. 215–232.
139. The relative optimism of Chinese analysts can also be attributed to the belief that any escalation to nuclear threats or use would be *intentional and belligerent*. Avery Goldstein, "First Things First," pp. 65–66.
140. Cunningham and Fravel, "Dangerous Confidence?" *International Security* 44, no. 2 (2019): pp. 104–105.
141. Lawrence Lewis and Anna Williams, *Impact of Unmanned Systems to Escalate Dynamics* (Summary) (Washington, DC: CNA, 2018), https://www.cna.org/CNA_files/PDF/Summary-Impact-of-Unmanned-Systems-to-Escalation-Dynamics.pdf.
142. Bernard Brodie, *Escalation and the Nuclear Option* (Princeton, NJ: Princeton University Press, 1966).
143. Chris Buckley, "Chinese Navy Returns Seized Underwater Drone to U.S.," *New York Times*, December 12, 2016, https://www.nytimes.com/2016/12/20/world/asia/china-returns-us-drone.html.
144. See Keir Lieber and Daryl Press, *The Myth of the Nuclear Revolution: Power Politics in the Atomic Age* (Ithaca, NY: Cornell University Press, 2020), pp. 108–110, 117–118.
145. The slow pace of expansion in Chinese and North Korean nuclear forces, together with the recent extension of the US-Russia New Strategic Arms Reductions Treaty through 2026, may obviate this problem in the short term.
146. A state might attempt to deflect scrutiny by claiming that because it could not foresee the actions of the algorithmic decision-making baked into weapon systems, it cannot be held responsible. This deflection would unlikely be accepted in international law, however. A state would be held responsible for the acts of the AIs in the same way as it is held responsible for the acts of (human) state agents. Deeks, Lubell, and Murray, "Machine Learning, Artificial Intelligence, and the Use of Force by States," pp. 23–24.
147. John M. Keynes, *The End of Laissez-Faire: The Economic Consequences of the Peace* (New York: BN Publishing, 2009).

2
Nuclear deterrence
New challenges for deterrence theory and practice

A growing number of great military powers are investing political capital and financial resources to develop the field of AI technology and AI-enhanced "autonomous"[1] weapon systems to derive the maximum potential military benefits—at a tactical, operational, and strategic level—that these systems offer.[2] As a result, the ubiquity of these new classes of advanced capabilities—and the incentives for militaries to adopt them—on the future battlefield is fast becoming a foregone certainty.[3] How might the rise of these capabilities weaken or strengthen deterrence? Given the recent genesis of AI and autonomy in a military context, and the rich body of work that describes these trends, this research is premised on the assumption that AI and autonomy technology will continue to be embraced—at different speeds and for different goals—by global militaries.[4] This chapter argues that AI and autonomy could decrease nuclear stability and increase the tendency for escalation to nuclear use, thereby undermining deterrence.

The chapter considers the potential implications of AI and autonomy for nuclear deterrence theory and practice. Using the rich deterrence literature as a point of departure, the research considers how advances in AI and autonomy might affect the manner, and means, by which nuclear-armed states seek to deter adversaries from embarking on a particular course of action through threats, denial, coercion, and compellence. Given the potentially transformative effects of AI and autonomy augmentation on a range of (nuclear and non-nuclear) strategic technologies, rethinking existing assumptions, theories, and permutations of deterrence—premised on human-rationality, perceptions, and nuanced signaling—is now needed.[5] The focus of this chapter is to examine the impact of the adoption of these technologies on nuclear deterrence. How might the introduction of non-human agents into a crisis or conflict between nuclear powers influence deterrence?

Because of the multifaceted possible intersections of AI with the nuclear enterprise, and *before* states reconfigure their nuclear doctrines and postures, research on this topic is important to anticipate (and mitigate)

AI and the Bomb. James Johnson, Oxford University Press. © James Johnson (2023).
DOI: 10.1093/oso/9780192858184.003.0003

the risk of misperception, miscalculation, and inadvertent escalation (see Chapter 3). The chapter considers how increasing complexity, speed, compressed decision-making, and human psychology associated with AI and autonomy might compound these risks, and thus impact deterrence theory and practice. How might states with different political systems, philosophies of military use, and deterrence policies view and likely respond to these dynamics? This cross-disciplinary lens broadens the conceptual field of deterrence research, providing a more fulsome empirical picture to test the AI deterrence theorizing presented in this chapter.

The text builds on the extensive body of work that considers nuclear deterrence theory and other theoretical frameworks to consider AI and autonomous systems' potential effects.[6] Classical nuclear deterrence approaches, premised on the assured threat of a retaliatory second strike (or mutually assured destruction (MAD)) is a fundamentally psychological-political-technical phenomenon that has worked to reduce the chances of deliberate use of nuclear weapons.[7] In other words, the potential (i.e., legal) use of nuclear weapons involves *both* technical calculations about the military effectiveness of a particular strategy, as well as more complex psychological and geopolitical predictions and intelligence estimates about the choices an adversary is likely to make in response to different modes of attack.[8] There is an extensive body of literature on the intersection of "emerging technology"[9] and the nuclear enterprise—and the potential for warfare more broadly defined.[10] The literature says very little about how existing concepts of escalation, nuclear terrorism, and classical deterrence theories might apply (or be tested) in the digital age—increasingly defined by developments in AI and autonomy—where perfect information and rational decision-making cannot be assumed.[11]

This remainder of the chapter is organized into four sections. The first identifies and contextualizes the core theories and concepts necessary to examine the implications of introducing AI and autonomous systems into the nuclear domain—namely, deterrence, escalation, and strategic stability. It describes the broadened notion of deterrence that emerged after the end of the Cold War, referred to in deterrence scholarship as the "fourth wave"—with insights from criminology, terrorism studies, and human psychology, which is a key focus of this chapter. Are existing theories of deterrence and related concepts still applicable in the era of AI and autonomy?

The second section applies this conceptual framework to examine the possible ways in which advances in AI and autonomy might undermine the central pillars that, for several decades, have undergirded nuclear deterrence and strategic stability. It unpacks the key technical and non-technical features of this emerging paradigm to elucidate how new AI-powered capabilities could affect the key components of strategic deterrence—especially Nuclear Command, Control, and Communication (NC3) systems and second-strike capabilities. Will AI-enabled capabilities strengthen or weaken deterrence?

The third section examines the growing multipolarity in international security competition—radically shifting the geostrategic balance—and considers how the confluence of this trend with the development of AI and autonomy might compound the problem of uncertainty caused by complexity and the speed of warfare, the entanglement of nuclear and conventional capacities, and the management of crisis escalation and signaling premised on the assumption of rationality—between nuclear-armed adversaries. It argues that although the explicit linkages between the structure of the international political system and developments in AI and autonomy are not clear,[12] shifts in the geopolitical balance will significantly influence how deterrence dynamics could play in future crises and conflict between peer and near-peer nuclear-powers.

The final section considers the implications of human and machine interactions for nuclear deterrence. How might these interactions spark inadvertent or accidental nuclear escalation to a strategic level? It highlights the potential for accidents and errors (both technical and human), misperception, and unintended consequences that result from new deterrence models that mix various levels of humans and machines, and new and legacy weapon systems in their interface with nuclear weapons. What trade-offs might arise from this synthesis? This analysis tentatively concludes that *ceteris paribus* greater degrees of automation, coupled with reduced human decision-making, will likely increase inadvertent escalation risk.

Deterrence theory in the digitized age: Concepts, assumptions, and paradoxes

How can we best conceptualize AI and autonomy in the context of nuclear weapons? While the definition of deterrence remains contested, at its core nuclear deterrence is concerned with seeking the means to induce caution in others by applying threats (implicit or explicit) of *punishment* (i.e., retaliation with nuclear weapons) to manipulate an adversary's behavior, or *denying* (i.e., removing the expected benefits of a particular behavior) the adversary the ability to realize their aggressive objectives in the first place.[13] Therefore, any behavior that results in caution or apprehension has possible deterrent effects—even when the behavior to be deterred has to be inferred.[14]

Deterrence and atomic weapons are not necessarily synonymous; an adversary can be deterred in various ways *below* the nuclear threshold.[15] A related concept is "compellence," which is about encouraging an adversary to pursue an action that they might have otherwise eschewed, rather than restraining their behavior with deterrence. However, once a situation escalates, much like the difference between defense and offense, the distinction can disappear—the coercion and counter-coercion mechanisms intrinsic to both deterrence and compellence often occur simultaneously.[16] The tension between the need to sustain deterrence and the simultaneous incentives to pursue coercive leverage lays at the heart of the credibility problem in nuclear deterrence theory.[17]

Whether nuclear deterrence—and deterrence generally—is effective and credible depends on three factors:[18] (1) rationality on both sides to ensure threats will suffice to shape the other's behavior;[19] (2) the adversary's perception of the defender's capacity and resolve (or will) to punish violations (or red-lines) or deny objectives; and (3) the deterrent threat must be communicated and understood by the adversary—defenders must also *reassure* aggressors that threats (or demands) will *not* be carried out in the event demands are met.[20] In sum, deterrence is about influencing an adversary's sense of risk, cost-benefit assessment, and decision-making outcomes. Deterrence requires a deep understanding of the other side's interests, priorities, strategic objectives, culture, identity, and above all, perceptions. How adversaries view each other's capabilities and intentions—especially the other side's philosophy of employment—will crucially influence these systems' deterrent effect and the potential risk of miscalculation and escalation if these assessments prove wrong.

Therefore, for deterrence to be effective, both sides need to understand how, why, and when adversaries adopt and employ deterrence strategies (or change them) and thus perceive their reality (i.e., value attached to their interests). This enables actors to discover ways to manipulate these constructions while simultaneously ensuring that allies' interests are addressed. Besides, these interests cannot be considered independent of a particular policy's deterrent effect, which will vary temporally.[21] Deterrence can fail if either side has divergent beliefs about the perceived cost of the punishment a state can inflict and the perceived probability that it will (or is able to) inflict it.[22]

Deterrence, and strategic bargaining, can fail between rational actors in asymmetric information situations (i.e., information about an actor's resolve); in particular, when incentives exist to manipulate this information, when credible commitment is problematic, and divisions exist on key policy issues.[23] While the existence of rational thinking—intrinsic to classical rational deterrence theorizing—is generally assumed to prove the likelihood of deterrence success; and juxtaposed, non-rational behavior by actors assumes a high-risk of deterrence failure, rational actors can challenge deterrence assumptions. For instance, as a result of misperception and other human cognitive fallibilities (discussed below), incomplete information (i.e., about a states' reputation costs or perception of the "balance of interests"), uncertainty about an adversary's future capabilities and intentions, time pressures and errors, or grandiose objectives can impel rational actors to fight losing wars to deter aggression.[24] Even in situations when an actor's threats are credible, and the balance of interests and reputational costs are known, fears about the actor's future intentions can create "rational" incentives to engage in—or even pre-emptively trigger—losing wars.[25] Thus, deterrence policies are counterproductive when they exacerbate an adversary's fear for their own security; sparking spirals of mistrust and fear associated with a deep security dilemma (see Chapter 4).[26]

The political psychology literature reveals that perceptions are heavily encumbered by inferences, reputation, interpretation, and cognitive-psychological theory.[27] Thus, although an actor's behavior might reveal or signal something important (resolve, deterrence, or reassurance), it is often "not clear exactly what is being revealed, and what others will think is being revealed."[28] For instance, a state's effort to reassure its allies that it will come to their defense may reduce the patron's sense of security if it infers that the threat is greater than it thought to the degree that the defender considers reassurances necessary.

Grounded in these core tenets and preconditions, nuclear deterrence has undergone several transformations since the Cold War. In light of changes to the geopolitical (from bipolarity to unipolarity and finally multipolarity), the technological landscape (missile defenses, hypersonic weapons, and AI and autonomy), and new security threats and domains (non-states, gray-zone conflict, space and cyberspace), political scientists have conceptualized "four waves" of deterrence theorizing.[29] The fourth wave followed the end of the Cold War and continues to the present day, coinciding with the "second nuclear age's" broader features, namely, multipolarity, asymmetric threats, non-state (especially rogue nation and terrorist) actors, and advanced strategic (nuclear and non-nuclear) weapons.[30] These "waves" progressively built upon the first wave's foundation, creating a more complex, nuanced, and applicable theory in light of new research, novel methodologies applied from other disciplines (military and non-military), and lessons from real-world events.[31] These methodologies were also accompanied by novel concepts, which have broadened the scope and practice of deterrence *inter alia*, conventional deterrence, extended deterrence (and its opposite "central deterrence"), inter-war deterrence, and, more recently, cross-domain, and cyberspace deterrence.[32]

Like research on cyber deterrence, early scholarship on deterrence theory and practice in the digital age has been predominately grounded in classical deterrence approaches—associated with the earlier waves in international relations (IR) deterrence theorizing rooted in known hierarchical relationships between actors and the principle of MAD.[33] A dearth of studies exists that explore the ways in which AI technology could intersect with deterrence theory and practice.[34] In this way, the research speaks to the nascent "fifth wave" of modern deterrence, representing a conceptual break from previous waves of classical deterrence theorizing (or post-classical deterrence) and non-human agents into deterrence.[35] Any discussion surrounding nascent emerging technology such as AI comes with an important caveat.

Since we have yet to see how AI might impact deterrence, coercion, escalation, and strategic stability in the real world, coupled with the nascent, uneven, and uncertain application of AI in the military sphere, the limitations of current "narrow" AI technology, and wider political, legal, and ethical issues to expound on—notwithstanding the valuable insights from experimental wargaming—the discourse is largely a theoretical and structured speculative endeavor.[36] Two other important concepts in the discussion about nuclear deterrence, nuclear-armed states, and AI and autonomy are *escalation*—especially inadvertent escalation, discussed in Chapter 3 and *strategic stability* (or lack thereof, instability) discussed in Chapter 1.

Escalation in the context of deterrence can be defined as an increase in the intensity or scope of a military situation that crosses a threshold(s) considered significant by one or more actors. Escalation occurs when at least one of the parties involved perceives (or misperceives) a significant qualitative shift in a situation.[37] These mechanisms can escalate a crisis or conflict between two or more nuclear-armed states into nuclear confrontation or cause a low-level conventional conflict to move up the "rungs" of the escalation ladder and cross the nuclear threshold—either intentionally, accidentally, or inadvertently (see Chapter 3).[38] The post-Cold War literature is rich in scholarship on how technologically complex nuclear systems can cause technical (and human-related) accidents and false alarms, which are considered particularly escalatory where one side lacks confidence in their retaliatory (or second-strike) capacity.[39] The fact that during the Cold War the perennial fear that an action or signal misinterpreted by the other—in the context of uncertainty and incomplete information associated with modern warfare—could trigger nuclear pre-emption is a useful point of departure to consider AI and autonomy.[40]

Accidental nuclear war—a nuclear confrontation without a deliberate and properly informed decision to use nuclear weapons on the part of the nuclear-armed state(s) involved—could be caused by a variety of accidents, most often encompassing a combination of human—and human-machine interaction failure—system errors, and procedural or organizational factors.[41] Moreover, despite paying lip-service to Machiavelli's Fortuna (the role of uncertainty, complexity, and chance in international politics), decision-makers underestimate the importance and frequency of accidents and randomness in these interactions.[42] AI-enhanced systems operating at higher speeds, levels of sophistication, and compressed decision-making timeframes will likely, therefore, further reduce the scope for de-escalating situations and contribute to future mishaps.[43] The rapid proliferation and ubiquity of advanced technologies like offensive-cyber, hypersonic weapons, and AI and autonomous weapons will make it increasingly difficult for states to mitigate this vulnerability without simultaneously improving their ability to strike first, thereby undermining the survivability of others' strategic forces.

Unraveling of deterrence in practice

In theory there is no difference between theory and practice. In practice there is—Yogi Berra[44]

How might AI and autonomy affect strategic deterrence's key components in ways that undermine states' second-strike capabilities? The size, mobility, hardened, and relatively hidden features of the superpowers' nuclear arsenals ensured the ability of states to withstand the first strike and deliver a retaliatory second strike, constituting the core pillars of the Cold War-era nuclear deterrence—known as the "nuclear revolution."[45] Like other technologies associated with the "computer revolution"—particularly big-data analytics, robotics, quantum computing, nanotechnology, and cyber capabilities—advances in AI and autonomy threaten to upend this fragile arrangement, *inter alia*, in several ways.[46]

Hunting for nuclear weapons in the digital age

How might AI-augmented intelligence gathering and analysis systems impact the survivability and credibility of states' nuclear deterrent forces? The integration of AI machine learning (ML) and big-data analytics can dramatically improve the ability of militaries to locate, track, target, and destroy a rival's nuclear deterrent forces—especially nuclear-armed submarines and mobile missile forces—and without the need to deploy nuclear weapons.[47] AI-enabled capabilities (e.g., space-based optical-electronic reconnaissance systems) that increase the vulnerability of second-strike capabilities (or are perceived to do so) heighten uncertainty and undermine deterrence—even if the state in possession of these counterforce capabilities did not intend to use them.[48] Russia, for example, is concerned that AI-enabled advances in US space-based optical-electronic reconnaissance systems might threaten Russia's land-based deterrent. Unlike the US, whose most reliable leg of the nuclear triad is its nuclear-powered ballistic missile submarines (SSBNs), Russia's road-mobile forces are the most survivable component of Russia's nuclear deterrent.[49] In short, the capabilities that AI might *enhance* (cyber-weapons, drones, precision-strike missiles, and hypersonic weapons), together with the ones it might *enable* (intelligence, surveillance, and reconnaissance (ISR), automatic target recognition (ATR), and autonomous sensor platforms) could make hunting for mobile nuclear arsenals faster, cheaper, and more effective than before.

AI-ML techniques could significantly improve existing machine vision and other signal processing applications, identify patterns from large datasets of signals and imagery, and enhance autonomy and sensor fusion applications. Taken together, strengthening ISR functionality, ATR, and terminal-guidance systems would have profound implications for strategic stability. Besides, AI used in conjunction with autonomous mobile sensor platforms might compound the threat posed to mobile intercontinental ballistic missiles (ICBM) launchers. Autonomous mobile sensors would only need to locate close to mobile ICBM launchers to be effective, and thus, as the "window of vulnerability" rapidly narrowed, and faced with the prospect of an imminent disarming strike, an adversary would be put under immense pressure to escalate. For instance, advances in deep learning (DL) can exponentially improve machine vision and other signal processing applications, which may overcome the main technical barriers for tracking and targeting adversaries' nuclear forces (i.e., sensing, image processing, and estimating weapon velocities and kill radius).[50] Some scholars argue that AI and autonomy could enable real-time tracking, shorten decision-cycles, and enable more accurate targeting—and reduce target selection errors—of an adversary's nuclear assets in ways that make counterforce operations more feasible.[51] Moreover, AI technology—especially that used to augment counterforce capabilities (see Chapter 4)—could put the defender at a distinct disadvantage: that is, giving the superior state an opportunity to escape the dynamics of perceived mutual and symmetric vulnerability—thus creating additional incentives to strike first (or pre-emptively) upon technologically superior military rivals.[52] Several technologies under development are designed explicitly for this purpose.[53] The less secure a nation considers its second-strike capabilities to be, therefore, the more likely it is to countenance the use of autonomous systems within its nuclear weapons complex to bolster the survivability of its strategic forces.

Given the tendency of Chinese and Russian strategists to extrapolate from current US capabilities malign intent—and to assume future capabilities will threaten their security—even modest and incremental improvements in AI techniques to integrate and synthesize data about the location of an adversary's mobile missiles could exacerbate pre-existing fears and distrust.[54] Irrespective of whether future breakthroughs in AI produce irrefutable evidence of a game-changing means of locating, targeting, and destroying mobile missile forces, Chinese and Russian perceptions of US intentions in the pursuit of these capabilities would, therefore, be far more salient.[55] Despite the US's reassurances, its adversaries would be unable to dismiss the possibility that military AI capabilities would not be used in future warfare to erode the survivability of their nuclear forces—a contingency for which the US has been preparing over several decades.[56]

Several observers posit that autonomous systems like the Defense Advanced Research Projects Agency (DARPA)'s *Sea Hunter*, by rendering the underwater domain "transparent," might erode the second-strike deterrence utility of stealthy SSBNs, triggering use-them-or-lose-them situations. Today, however, the technical feasibility of this hypothesis remains highly contested.[57] On the one hand, several experts posit that emerging technologies such as AI, quantum communications, and big-data analytics will empower new iterations of highly portable sensing, communications, and signal-processing platforms that could render at-sea nuclear deterrence all but obsolete.[58] On the other hand, others consider this hypothesis technically and operationally premature for several reasons, not least the notoriously complex and dynamic underwater conditions that hamper underwater targeting, and the need for increased power for autonomous systems to operate extended ranges.[59]

AI-cyber threats to nuclear systems

How might AI-infused cyber capabilities be used to subvert or otherwise compromise states' control over their nuclear systems? Today, it is thought possible that a cyber-attack (spoofing, hacking, manipulation, data-poisoning, or digital jamming) could infiltrate a nuclear weapon system, threaten the integrity of its communications, and ultimately (and possibly unbeknown to its target) gain control of its (potentially dual-use) C2 systems. For instance, a non-state third-party hacker might "break into, interfere with, or sabotage nuclear C2 facilities; spoof or compromise early-warning systems or components of the nuclear firing chain; or in a worst-case scenario even cause a nuclear explosion or launch."[60] For example, to erode the confidence of an adversary in its nuclear deterrent, State A plants malware or ransomware on State B's nuclear delivery systems, exploiting vulnerabilities in B's supply chains—akin to the "colossal" Kaseya ransomware supply-chain attack in 2021.[61] Signaling resolve as a crisis unfolds, A informs B aware of the penetration, and that latent infiltration preparations lay in wake. B, unable to determine the precise nature or extent of these vulnerabilities, responds to "use-it-or-lose-it" pressures by pre-emptively escalating the situation.

Because of the intense time pressures that would loom large with the decision to use nuclear weapons—especially where a state maintains a launch-on-warning posture—AI-enhanced cyber-attacks against nuclear systems would be almost impossible to detect and authenticate. Further, warning signals would be difficult to authenticate, let alone attribute, to initiate a nuclear strike within a short timeframe. A shared concern of China, the US, and Russia—albeit with varying degrees of sensitivity—are the potential threats posed by AI-augmented cyber warfare that might impel states to adopt (or rely more heavily upon) a launch-on-warning nuclear posture or a policy of pre-emption during a crisis.[62] Cognizant of the potential vulnerabilities of NC3 infrastructure to cyber-attacks, nuclear-armed states (including Russia) under the Nuclear Non-Proliferation Treaty have begun to incorporate this threat into their deterrence policies.[63] In short, conventional strikes against an adversary's nuclear deterrent forces (especially NC3) combining AI-enabled ISR and cyber capabilities would likely amplify the potentially destabilizing impact of such an operation.

Advances in AI could also exacerbate this cybersecurity challenge by enabling improvements to the cyber offense.[64] By automating advanced persistent threat (APT) operations, ML and AI could dramatically reduce the extensive resources and skill required to execute APT operations (or "hunting for weaknesses"), especially against hardened nuclear targets. AI-augmented cyber-tools' machine speed could enable an attacker to exploit a narrow window of opportunity to penetrate an adversary's cyber defenses or use APT tools to find new vulnerabilities faster and more easily than before. As the former US Chairman of the Joint Chiefs, General Joseph Dunford, recently warned, "the *accelerated speed of war* ensures the ability to recover from early missteps is greatly reduced" (emphasis added).[65] According to scholar Robert Jervis, "a great deal of the information received at the start of a crisis is wrong ... although waiting for more information does not guarantee a good understanding of the situation, quick moves are more likely to be based on false understandings."[66]

For example, when docked for maintenance, air-gapped SSBNs, considered secure when submerged, could become increasingly vulnerable to a new generation of low-cost and highly automated APT cyber-attacks.[67] An attacker might also use ML tools to target autonomous dual-use early-warning systems with "weaponized software" (i.e., hacking, subverting, spoofing, or tricking), causing random and potentially undetectable errors, malfunctions, and behavioral manipulation to these networks.[68] An attacker may, for instance, poison a data-set to inhibit an algorithm from learning specific patterns, or insert a secret backdoor that can trick the system in the future.[69]

Chinese analysts would view cyber-infiltrations on China's NC3 systems as highly escalatory, even if the perpetrator's goals were limited to collecting information (i.e., espionage) about cyber threats to prevent a future attack.[70] By contrast, Russian analysts tend to view Russia's NC3 network as relatively isolated and insulated from cyber-attacks.[71] However, both Chinese and Russian analysts worry about the vulnerability of their NC3 systems to fast-paced and stealthy conventional counterforce operations augmented by AI technology. The discovery of an adversary's attempt (successful or otherwise) to degrade a state's nuclear network would heighten mistrust and tension in future nuclear crises.[72]

Uncertainty about the efficacy of AI-augmented cyber capabilities during a crisis or conflict would likely reduce *both* sides' risk tolerance, increasing the incentive to strike pre-emptively as a hedging strategy.[73] For instance, a state might strike pre-emptively in response to information mined from AI-augmented ISR systems that an adversary was planning a surprise attack. During crisis conditions, an offensive AI cyber-tool that succeeds in compromising an adversary's nuclear weapon systems—resulting in an "asymmetric information" situation—may cause either or both sides to overstate (or understate) its retaliatory capabilities, thus making them more inclined to act in a risky and escalatory fashion.[74] In short, in a competitive strategic environment where states are inclined to assume the worst of others' intentions, one state's efforts to enhance its strategic forces' survivability may be viewed by others as a threat to their nuclear retaliatory capability—or as a classic "security dilemma" (see Chapter 4).[75]

Drone swarming under the nuclear shadow

How might AI-augmented drones swarming and hypersonic weapons complicate missile defense, undermine states' nuclear deterrent forces, and increase the risk of escalation? Drones (especially micro-drones[76]) used in swarms are conceptually well-suited to conduct pre-emptive attacks and nuclear-ISR missions against an adversary's nuclear mobile missile launchers and SSBNs and their enabling facilities (e.g., early-warning systems, antennas, sensors, and air intakes).[77] In short, the ability of future iterations of AI-ML—with expanded and dispersed mining data pools—infused drone swarming technology to locate, track, and target strategic missiles (i.e., mobile ICBM launchers in underground silos and onboard stealth aircraft or SSBNs) is set to grow.[78]

The following four scenarios illustrate the possible strategic operations that AI-augmented drone swarms would execute.[79] First, drone swarms could be deployed to conduct ISR operations to locate and track dispersed (nuclear and conventional) mobile missile launchers and their attendant dual-use command, control, communications, and intelligence (C3I) systems.[80] Specifically, swarms incorporating AI-infused ISR, autonomous sensor platforms, ATR, and data analysis systems may enhance sensor drones' effectiveness and speed to locate mobile missiles and evade enemy defenses.

Second, swarming may enhance legacy conventional and nuclear weapon delivery systems (e.g., ICBMs and submarine launched ballistic missiles), potentially incorporating hypersonic variants. At least two nuclear-armed states have developed unmanned air and underwater vehicle prototypes with nuclear delivery optionality.[81] AI applications will likely enhance the delivery system targeting, tracking, and improving drone swarms' survivability against the current generation of missile defenses.

Third, swarming tactics could bolster a states' ability to disable or suppress an adversary's defenses, clearing the path for a disarming attack.[82] Drone swarms might be armed with cyber or electronic warfare capabilities—in addition to anti-ship, anti-radiation, or regular cruise and ballistic missiles—to interfere with or destroy an adversary's early-warning detection and C3I systems in advance of a broader offensive campaign.[83]

Finally, in the maritime domain, unmanned underwater (UUVs), aerial (UAVs), and surface vehicles (USVs), supported by AI-enabled intra-swarm communication and ISR systems, may be deployed in *both* offensive and defensive anti-submarine warfare operations to saturate an enemy's defenses and to locate, disable, and destroy its nuclear-armed or non-nuclear attack submarines.[84]

Perceived as a relatively low-risk *force majeure* with ambiguous rules of engagement and absent a robust normative and legal framework,[85] lethal and non-lethal autonomous weapons will likely become an increasingly attractive asymmetric capability to erode a superior adversary's deterrence and resolve.[86] Notwithstanding the remaining technical challenges—especially the demand for power—swarms of robotic systems fused with AI-ML techniques may presage a powerful interplay of increased range, accuracy, mass, coordination, intelligence, and speed in a future conflict.[87] How might new technologies undermine deterrence between nuclear-armed adversaries in conflict-prone—and possibly asymmetric—situations?

Deterrence in digital multipolarity

Nuclear multipolarity will likely challenge several long-held nuclear deterrence assumptions—especially the probability that actors can be coerced to behave in particular ways premised on the assumption of rationality—which significantly impair states' ability to manage escalation and interpret signaling (i.e., deterrence and resolve); thereby increasing the likelihood of impregnating misperception and miscalculation into fragile strategic dyads—and possibly triads—of nuclear-armed states.[88] In a multipolar world nuclear order with competing and contested strategic dyads—India-Pakistan, US-Russia, US-North Korea, US-China, and perhaps India-China—with weak mechanisms for de-escalation, doctrinal opacity, and questionable attitudes towards nuclear restraint, the potential risk of nuclear use increases.[89] A 2019 report by the UK's House of Lords, for instance, concludes that "the risk of the use of nuclear weapons has increased, in the context of rising inter-state competition, *a more multipolar world*, and the development of *new capabilities and technologies*" (emphasis added).[90]

Nuclear multipolarity—associated with the "second nuclear age"—is important because each state will likely choose different responses to the new choices emerging in the digital age.[91] State actors making decisions under the nuclear shadow will be more inclined to assume the worst of others' intentions, especially in situations where the legitimacy of the status quo is contested (e.g., maritime Asia). Motivated states such as China and Russia might eschew the limitations of AI, i.e., brittleness, "explainability," bias, and vulnerability; thereby compromising safety and verification standards to secure or capture the first-mover advantages—or to offset a perceived strategic disadvantage at a conventional level—vis-à-vis an adversary on the future digitized battlefield.[92] According to Kenneth Waltz, while nuclear weapons served as a stabilizing force during the Cold War-era, "increased numbers of actors increase levels of systemic uncertainty ... rising uncertainty heightens potential miscommunication and conflict. Bipolarity [compared to multipolarity] is, therefore, the most stable form of international power distribution."[93]

During a crisis, when prudent and careful planning can run aground in complex situations, coupled with an abundance of ambiguous information (about resolve and power), misperceptions and miscalculation can generate temptations for pre-emption.[94] Against today's geopolitical backdrop, therefore, the use of autonomous weapon systems (perceived as low-risk and low-cost) during tense and complex adversarial environments—with ambiguous rules of engagement such as anti-access/area denial zones—will become an increasingly enticing asymmetric option to undermine an adversary's military readiness, deterrence, and resolve. According to John Mearsheimer, "as long as the system remains anarchic, states will be tempted to use force to alter an unacceptable status quo."[95] The other side could easily perceive one state's efforts to enhance its strategic forces' survivability with state-of-the-art dual-use technology like AI (see Chapter 4) as a potential threat to its ability to survive and respond to a nuclear first strike.[96]

How might regime type or political systems affect deterrence dynamics? Authoritarian states may perceive an adversary's intentions very differently from a democratic one.[97] The belief that a regime's survival or political legitimacy is threatened might cause leaders to consider worst-case scenario judgments and behave in a manner predicted by offensive realist scholars.[98] Conversely, non-democratic leaders operating in closed political systems such as China, Russia, or North Korea may exhibit a higher degree of confidence (or overconfidence) in their ability to respond to perceived threats in world politics (see Chapter 5).[99] Bias assessments from a non-democratic regime's (or "Stasi" type) intelligence services might reinforce a leader's faith—or a false sense of security—in their diplomatic skill and maneuverability.[100] Without institutionalized structures (i.e., a general staff system) connecting the intelligence services with the military and broader political context, decisions will likely be made in vacuums, with minimal checks and balances on the political leadership (or supreme leader), and a reduction in "bottom-up" (or a "fact searching" organizational culture) information flow—because of the fear of contradicting the leadership.[101] Social media—and other AI-enhanced information tools—which supply decision-makers with a continuous flow of near-real-time information might also complicate the practice of deterrence and escalation management before and during future crises (see Chapter 3).[102]

Authoritarian regimes whose political legitimacy are conditioned or legitimized by the general acceptance of official narratives and dogma—to extend influence, manage expectations, and control the discursive environment—tend to become empowered when people's trust in truth (i.e., faith in what they see and hear) is undermined. Though authoritarian regimes might follow a rational deterrence logic and pattern of behavior in strategic decision-making, this "logic" will inevitably reflect the myriad factors that can influence a leader's threat perceptions including, *inter alia*, leaders' personalities, beliefs, experiences, and the importance attached to reputation and status—tethered to its (and possibly distorted or misinformed) version of events.[103] Thus, confounding the expectations and calculations (i.e., notions of deterrence and escalation thresholds) of adversaries and increasing accidental and inadvertent escalation risk (see Chapter 3).[104] The relative importance of these factors upon states' resolve under crisis conditions in context-bound situations is, however, difficult to answer with any certainty.[105]

A regime that views its second-strike capabilities—especially its NC3 systems—as vulnerable or insecure (North Korea, Pakistan, or perhaps China) may be more inclined to automate its nuclear forces and launch postures. An insecure nuclear-armed regime in an asymmetric dyad—assuming adequate technical, economic, organizational, and political resources—could do, *inter alia*, one or a combination of the following: (1) alter its nuclear force structure or doctrine (i.e., adopt a launch-on-warning posture, or reject a no-first-use commitment);[106] (2) increase the alert status of its nuclear arsenals;[107] (3) modernize or expand its nuclear capabilities (both strategic and tactical nuclear weapons);[108] (4) develop AI capabilities to bolster deterrence;[109] or (5) use AI technology to automate its second-strike nuclear capability (i.e., launch policy and NC3 systems). And, *in extremis*, to deter a decapitating strike during an existential crisis, pre-delegate launch decisions to machines.[110] In sum, non-democratic nuclear states with centralized Command and Control (C2) structures, whose leaders are less confident in the survivability of their nuclear arsenal, and whose legitimacy and stability are tethered to strategic narratives, would likely be more persuaded by the merits of automation, and less concerned about the potential risks—least of all the ethical, human cognitive, or moral challenges—associated with this decision.

Despite official Chinese statements supporting the regulation of military AI by global militaries, for instance, much of China's AI-related initiatives (e.g., the use of data for social surveillance to distil a social-credit scoring system and ubiquitous facial recognition policies) focus on the impact on social stability and, in particular, efforts to insulate the legitimacy of the regime against potential internal threats.[111] By contrast, the political processes, accountability (especially elected leaders and head of state vis-à-vis public opinion), nuclear-launch protocols, nuclear strategy and doctrine, mature civil-military relations, and shared values between allies (i.e., the US and its NATO allies), in democratic societies should make them less predisposed—or at least more reticent and encumbered—in use of AI in the nuclear domain.[112] In short, while technological factors involved in NC3 systems are of great importance, political, psychological, strategic-cultural, doctrinal, and organizational considerations are also central to understanding the nature and key to the mitigation of deterrence failure and escalation risk.

It is politics and the breakdown of human bargaining that ultimately leads to conflict; thus AI and other kinds of advanced technologies (drones, deepfakes, and cyber capabilities) can be viewed as the tools (or independent variables) that can complicate and exacerbate the escalation dynamics determined by human choices, biases, and interactions—both human-to-human and human-to-machine.[113] The amplification of false alarms or the creation of false signals (false positives and false negatives) by social media might also disrupt critical communication channels between commanders and their political leadership and between allies and adversaries during crisis or conflict—raising questions about the reliability of nuclear deterrence (see Chapter 3). In contested information environments, errors in early-warning systems and decision-making processes are characterized by a lack of data, ambiguous indicators, mixed signals, and conflicting sensor data inputs. Amidst this complexity and ambiguity, false signals frequently occur (and are even expected), not least because sensor systems may not be cross calibrated to provide cross-checking confirmation.[114]

Automating strategic decision-making: A double-edged sword for deterrence?

Today, the potential tactical and operational (i.e., "the ways wars are fought") impact of AI is qualitatively axiomatic; its effect at a strategic level (i.e., actions that affect balance of power and great-power competition), however, remains uncertain. Some scholars argue that the adoption of AI at a tactical and operational level could have significant strategic implications—for example, on deterrence policy, escalation mechanisms (see Chapter 3), and strategic stability between great powers.[115] AI systems that are programed to pursue tactical and operational advantages aggressively, for example, might misperceive (or simply ignore) an adversary's bid to signal to resolve (i.e., to demonstrate restraint or de-escalate a situation) as a prelude to an imminent attack, and might increase the risks of inadvertent escalation and first-strike instability.[116]

Data limitations, coupled with constraints on the ability of AI algorithms to capture the nuanced, dynamic, subjective accurate, and changeable nature of human commanders—or "theory-of-the-mind" functions—will mean that for the foreseeable future strategic decision-making will remain a fundamentally human endeavor—albeit imbued with increasing degrees of interaction with intelligent machines.[117] Thus, AI will continue to include some human agency—especially in collaboration with machines—for managing the attendant issues associated with technological complexity and interdependence, avoiding, for now at least, the risks associated with pre-delegating the use of military force. With the innovation of AI-infused battlefield digital assistants, however, the human is often moving to a supervisory role as the physical translation of commanders' intentions (or "mission command") is being performed by non-human agents, with a potential reduction in a human sense of agency and thus attention and vigilance.[118]

While human agency should ensure that the role of AI in the nuclear domain is confined to a predominately ancillary one, through the discharge of its "support role" at a tactical level (data collection and analysis, situational awareness, stockpile management, and decision-making support systems), it may still—and possibly unbeknownst to commanders—influence strategic decisions that involve nuclear weapons.[119] In other words, the distinction between AI's impact at a tactical and strategic level is *not a binary one*.[120] Thus, irrespective of whether human operators remain "in the loop" or "out of the loop" (see Figure 2.2) the complexities of human-to-machine interaction intrinsic to AI-augmented applications should be taken into consideration. Technology designed ostensibly to augment autonomous weapon systems to fulfill tactical and operational missions such as ISR remote sensing, target recognition, and battlefield situation awareness—because of its potential effect on surprise attacks and on mutual strategic vulnerability—could nonetheless inform and shape strategic calculations.[121] In short, escalation at the tactical level could easily have *strategic effects*—inviting instability exacerbated by misperception, misinterpretation, and miscalculation (see Chapters 3 and 4).

On the one hand, future AI-augmented C2 support tools may overcome many of the psychological shortcomings inherent to human strategic decision-making during wartime (e.g., susceptibility to invest in sunk costs, skewed risk judgment, heuristics, cognitive bias, and groupthink) with potentially stabilizing effects. Besides, faster and more reliable AI applications could also enable commanders to make more informed decisions during a crisis, improve the safety and reliability of nuclear support systems, strengthen the cyber defenses of C2 networks, enhance battlefield situational awareness, and reduce the risk of human error caused by fatigue and repetitive tasks.[122] On the other hand, AI systems that allow commanders to predict the potential production, commissioning, deployment, and, ultimately, launch of nuclear weapons by adversaries will likely lead to unpredictable system behavior and outcomes, which, *in extremis*, could undermine first-strike stability—the premise of MAD—thus making nuclear wars winnable.[123]

The US 2018 Nuclear Posture Review, for example, explicitly states that the DoD would pursue design support technologies such as ML to facilitate more effective and faster strategic decision-making.[124] China have also begun to research the use of big-data and deep-learning AI techniques to enhance the processing speed and intelligence analysis of satellite images, support its early-warning capabilities, and, ultimately, enable a "prediction revolution" in future warfare.[125] Chinese researchers have also applied AI to wargaming and military simulations, and have researched AI-enabled data retrieval and analysis from remote sensing satellites, to generate data and insights that may be used to enhance Chinese early-warning systems, situational awareness, and targeting.[126]

Under crisis conditions, the deterrent effect of AI is predicated on the perceived risks associated with a particular capability it enables or enhances. The higher the uncertainty generated by a capacity, the more likely deploying AI-augmented capabilities in a crisis might encourage an adversary to act more cautiously and, thus, in turn, bolster stability. Counter-intuitively, therefore, states may view the expanded automation of their NC3 systems to manage escalation and strengthen deterrence, signaling to an adversary that any attack (or the threat of one) might trigger nuclear escalation. Put another way, if a nuclear-armed state used automation to reduce its flexibility during a crisis, and without the ability to signal this to an adversary, it would be akin to Herman Kahn's notion of "being drunk, blind, and without a steering wheel" in a game of chicken.[127] Thus, even if an actor is unable to credibly threaten to use nuclear weapons in a crisis, it can still take steps that raise the risk that a crisis might inadvertently or accidentally escalate to nuclear confrontation. Thomas Schelling noted, "another paradox of deterrence is that it does not always help to be, or to be believed to be, fully rational, cool-headed, and in control of oneself or of one's country."[128]

A prima facie argument exists, therefore, that fusing NC3 systems with AI-augmented automated response mechanisms—akin to an enhanced version of Russia's primitive "semi-autonomous"[129] (i.e., a human's finger was ultimately on the nuclear button) *Perimeter* or "Dead Hand" (see Figure 2.1)—might resolve the logical paradox inherent with rational-based classical deterrence.[130] That is, predicated on the notion of mutual destruction and the will to retaliate, thus ensuring mutual deterrence and improving stability.[131] Using game theoretic logic, Schelling described a condition in which neither side can credibly threaten deliberate nuclear attack, but where one side may credibly make "threats which leave something to chance"—how and why actors can unintentionally stumble into war and strategically seek to manipulate risk to achieve advantage.[132] In this situation, brinksmanship—competition in risk-taking or manipulating risk—may be used by the actor with higher stakes in the contest to out-bargain the other by running greater risks of inadvertent or accidental escalation—risks that are complicated further in asymmetric regional contests involving multiple players with mutually exclusive stakes in the game (see Chapters 3 and 4).[133]

This perceived deterrent effect, coupled with the uncertainty caused by the introduction of AI into a situation, might incentivize states facing a militarily superior adversary to delegate decisions to machines (i.e., fully autonomous mode) to signal resolve during a crisis.[134] Because of the difficulty of demonstrating a posture like this *before* a crisis or conflict, this implicit threat—akin to the Dr. Strangelove's doomsday machine parable—may equally worsen crisis instability.[135] Moreover, the confusion and uncertainty that would result from mixing various (and potentially unknown) levels of human-machine interactions and AI is reacting to events—such as signaling and low-level

To launch any retaliatory strike, the system would have had to run through an if/then propositional formula:*
IF the system was turned on
AND
IF a nuclear weapon had hit Soviet soil
AND
IF there was no communications link to the war room of the Soviet General Staff,
THEN the system would initiate the process of launching a retaliatory strike.

*If command and control communications links survived, *Perimetr* would assume that high-level decision-makers who could order an attack were still alive and remained in full control of the decision-making process. However, if the link was lost, then the system would conclude that these decision-makers did not survive and transfer the decision authority to whoever was then manning the system deep inside a protected bunker. See David E. Hoffman, *The Dead Hand: The Untold Story of the Cold War Arms Race and Its Dangerous Legacy* (Anchor Books: New York, 2009), chapter 6.

Figure 2.1 Russia's Perimeter or "Dead hand" launch formula

conflict—in non-human ways (using force where a human commander would not have), and at machine speed, could dramatically increase inadvertent risk.

The recent defeat of a human pilot by an AI system in a DARPA-hosted Alpha Dogfight Challenge demonstrated how AIs performing in complex physics in a dynamic (albeit virtual) environment can compress the observe, orient, decide, and act (OODA) decision-making loop—used by combatants to gain the tactical upper hand by generating confusion and disorder among adversaries—and apply non-conventional (or non-human) tactics in a high-stakes game of human-to-machine chicken.[136] According to Lieutenant General Dennis Crall, who heads the Joint All Domain Command and Control initiative for the US DoD Joint Chiefs, AI on the battlefield will mean "the *time that we had to make these decisions has all but evaporated*, that the real battlefield calculus has changed. So, *our OODA loop ... really has to change significantly*" (emphasis added).[137]

Deterrence and human-to-machine interaction

How might the introduction of intelligent machines affect human-to-human deterrence? Is it possible for humans and machines to combine their respective talents and work together as a symbiotic whole? Recent experimental wargaming hosted by the RAND Corporation, which explored the effects of mixing various levels of human and machine configurations on escalatory dynamics—signaling, decision-making, and de-escalation—during a crisis revealed some interesting preliminary findings.[138] The tentative results demonstrated that where high levels of autonomy coincide with primarily human decision-making (known as "humans in the loop"), escalation risk is generally lower. This hypothesis was attributed to the fact that human involvement in decisions allowed more time to de-escalate (e.g., devise off-ramps), and that humans are likely to have a better understanding of signaling (i.e., resolve, deterrent threat, desire to de-escalate, or reassurance) compared to AI. In short, AI would likely be worse (or at least less reliable/predictable) than humans at understanding signaling involved in deterrence, particularly signaling involving resolve and de-escalation.[139]

Conversely, and more speculatively, when decisions are primarily made by machines and combined with high levels of autonomy (known as "humans out of the loop"), escalation risk is higher—but because of the lower human-risk the perceived costs of miscalculation are lower. The potential deterrent effect of this futuristic configuration of machine vs. machine interaction (or

"AI-enhanced general intelligence") is confounding.[140] For instance, would an adversary calculate risks differently due to the presence of a non-human agent on the opposing side? Traditional IR deterrence theory—intertwining human perception, cognition, concepts, beliefs, judgments, emotion, intuition, signaling intent, and rationality—would be effectively redundant. AI researchers have already made significant progress in the synthetic modeling of human belief mechanisms to predict an adversary's intentions and behavior more accurately in simulated battlefield situations.[141] Many AI algorithms used by (semi)autonomous systems today are based on game and decision theoretic rational decision-making,[142] which is related to a broader field of "rational choice theory," which includes neoclassical behavioral economics, decision theory, and analytical philosophy.[143] By replacing human agency and rationality with complex and opaque (or "black box") algorithmic imitations of human rationality, AI technology risks obfuscating and upending the traditional deterrence mechanisms, norms, and postures.[144]

On the one hand, AIs—embedded with algorithms that enable rational decision-making—that assume the cognitive load for judgment, coordination, and control during a crisis or conflict might reduce the traditional risks associated with accidents in human-machine interactions.[145] On the other hand, the total absence of a normative deterrence framework—in particular, to signal to resolve to an adversary while simultaneously seeking to de-escalate a situation—may compress (or remove entirely) the various rungs of the inherently psychological escalation ladder framework (see Chapter 3). This could increase inadvertent escalation risk and complicate de-escalation and conflict termination—especially between asymmetric dyads where incentives to strike pre-emptively to achieve escalation dominance exist.[146] These risks, to be sure, would be compounded if the role of increasingly sophisticated AIs extended all the way from the early-warning and detection to the judgment and launch decision-making phase.[147]

Figure 2.2 illustrates the theoretical delineation between the use of AI technology to *make* a decision and to *inform* a decision. These lines would quickly blur, however, if commanders became overly reliant on machine speed, real-time battlefield vision, and predictions—taking humans *de facto* out of the loop even in the context of a deceptively alluring but vague policy framework designed to ensure "meaningful human control."[148] Even if the final war-making decision rests with human decision-makers, the degree to which human control is "meaningful" will be determined by (a) the centrality of AI technology to the human decision-making process and (b) how much deference and authoritativeness AI's decisions are imbued with. In short, the

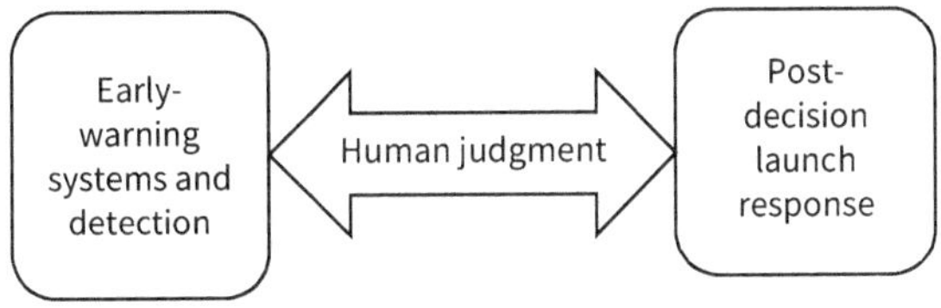

Figure 2.2 Human decision-making and judgment ("human in the loop") vs. machine autonomy ("human out of the loop")

delegation of the decision-making process (to inform *and* make decisions) to machines is not a binary choice but rather a continuum between the two extremes—human decision-making and judgment and machine autonomy at each stage of the kill-chain. However, and irrespective of whether humans keep their finger on the red button, the higher the centrality or (over)reliance upon machines' (known as "automation bias") decision-making process to *inform* a decision will increase AI's influence on the range and quality of options offered to planners at each stage of the decision-making chain.

The potentially escalatory effects of AI's tactical optimization programming would likely be compounded by differences in adversaries' goal setting (an AI's priorities, value alignment, control, and off-ramps),[149] C2 organization (centralized vs. decentralized), and the configuration of their human-machine interactions. Specifically, machine decision-making—designed to exploit the tactical and operational advantages in a situation—may lack the "theory of the mind" in *a priori* situations with their interactions with humans.[150] AIs analyzing a situation need to understand observable factors such as troop movements and weapon stockpiling; they must also grasp the complex, unpredictable, and highly nuanced nature of human psychological interactions in domestic and IR. The use of machines to predict the probability that an adversary will use force based on their signaling and behavior is, therefore, inherently difficult. A state might, for instance, use brinksmanship or other aggressive posturing behavior (e.g., putting its nuclear weapons on high alert or moving its warheads out of their garrisons) to signal that it is *prepared* to use force or to persuade the other side to back down, while not actually intending to escalate a situation.[151]

Theoretically, an AI algorithm optimized to pursue pre-programed goals—trained on historical combat scenarios, wargaming analytics, and intelligence datasets—could misinterpret an adversary simultaneously signaling resolve who is seeking to avoid conflict or de-escalate a situation. Though

Russia, for example, is actively developing AI-enabled systems to support weapons management, strategic forecasting, and decision-making—to ameliorate some of the natural, physical, and psychological constraints on human combatants—analysts remain concerned about the potential ramifications of future AI-infused military systems that can autonomously set goals, eliminating the role of human agency altogether.[152]

Absent reliable means to determine an actor's intentions—and making decisions using simplified value-maximizing processes—AI systems may convey undesirable and unintended (by human commanders) signals to the enemy, thus complicating the delicate balance between an actor's willingness to escalate a situation as a last resort and keeping the option open to step back from the brink. In short, an overreliance on AI to inform strategic decisions by relegating the role of humans in reviewing, scrutinizing, and invalidating the assumptions about an adversary's intentions and behavior—which undergirds deterrence theory—would risk deterrence failure and (especially inadvertent) escalation. An open question remains about how the various parts of hypothetical human-to-machine synergetic would work in practice include: would the humans become too reliant on the AI, believing it to be superior at its job? Or might commanders distrust AI's recommendations because of its "black box" machine fuzzy logic and non-human nature?

Conclusion

This chapter argues that AI and autonomy could affect nuclear deterrence in two ways: decreasing stability in nuclear multipolarity and increasing the tendency for (especially inadvertent) escalation to nuclear use. The chapter found that existing classical deterrence theories rooted in rational-based human actor assumptions are no longer applicable in the light of recent developments in AI and autonomy. It builds on the nascent "fifth wave" of modern post-classical deterrence theorizing, which considers the potential implications of introducing non-human agents—and the corresponding disengagement or even removal of human agents—from strategic interactions.

Additional early insights from the theoretical and empirical research for nuclear deterrence in the age of intelligence machines includes the following. First, AI's impact on nuclear deterrence will be as important (if not more so) for how its effect on human perception, cognition, and human-to-machine interaction, rather than the immediate result of the technology per se.

Second, how AI-enabled systems are designed, adapted, and assimilated into doctrine and force structure by states may lead to technological displacement or deflection that might undermine presumed deterrence successes—and thus increase the risk of misperception, miscalculation, and inadvertent or accidental escalation (see Chapter 3). Third, and closely related, determining with any fidelity and measuring "deterrence success" in the use of AI applications is a complicated endeavor, requiring counterfactual hypothesis testing that definitively connects AI itself to behavior that has yet to occur (see Chapter 5). Experimental wargaming—which tests various hypotheses and variables in different combat scenarios, using multiple potential applications of AI-augmentation, and from the perspective of various adversaries—and other empirical case studies might be helpful here.[153] Fourth, ethics, norms, and international law will play an important role in AI and nuclear deterrence (and in the role of AI in warfare more generally), which will shape and inform the technology's future use and utility in the nuclear deterrence architecture—and, ultimately, its justification and applicability.

The employment of AI and autonomy in the nuclear enterprise entails a multitude of trade-offs and open questions that should persuade designers, policymakers, and operators to explore ways in which new AI-powered capabilities might strengthen or complicate deterrence in the fragile nuclear "balance of terror,"[154] how adversaries think about these developments, and the effect of synthesizing legacy nuclear systems (especially NC3) with AI technology. Examples of these trade-offs, *inter alia*, include the following.[155] First, ensuring adequate control, safety, and supervision of AIs to mitigate escalation risk, while exploiting the potential benefits of increased lethality, scale, and speed afforded by this technology. Second, assimilating AI technology into force structure and doctrine, while mitigating the dangers posed by the speed and reduced levels of control of nuclear weapons. Third, instilling AI-augmented C2 systems with sufficient knowledge of the laws of engagement—and other legal and normative war-faring frameworks—without degrading AI's capacity to predict and respond to threats—especially confrontations that arise from introducing new capabilities into a situation (i.e., vertical escalation).[156] Finally, and related, striking the right balance between communicating, signaling, or demonstrating AI-enabled capabilities to adversaries (to signal resolve, or increase transparency to reduce ambiguity, miscalculation, and accidents), while preserving the technological upper hand.

Objectively appraising and reconciling these trade-offs will likely be complicated by people's tendency to avoid making decisions when faced with incommensurable problems containing moral and ethical choices.[157] A reticence to engage in these issues or, worse, downplay the potential risks associated with AI and autonomy would make it more challenging to alter incentives to enhance strategic stability and shape deterrence and escalation as the technology matures, and the military use of these systems inevitably increases. Cognizant that some states have or plan to deploy AI systems in their nuclear deterrence structures (e.g., high-precision missile systems, missile defenses, cyber offense, electronic-warfare, and physical security),[158] experts generally agree that AI requires further experimentation, testing, and verification before being integrated into nuclear support systems.

How might militaries develop and deploy AI to steer it towards ensuring mutual deterrence and enhancing strategic stability? To improve strategic stability in an era of rapid technological change, great-power strategic competition, and nuclear multipolarity, the formulation of future arms control frameworks must reflect the shifting perspectives described in this study. How might the existing nuclear arms control regime be reconfigured and broadened to incorporate emerging technologies like AI?[159] The dual-use and diffused nature of AI (see Chapter 4) compared to nuclear technology will, however, make arms control efforts particularly problematic (see the conclusion). Moreover, when nuclear and non-nuclear capabilities and war-faring are blurred, strategic competition and arms racing are more likely to emerge, complicating arms control efforts.[160] In short, legacy arms control frameworks, norms, and even the notion of strategic stability itself will increasingly struggle to assimilate and respond to these fluid and interconnected trends—this analysis is elaborated upon in the conclusion.

Governments should also explore ways to increase transparency and accountability for AI and national security, such as addressing the implications of deepfakes and lethal autonomous weapons.[161] To counter the threat posed by non-state actors using AI-enabled tools such as deepfakes to manipulate, deceive, or otherwise interfere with strategic decision-making systems in misinformation attacks, governments should coordinate with both allies and adversaries—continuing to harden NC3 systems and processes against cyber-attacks.[162] Other measures that may also improve stability, among other things, include: reducing the number of nuclear weapons; taking arsenals off high alert (or launch-on-warning) status; separating warheads from delivery systems (or de-matting warheads); shifting to a deterrent-only (or minimum deterrence) force posture, and adopting a no-first-use declaratory policy—as China and India do today.[163] How might AI enhanced weapons increase the risk of inadvertent nuclear war? The next chapter picks up this challenge.

Notes

1. "Autonomous" systems, by ingesting vast datasets, can learn and adapt to dynamic (or non-deterministic) environments, evolving as the environment around it changes faster (and eventually more reliably) than what would be reasonable for a human. In contrast, "automated systems" typically run within a well-defined set of parameters and are very restricted in what tasks they can perform, and the decisions made or actions taken by these systems are based on predefined heuristics. AI and ML will have an increasing role in automated and autonomous systems. For automated systems, ML will enable them to be more resilient and efficient, and for autonomous systems AI will bring about more reliable and sophisticated decision-making and the ability to handle greater complexity. See Scott Matterson, "Autonomous Versus Automated: What Each Means and Why it Matters," *TechRepublic*, June 7, 2021, https://www.techrepublic.com/article/autonomous-versus-automated-what-each-means-and-why-it-matters/.
2. Sections of this chapter are derived in part from James Johnson, "Deterrence in the Age of Artificial Intelligence & Autonomy: A Paradigm Shift in Nuclear Deterrence Theory and Practice?" *Defense & Security Analysis* 36, no. 4 (2019): pp. 422–448. Copyright Taylor & Francis, available online: doi: 10.1080/14751798.2020.1857911.
3. For example, see Michael C. Horowitz, Sarah E. Kreps, and Matthew Fuhrmann, "Separating Fact from Fiction in the Debate over Drone Proliferation," *International Security* 41, no. 2 (Fall 2016): pp. 7–42; Andrew Ilachinski, *Artificial Intelligence & Autonomy: Opportunities and Challenges* (Arlington, VA: CNA, 2017); Mary L. Cummings, *Artificial Intelligence and the Future of Warfare* (London: Chatham House, The Royal Institute of International Affairs, January 2017); and Elsa B. Kania, *Battlefield Singularity: Artificial Intelligence, Military Revolution, and China's Future Military Power* (Washington, DC: Center for a New American Security, 2017).
4. See Michael C. Horowitz, "Artificial Intelligence, International Competition, and the Balance of Power," *Texas National Security Review* 1, no. 3 (2018): pp. 37–57; and Itai Barsade and Michael C. Horowitz, "Artificial Intelligence Beyond the Superpowers," *Bulletin of the Atomic Scientists*, August 16, 2018, https://thebulletin.org/2018/08/the-ai-arms-race-and-the-rest-of-the-world/.
5. Notable studies that consider the implications of AI for deterrence and nuclear stability include: Yuna Huh Wong et al., *Deterrence in the Age of Thinking Machines* (Santa Monica, CA: RAND Corporation, 2020), doi: 10.7249/RR2797; Edward Geist and Andrew Lohn, *How Might Artificial Intelligence Affect the Risk of Nuclear War?* (Santa Monica, CA: RAND Corporation, 2018); Michael Horowitz, Paul Scharre, and Alex Velez-Green, *A Stable Nuclear Future? The Impact of Automation, Autonomy, and Artificial Intelligence* (Philadelphia, PA: University of Pennsylvania, 2017); and James Johnson, "Artificial Intelligence in Nuclear Warfare: A Perfect Storm of Instability?" *The Washington Quarterly* 43, no. 2 (2020): pp. 197-211.
6. The literature is expansive; notable studies include, *inter alia*, Lawrence Freedman, *Deterrence* (New York: Polity, 2004); Lawrence Freedman, *The Evolution of Nuclear Strategy* (Basingstoke: Palgrave Macmillan, 2003); Thomas Schelling, *Arms and Influence* (New Haven and London: Yale University Press, 1966); Glen H. Snyder, *Deterrence and Defense: Towards a Theory of National Security* (Princeton, NJ: Princeton University Press, 1961); Patrick M. Morgan, *Deterrence Now* (Cambridge: Cambridge University Press, 2003);

and Michael Quinlan, *Thinking About Nuclear Weapons: Principles, Problems, Prospects* (Oxford: Oxford, University Press 2009).

7. The origins of classical IR deterrence theory can be traced to the aftermath of World War I and can be categorized into several broad threads, including structural deterrence theory and decision-theoretic deterrence theory. See Stephen L. Quackenbusch and Frank C. Zagare, *Modern Deterrence Theory: Research Trends, Policy Debates, and Methodological Controversies, Oxford Handbooks* (Oxford: Oxford University Press, 2016).
8. The proper legal employment of nuclear weapons depends on intelligence estimates and strategic judgment. The US DoD Law of War Manual notes, for instance, in the context of assessing military advantage, that decisions about application of the law of armed conflict "may require knowing the broader strategy being employed by the attacking party or knowing intelligence information about the strategic and operational context in which the attack takes place." US DoD Nuclear Employment Strategy; and Department of Defense Law of War Manual (Washington, DC: General Counsel of the Department of Defense, June 2015), p. 243.
9. The term "emerging technologies" lacks definitional clarity and is thus excessively broad and heterogenous—policymakers tend to treat them as a broad risk category. In 2020, NATO defined "emerging technologies" as "those technologies or scientific discoveries that are expected to reach maturity in the period 2020–[2030]; and are not yet widely in use or whose effects on defence [and] security are not entirely clear." NATO Science & Technology Organization, "Science & Technology Trends 2020–2040: Exploring the S&T Edge," March 2020, https://www.nato.int/nato_static_fl2014/assets/pdf/2020/4/pdf/190422-ST_Tech_Trends_Report_2020-2040.pdf.
10. For scholarship on the intersection of technology with the potential for war, see James D. Fearon, "Cooperation, Conflict, and the Costs of Anarchy," *International Organization* 72, no. 3 (2018): pp. 523–559; for escalation dynamics, see Caitlin Talmadge, "Emerging Technology and Intra-War Escalation Risks: Evidence from the Cold War, Implications for Today," *Journal of Strategic Studies* 42, no. 6 (2019): pp. 864–887; for arms racing and strategic stability, see Todd S. Sechser, Neil Narang, and Caitlin Talmadge, "Emerging Technologies and Strategic Stability in Peacetime, Crisis, and War," *Journal of Strategic Studies* 42, no. 6 (2019): pp. 727–735; and for deterrence and the offense-defense balance, see Ben Garfinkel and Allan Dafoe, "How Does the Offense-Defense Balance Scale?" *Journal of Strategic Studies* 42, no. 6 (2019): pp. 736–763; and Joseph S. Nye, "Deterrence and Dissuasion in Cyberspace," *International Security* 41, no. 3 (2017): pp. 44–71.
11. Notable exceptions include Herbert Lin, "Escalation Dynamics and Conflict Termination in Cyberspace," *Strategic Studies Quarterly* 6, no. 3 (Fall 2012): pp. 46–70; and Erik Gartzke and Jon R. Lindsay, "Thermonuclear Cyberwar," *Journal of Cybersecurity* 3, no. 1 (February 2017): pp. 37–48.
12. James Johnson, "Artificial Intelligence & Future Warfare: Implications for International Security," *Defense & Security Analysis* 35, no. 2 (2019): pp. 147–169.
13. For example, see Bernard Brodie, *Escalation and the Nuclear Option* (Princeton, NJ: Princeton University Press 1965); Snyder, *Deterrence and Defense: Towards a Theory of National Security*; Thomas Schelling, *The Strategy of Conflict* (Cambridge, MA: Harvard University Press, 1960), and Schelling, *Arms and Influence.*
14. Freedman, *Deterrence*, p. 122.

15. During the Cold War, deterrence was viewed almost solely in terms of nuclear weapons. However, nuclear-armed states have also sought to deter—both state and non-state—aggression and developed advanced conventional military capabilities to underwrite their "conventional deterrence" postures. Michael S. Gerson, "Conventional Deterrence in the Second Nuclear Age," *Parameters* (Autumn 2009): pp. 32–48.
16. Freedman, *Deterrence*, pp. 110–111.
17. See Robert Powell, "Nuclear Deterrence Theory, Nuclear Proliferation, and National Missile Defense," *International Security* 27, no. 4 (Spring 2003): pp. 89–91.
18. Alex S. Wilner, "US Cyber Deterrence: Practice Guiding Theory," *Journal of Strategic Studies* 43, no. 2 (2020): pp. 245–280.
19. Some have argued that policymakers, by confusing "rational" with "reasonableness," means that when adversaries fail to act as intended—or as a particular theory predicts—they appear irrational. An actor can be "rational" within their belief structure or theoretical framework of understanding. Kenneth Payne, *The Fallacies of Cold War Deterrence and a New Direction* (Lexington, KY: University of Kentucky, 2001), pp. 7–15.
20. For example, President John F. Kennedy's public pledge in 1962 not to invade Cuba—assuring the Soviets that capitulation would not invite future demand—helped resolve the missile crisis.
21. A deterrent effect can result from military threats and non-military forces such as economic, diplomatic, and political inducements and assurances. Freedman, *Deterrence*, p. 59.
22. Robert Jervis, *How Statesmen Think: The Psychology of International Politics* (Princeton, NJ: Princeton University Press, 2017), p. 192.
23. James Fearon, "Rationalist Explanations for War," *International Organization* 49, no. 3 (1995): pp. 379–414.
24. Jeffrey W. Knopf, "The Fourth Wave in Deterrence Research," *Contemporary Security Policy* 31, no. 1 (2010): pp. 1–33.
25. Todd S. Sechser, "Goliath's Curse: Coercive Threats and Asymmetric Power," *International Organization* 64, no. 4 (Fall 2010): pp. 627–660; and Knopf, "The Fourth Wave in Deterrence Research," p. 646.
26. Richard N. Lebow, "Conclusions," in Robert Jervis, Richard N. Lebow, and Janice G. Stein, *Psychology and Deterrence* (New York: Johns Hopkins University Press, 1985), p. 226.
27. Jonathan Mercer, *Reputation and International Politics* (Ithaca, NY: Cornell University Press, 1996); Robert Jervis, *Perception and Misperception in International Politics* (Princeton, NJ: Princeton University Press, 1976); and Ted Horf, *Peripheral Visions: Deterrence Theory and American Foreign Policy in the Third World* (Ann Arbor, MI: University of Michigan Press, 1994).
28. Horf, *Peripheral Visions*, p. 111.
29. Sechser, "Goliath's Curse: Coercive Threats and Asymmetric Power"; and Knopf, "The Fourth Wave in Deterrence Research," pp. 1–33.
30. Thomas Rid, "Deterrence Beyond the State: The Israeli Experience," *Contemporary Security Policy* 33, no. 1 (2012): pp. 124–147; and Amir Lupovici, "The Emerging Fourth Wave of Deterrence Theory," *International Studies Quarterly* 54 (2010): pp. 705–732.
31. While this chapter focuses on how traditional IR intersects with AI deterrence, deterrence theory also permeates a broad spectrum of human behavior, including criminology, terrorism studies, cybersecurity, and cognitive-psychological disciplines. For example, see

Anthony Braga and David Weisburd, "The Effects of Focused Deterrence Strategies on Crime," *Journal of Research in Crime and Delinquency* 49 (2012), pp. 1–65; Rob Guerette and Kate Bowers, "Assessing the Extent of Crime Displacement and Diffusion of Benefits," *Criminology* 47, no. 4 (2009), pp. 1331–1368; Trevor Bennett, Katy Holloway, and David Farrington, "Does Neighborhood Watch Reduce Crime?" *Journal of Experimental Criminology* 2, no. 4 (2006), pp. 437–458; and Wilner, "US Cyber Deterrence: Practice Guiding Theory."

32. See Gerson, "Conventional Deterrence in the Second Nuclear Age," pp. 32–48; Daniel Sobelman, "Learning to Deter," *International Security* 41, no. 3 (2016/17), pp. 151–196; and Linton Brooks and Mira Rapp-Hooper, "Extended Deterrence, Assurance, and Reassurance in the Pacific During the Second Nuclear Age," in *Strategic Asia 2013–14: Asia in the Second Nuclear Age*, edited by Ashley J. Tellis, Abraham M. Denmark, and Travis Tanner (Seattle, Washington: National Bureau of Asian Research, 2013), pp. 266–300.
33. See Wilner, "US Cyber Deterrence: Practice Guiding Theory," pp. 7–8; and Andrew Futter, *Hacking the Bomb: Cyber Threats and Nuclear Weapons* (Washington, DC: Georgetown University Press, 2018).
34. Notable examples include Wong et al., *Deterrence in the Age of Thinking Machines*; Horowitz, "When Speed Kills: Lethal Autonomous Weapon Systems, Deterrence, and Stability," *Journal of Strategic Studies* 42, no. 6 (2019): pp. 764–788; and Johnson, "Deterrence in the Age of Artificial Intelligence & Autonomy."
35. Tim Prior, "Resilience: The 'Fifth Wave' in the Evolution of Deterrence," in *Strategic Trends 2018*, edited by Oliver Thränert and Martin Zapfe (Zurich: Center for Security Studies, ETH Zurich, 2018), pp. 63–80; Beyza Unal, Yasmin Afina, and Patricia Lewis, "Perspectives on Nuclear Deterrence in the 21st Century," International Security Program Research Paper, Chatham House (April 2020); and Stephen L. Quackenbush and Frank C. Zagare, "Modern Deterrence Theory: Research Trends, Policy Debates, and Methodological Controversies," *Oxford Handbooks Online*, May 2016, doi: 10.1093/oxfordhb/9780199935307.013.39.
36. Reid B.C. Pauly, "Would US Leaders Push the Button? Wargames and the Sources of Nuclear Restraint," *International Security* 43, no. 2 (2018): pp. 151–192; Jacquelyn G. Schneider, "Cyber Attacks on Critical Infrastructure: Insights from War Gaming," *War on the Rocks*, July 26, 2017; and Erik Lin-Greenberg, "Wargame of Drones: Remotely Piloted Aircraft and Crisis Escalation" August 22, 2020; *SSRN*, https://ssrn.com/abstract=3288988.
37. Forrest E. Morgan et al., *Dangerous Thresholds: Managing Escalation in the 21st Century* (Santa Monica, CA: RAND Corporation, 2008), p. 8.
38. Herman Kahn, *On Escalation: Metaphors and Scenarios* (New York: Praeger, 1965).
39. Charles Perrow, *Normal Accidents: Living with High-Risk Technologies* (New York: Basic Books, 1984); Scott D. Sagan, *The Limits of Safety: Organizations, Accidents, and Nuclear Weapons* (Princeton, NJ: Princeton University Press, 1993); and Schelling, *Arms and Influence*, pp. 227–228.
40. Stephen J. Cimbala, *The Dead Volcano: The Background and Effects of Nuclear War Complacency* (Westport, CT: Praeger, 2002), p. 147.
41. Michael D. Intriligator and Dagobert L. Brito, "Minimizing the Risks of Accidental Nuclear War: An Agenda for Action," in *Inadvertent Nuclear War*, edited by Hakan

Wilberg, Ib Damgaard Petersen, and Paul Smoker (New York: Pergammon Press, 1993), pp. 221–237.

42. Jervis, *Perception and Misperception in International Politics*, Ch. 3.
43. Military autonomous systems, and inadvertent risks associated with these systems, is not a new phenomenon. Examples of existing autonomous systems include landmines, torpedoes, the *Aegis* and *Patriot* missile defense systems, and the close-in anti-ship missile *Phalanx* system—a defensive weapon system which has been deployed for over thirty years. For analysis of errors involving these systems, see Wong et al., *Deterrence in the Age of Thinking Machines*, pp. 71–72.
44. David Mikkelson, "Did Yogi Berra Eloquently Explain the Difference between Practise and Theory?" *Snopes*, April 29, 2015, https://www.snopes.com/fact-check/practice-and-theory/.
45. Jervis, *The Meaning of the Nuclear Revolution*.
46. See Keir A. Lieber and Darryl G. Press, "The New Era of Counterforce: Technological Change and the Future of Nuclear Deterrence," *International Security* 41, no. 4 (Spring 2017): pp. 9–49; and Paul Bracken, "The Cyber Threat to Nuclear Stability," *Orbis* 60, no. 2 (2016): pp. 188–203.
47. Keir A. Lieber and Daryl G. Press, "Why States Won't Give Nuclear Weapons to Terrorists," *International Security* 38, no. 1 (2013): pp. 80–104.
48. The technical feasibility of this hypothesis—especially in the case of nuclear-powered submarines on extended deterrence patrols—remains highly contested. See Sebastian Brixey-Williams, "Will the Atlantic Become Transparent?" 2nd ed. *British Pugwash*, November 2016; and Owen R. Cote Jr., "Invisible Nuclear-Armed Submarines, or Transparent Oceans? Are Ballistic Missile Submarines Still the Best Deterrent for the United States?" *Bulletin of the Atomic Scientists*, 75, no. 1 (2019): pp. 30–35.
49. This concern arguably drives Russia's launch-on-warning nuclear posture. See Jeffrey Edmonds et al., "Artificial Intelligence and Autonomy in Russia," *Center for Naval Analysis* (May 2021), https://www.cna.org/CNA_files/centers/CNA/sppp/rsp/russia-ai/Russia Artificial-Intelligence-Autonomy-Putin-Military.pdf.
50. Defense experts estimate that while AI will solve some of these challenges in the near term, the many technical problems associated with tracking and targeting mobile missiles are unlikely to be overcome within the next two decades. Geist and Lohn, *How Might Artificial Intelligence Affect the Risk of Nuclear War?* p. 16.
51. Lieber and Press, "The New Era of Counterforce."
52. For recent literature that considers the effects of advanced technological conventional counterforce capabilities on strategic nuclear competition and escalation risks, see James M. Acton, "Escalation through Entanglement," *International Security* 43, no. 1 (Summer 2018): pp. 56–99; Stephen J. Cimbala, "Nuclear Crisis Management and 'Cyberwar': Phishing for Trouble?" *Strategic Studies Quarterly* (Spring 2011): pp. 117–128; and Erik Gartzke and Jon R. Lindsay, "Thermonuclear Cyberwar," *Journal of Cybersecurity*, 3, no. 1 (March 2017): pp. 37–48.
53. US Defense Advanced Research Projects Agency (DARPA), "ACTUV Sea Hunter Prototype Transitions to the US Office of Naval Research for Further Development," January 30, 2018.
54. Tong Zhao and Li Bin, "The Underappreciated Risks of Entanglement: A Chinese Perspective," in *Entanglement: Russian and Chinese Perspectives on Non-Nuclear Weapons*

and Nuclear Risks, edited by James M. Acton et al. (Washington, DC: Carnegie Endowment for International Peace, 2017), pp. 47–75.

55. In an asymmetric dyad, a weaker state's perception about the future (i.e., relative power and resolve) is likely to diverge with more powerful rivals, because stronger states have incentives to understate the future while weaker ones have incentives to overstate it—the existence of uncertainty in asymmetric relationship can lead to the failure of bargaining and deterrence. Erik Gartzke, "War is in the Error Term," *International Organization* 53 no. 3 (Summer, 1999): p. 584.
56. Caitlin Talmadge, "Would China Go Nuclear? Assessing the Risk of Chinese Nuclear Escalation in a Conventional War with the United States," *International Security* 41, no. 4 (Spring 2017): pp. 50–92.
57. Bradley Martin et al., *Advancing Autonomous Systems: An Analysis of Current and Future Technology for Unmanned Maritime Vehicles* (Santa Monica, CA: RAND Corporation, 2019).
58. See Cote Jr., "Invisible Nuclear-Armed Submarines, or Transparent Oceans?" pp. 30–35.
59. See Jonathan Gates, "Is the SSBN Deterrent Vulnerable to Autonomous Drones?" *The RUSI Journal* 161, no. 6 (2016): pp. 28–35.
60. Futter, *Hacking the Bomb*, p. 164.
61. Joe Tidy, "US Companies Hit by 'Colossal' Cyber-Attack," *BBC News*, July 3, 2021, https://www.bbc.com/news/world-us-canada-57703836.
62. China and Russia have already taken steps towards placing their retaliatory forces on higher alert, and the fear of a US cyber-attack against their nuclear deterrent forces could prompt them to accelerate efforts to place more of their nuclear weapons on hair-trigger alert. Gregory Kulacki, "China's Military Calls for Putting its Nuclear Forces on Alert," *Union of Concerned Scientists*, January 2016, pp. 1–9; and Keir Lieber and Daryl Press, *The Myth of the Nuclear Revolution: Power Politics in the Atomic Age* (Ithaca, NY: Cornell University Press, 2020) pp. 108–110, 117–118; and See Jeffrey Edmonds et al., "Artificial Intelligence and Autonomy in Russia," *Center for Naval Analysis* (May 2021), https://www.cna.org/CNA_files/centers/CNA/sppp/rsp/russia-ai/Russia-Artificial-Intelligence-Autonomy-Putin-Military.pdf.
63. Dmitry Stefanovich, "Russia's Basic Principles and the Cyber-Nuclear Nexus," *European Leadership Network*, July 14, 2020, https://www.europeanleadershipnetwork.org/commentary/russiasbasic-principles-and-the-cyber-nuclear-nexus.
64. See James Johnson, "The AI-Cyber Nexus: Implications for Military Escalation, Deterrence, and Strategic Stability," *Journal of Cyber Policy* 4, no. 3 (2019): pp. 442–460, doi: 10.1080/23738871.2019.1701693.
65. Joseph F. Dunford, speech quoted at Jim Garamone, "Dunford: Speed of Military Decision-Making Must Exceed Speed of War," US Department of Defense, January 31, 2017.
66. Robert Jervis, "Some Thoughts on Deterrence in the Cyber Era," *Journal of Information Warfare* 15, no. 2 (2016): pp. 68–69.
67. Paul Ingram, "Hacking UK Trident: A Growing Threat," *BASIC*, May 31, 2017.
68. ML systems rely on high-quality datasets to train their algorithms; poisoning this data could lead these systems to perform in unanticipated and potentially undetectable ways.
69. Eugene Bagdasaryan et al., "How to Backdoor Federated Learning," (preprint August 6, 2018), arXiv:1807.00459.

70. Acton et al., *Entanglement: Russian and Chinese Perspectives on Non-Nuclear Weapons and Nuclear Risks*, p. 81.
71. Ibid.
72. Gartzke and Lindsay, "Thermonuclear Cyberwar," pp. 37–48.
73. For analysis on how AI-augmented cyber capabilities might affect nuclear deterrence, escalation, and stability, see Johnson, "The AI-Cyber Nexus."
74. See Barton Whaley, "Covert Rearmament in Germany, 1919–1939: Deception and Misperception," *Journal of Strategic Studies* 5, no. 1 (March 1982): pp. 3–39; John J. Mearsheimer, *Why Leaders Lie: The Truth about Lying in International Politics* (Oxford: Oxford University Press, 2013); and Robert L. Jervis, *The Logic of Images in International Relations* (New York: Columbia University Press, 1989).
75. Schelling and Halperin, *Strategy and Arms Control*, p. 30.
76. Micro-drone UAVs range from small flying insects to palm-sized devices equivalent to small birds and can carry payloads, including small guns and explosives and unconventional munitions such as poisons and nerve agents. See Berenice Baker, "Dogfighting Drones—Swarms of Unmanned Battle-Bots Take to the Skies," Airforce-technology.com, July 23, 2013, www.airforcetechnology.com/features/featuredogfight-drones-unmanned-battle-bot-swarms/.
77. The US Office of Naval Research, for example, envisions currently experimenting with swarms of unmanned surface vehicles to form a defensive perimeter around larger ships and surround enemy ships as part of the counterterrorism effort. Thomas Claburn, "Navy Tests Swarming Autonomous Boats," *InformationWeek*, November 7, 2014, https://www.informationweek.com/messages.asp?piddl_msgthreadid=21052&piddl_msgorder=thrd.
78. Elias Groll, "How AI Could Destabilize Nuclear Deterrence," *Foreign Policy*, April 24, 2018, https://foreignpolicy.com/2018/04/24/how-ai-could-destabilize-nuclear-deterrence/.
79. The value of drones in these scenarios does not mean that they are the *only* or necessarily most effective way to fulfill these missions. Gates, "Is the SSBN Deterrent Vulnerable to Autonomous Drones?"
80. The US, Russia, South Korea, and China are also actively pursuing drone swarm technology programs.
81. In 2015, Russia revealed a large nuclear-armed UUV delivery vehicle, *Poseidon* (also known as Status-6). The US is also developing an "optionally manned" nuclear-capable long-range bomber, the B-21 *Raider*, carrying nuclear payloads. Ria Novosti, "Russia Could Deploy Unmanned Bomber After 2040—Air Force," *GlobalSecurity.org*, February 8, 2012, and Robert M. Gates, "Statement on Department Budget and Efficiencies" (US Department of Defense, January 6, 2011).
82. Mike Pietrucha, "The Need for SEAD Part 1: The Nature of SEAD," *War on the Rocks*, May 17, 2016.
83. Conversely, drone swarms might enhance states' missile defenses as countervails to this type of offensive threat. Polat Cevik et al., "The Small and Silent Force Multiplier: A Swarm UAV-Electronic Attack," *Journal of Intelligent and Robotic Systems* 70 (April 2013): pp. 595–608.
84. For example, DARPA's Anti-Submarine Warfare Continuous Trail Unmanned Vessel program, designed to track quiet diesel-electric submarines with USVs from the surface.

85. As technology advances—particularly AI—an increasing number of states will likely consider developing and operating lethal autonomous weapon systems (LAWS). The international community considers the implications of LAWS in conversations held under the auspices of the United Nations Convention on Certain Conventional Weapons, a multilateral arms control agreement to which the US became a party in 1982. The US and Russia (albeit for different reasons) have consistently opposed banning LAWS. China supports a ban on the use, but not development, of LAWS, which it defines as indiscriminate—lethal systems that do not have any human oversight and cannot be terminated. However, some have argued that China is maintaining "strategic ambiguity" about its position on LAWS. See Kelly M. Sayler and Michael Moddie, "International Discussions Concerning Lethal Autonomous Weapon Systems," *Congressional Research Service* (In Focus), October 15, 2020, https://fas.org/sgp/crs/weapons/IF11294.pdf.
86. Paul Scharre, *Autonomous Weapons and Operational Risk: Ethical Autonomy Project* (Washington, DC: Center for a New American Security, February 2016).
87. For a technical study on drone swarming technology and tactical deployment challenges for autonomous weapon systems, see Arthur M. Holland, *Known Unknowns: Data Issues & Military Autonomous Systems* (Geneva: UNIDIR, 2021); and Paddy Walker, "Leadership Challenges from the Deployment of Lethal Autonomous Weapon Systems," *The RUSI Journal* (2021) doi: 10.1080/03071847.2021.1915702.
88. Robert Einhorn and W.P.S Sidhu, *The Strategic Chain: Linking Pakistan, India, China, and the United States, Arms Control and Non-Proliferation Series Paper 14* (Washington, DC: Brookings Institution, March 2017).
89. Paul Bracken, *The Second Nuclear Age: Strategy, Danger, and the New Power Politics* (New York: Times Books, 2012); and Michael Krepon, "Can Deterrence Ever Be Stable?" *Survival* 57, no. 3 (2015): pp. 111–132.
90. "Rising Nuclear Risk, Disarmament, and the Nuclear Non-Proliferation Treaty," *UK House of Lords Select Committee on International Relations*, April 24, 2019, https://publications.parliament.uk/pa/ld201719/ldselect/ldintrel/338/33803.htm.
91. For literature on the "Second Nuclear Age" see Bracken, *The Second Nuclear Age*; Colin S. Gray, *The Second Nuclear Age* (Boulder, CO: Lynne Rienner, 1999); and Keith Payne, *Deterrence in the Second Nuclear Age* (Washington, DC: Georgetown University Press, 1996).
92. For China's approach to military AI, see Elsa B. Kania, "AI Weapons in China's Military Innovation," Global China, Brookings Institution, April 2020, https://www.brookings.edu/research/ai-weapons-in-chinas-military-innovation/; and for Russia's approach, see "Artificial Intelligence in Russia," Issue 12, *The Russia Studies Program, Center for Navy Analysis*, October 2020.
93. Kenneth N. Waltz, *Theory of International Politics* (Reading, MA: Addison-Wesley, 1979), p. 168.
94. Game-theoretic analysis suggests that while significant differences exist in crises, they only need to raise the specter of military conflict to generate reputations for resolve—actors who can signal that they place a low value on military costs during a crisis (or cost-tolerance) will likely obtain improved settlement terms. James. D. Fearon, "Signaling Foreign Policy Interests: Tying Hands Versus Sinking Costs," *Journal of Conflict Resolution* 41, no. 1 (1997): pp. 68–90.

95. John J. Mearsheimer, *Conventional Deterrence* (Ithaca, NY: Cornell University Press, 1984), p. 210.
96. Thomas J. Christensen, "The Meaning of the Nuclear Evolution: China's Strategic Modernization and U.S.-China Security Relations," *Journal of Strategic Studies* 35, no. 4 (August 2012): pp. 467–471.
97. James Johnson, "Delegating Strategic Decision-Making to Machines: Dr. Strangelove Redux?" *Journal of Strategic Studies* (2020) doi: 10.1080/01402390.2020.1759038.
98. John J. Mearsheimer, "The Gathering Storm: China's Challenge to US Power in Asia," *The Chinese Journal of International Politics* 3, no. 4 (Winter 2010): pp. 381–396.
99. For example, see Yang Yaohui, "A Vision of a New Kind of Combat Systems," *PLA Daily*, June 20, 2020; and Yang Feilong and Li Shijiang, "Cognitive Warfare: Dominating the Era of Intelligence," *PLA Daily*, March 19, 2020.
100. Keren Yarhi-Milo, *Knowing the Adversary* (Princeton NJ: Princeton University Press, 2014), p. 250.
101. David E. Apter, *The Politics of Modernization* (Chicago, IL: Chicago University Press, 1967).
102. See Michael J. Mazarr et al., *The Emerging Risk of Virtual Societal Warfare: Social Manipulation in a Changing Information Environment* (Santa Monica, CA: RAND Corporation, 2019).
103. Allan Dafoe, Jonathan Renshon, and Paul Huth, "Reputation and Status as Motives for War," *Annual Review of Political Science* 17, no. 1 (2014): pp. 371–393.
104. Lloyd R. Leavitt and Paul Bracken, "Nuclear Proliferation: Neither Safe nor Stable," in *Inadvertent Nuclear War*, edited by Hakan Wilberg, Ib Damgaard Petersen, and Paul Smoker (New York: Pergammon Press, 1993), pp. 203–217, pp. 215–216.
105. Jervis, *How Statesmen Think*, pp. 196–197.
106. Reports indicate that China and India have considered making changes to their respective no-first-use pledges. Zhenqing Pan, "A study of China's No-First-Use Policy on Nuclear Weapons," *Journal of Peace and Nuclear Disarmament* 1, no. 1 (2018): pp. 115–136; and Harsh V. Pant and Yogesh Joshi, "Nuclear Re-Think: A Change in India's Nuclear Doctrine Has Implications on Cost & War Strategy," *Economic Times* (New Delhi), August 17, 2019.
107. China, India, and Pakistan have reportedly shifted their policy of keeping nuclear warheads separate from delivery systems—in the context of expanding their nuclear submarine deterrence capacity.
108. Lauren J. Borja and MV Ramana, "Command and Control of India's Nuclear Arsenal," *Journal for Peace and Nuclear Disarmament* 3, no, 1 (2020): pp. 1–20; and Tim Craig and Karen De Young, "Pakistan is Eyeing Sea-Based and Short-Range Nuclear Weapons, Analysts Say," *Washington Post*, September 21, 2014, https://www.washingtonpost.com/world/asia_pacific/pakistan-is-eyeing-sea-based-and-short-range-nuclear-weapons-analysts-say/2014/09/20/1bd9436a-11bb-11e4-8936-26932bcfd6ed_story.html.
109. Nuclear powers, including Russia, the United Kingdom, China, France, India, and the US, have all released official policy statements that explicitly underscore the importance of AI technologies as key military enablers in the multipolar world order. Vincent Boulanin, ed., *Artificial Intelligence, Strategic Stability and Nuclear* (Stockholm: SIPRI Publications, June 2020), p. 110.
110. For example, reports suggest that President Putin has considered reactivating Russia's controversial *Perimeter* (or "Dead-Hand") system. President of Russia, "Meeting of the

Valdai International Discussion Club," October 18, 2018, http://en.kremlin.ru/events/president/news/58848.

111. For example, see Yuan Yi, "The Development of Military Intelligentization Calls for Related International Rules," *PLA Daily*, October 16, 2019, http://military.workercn.cn/32824/201910/16/191016085645085.shtml; and Lin Juanjuan, Zhang Yuantao, and Wang Wei, "Military Intelligence is Profoundly Affecting Future Operations," *Ministry of National Defense of the People's Republic of China*, September 10, 2019.
112. Brian W. Everstine, "DOD AI Leader Wants Closer Collaboration With NATO," *Airforce Magazine*, January 15, 2020.
113. Adam P. Liff, "Cyberwar: A New 'Absolute Weapon'? The Proliferation of Cyberwarfare Capabilities and Interstate War," *Journal of Strategic Studies*, 35, no. 3 (2012): pp. 401–428.
114. The use of false information, false signals, and elaborate spoofing between adversaries are not entirely new. Roberta Wohlstetter, *Pearl Harbor: Warning and Decision* (Stanford, CA: Stanford University Press, 1962), pp. 393–394; Liff, "Cyberwar: A New 'Absolute Weapon'?"
115. See Zachary Davis, "Artificial Intelligence in the Battlefield," PRISM 8, no. 2 (2019): pp. 114–131; and Johnson, "Delegating Strategic Decision-Making to Machines."
116. Johnson, "Delegating Strategic Decision-Making to Machines," pp. 60–61.
117. Kenneth Payne, "Fighting on Emotion and Conflict Termination," *Cambridge Review of International Affairs* 28, no. 3 (August 2015): pp. 480–497.
118. Recent research found that a fully autonomous system would reduce the sense of agency not only because the user is not actively involved in a task, but also because the system appears as an intentional entity taking autonomous decisions and possesses agency attributes. See Hannah Limerick, David Coyle, and James W. Moore, "The Experience of Agency in Human-Computer Interactions: A Review," *Frontiers in Human Neuroscience* 8 (2014): p. 643.
119. The determination of human-to-machine interaction and degree of "autonomy" of a particular capability (and thus meaningful human control) is complicated by the propensity of defense contractors to "play up the sophistication and autonomy of their products in marketing and downplay them when scrutinized by international bodies such as the United Nations." Quoted in the *UK House of Lords Select Committee on Artificial Intelligence*, "AI in the UK: Ready, Willing and Able?" *Report of Session 2017–19* (London: House of Lords, April 16, 2018), https://publications.parliament.uk/pa/ld201719/ldselect/ldai/100/100.pdf, p.26.
120. Kenneth Payne, *Strategy, Evolution, and War: From Apes to Artificial Intelligence* (Washington, DC: Georgetown University Press), p. 183.
121. Ibid.
122. Rebecca Hersman et al., "Under the Nuclear Shadow: Situational Awareness Technology and Crisis Decision-making," *Center for Strategic and International Studies*, March 18, 2020, https://ontheradar.csis.org/analysis/final-report/.
123. Keir A. Lieber and Daryl G. Press, "The End of MAD: The Nuclear Dimension of US Primacy," *International Security* 30, no. 4 (Spring 2006): pp. 7–44.
124. US Office of the Secretary of Defense, *Nuclear Posture Review* (Washington DC: Department of Defense, February 2018), pp. 57–58.
125. Jia Daojin and Zhou Hongmei, "The Future 20–30 Years Will Initiate Military Transformation," *China Military Online*, June 2, 2016.

126. From the open sources, no unambiguous evidence has emerged to suggest that China plans to use AI to augment its NC3 systems. Fiona S. Cunningham and Taylor M. Fravel, "Dangerous Confidence? Chinese Views on Nuclear Escalation," *International Security* 44, no. 2 (2019): pp. 106–108.
127. Herman Kahn, *On Escalation: Metaphors and Scenarios* (New York: Prager, 1965), p. 11.
128. Schelling, *Arms and Influence*, p. 37.
129. "Semi-autonomous" (or "fire and forget" and "launch and leave" weapons), which are commonplace in contemporary warfare systems, should be distinguished from "fully autonomous" ones. The former engages specific targets or categories of targets that a human operator pre-selects and then attacks without human involvement (e.g., US AGM-130 and AGM-65 missiles used for attacking ground targets). See US Airforce, Factsheets (Weapons), http://www.af.mil/information/factsheets/index.asp.
130. If command and control communications links survived, *Perimeter* would assume that high-level decision-makers who could order an attack were still alive and remained in full control of the decision-making process. However, if the link was lost, then the system would conclude that these decision-makers did not survive and transfer the decision authority to whoever was then manning the system deep inside a protected bunker. See David E. Hoffman, *The Dead Hand: The Untold Story of the Cold War Arms Race and Its Dangerous Legacy* (Anchor Books: New York, 2009), Ch. 6.
131. In 1985 the Soviet Union deployed a semi-automated retaliation system—known in Western literature as the Dead Hand and as *Perimeter* in Russian sources—the system was designed to ensure a Soviet response to a nuclear strike from the US. In 2011, the Commander-in-Chief of the Russian SRF, General S. Karakayev, confirmed in an interview with one of the prominent Russian newspapers that *Perimeter* exists and continues to be on combat duty. The system's specific capabilities are unknown, however. See Ryabikhin Leonid, "Russia's NC3 and Early Warning Systems," *Tech4GS* July 11, 2019, https://www.tech4gs.org/nc3-systems-and-strategic-stability-a-global-overview.html; Eric Schlosser, *Command and Control* (New York: Penguin Group, 2014); and Richard Rhodes, *Arsenals of Folly: The Making of the Nuclear Arms Race* (London: Simon & Schuster, 2008).
132. From the perspective of game theory, deterrence equals one player threatening another player with the goal of preventing them from conducting an aggressive action that it has not yet taken. A "game" generally involves players, strategies, payoffs, and an information structure. The most well-known games are non-cooperative two-player zero-sum games such as the "prisoner's dilemma." See Schelling, *The Strategy of Conflict*, p. 187.
133. Robert Powell, "Nuclear Deterrence Theory, Nuclear Proliferation, and National Missile Defense," *International Security* 27, no. 4 (Spring 2003): pp. 86–118; and Francis J. Gavin, "Strategies of Inhibition: U.S. Grand Strategy, the Nuclear Revolution, and Nonproliferation," *International Security* 40, no. 1 (Summer 2017): pp. 9–46.
134. By contrast, a superior military power that automates combat to AIs would appear to have relatively less to lose in a military contest, thus reducing its ability to demonstrate resolve. See Erik Gartzke, "Blood and Robots: How Remotely Piloted Vehicles and Related Technologies Affect the Politics of Violence," *Journal of Strategic Studies* (2019) doi: 10.1080/01402390.2019.1643329. Wong et al., *Deterrence in the Age of Thinking Machines*.
135. Johnson, "Delegating Strategic Decision-Making to Machines."

136. In this simulated challenge, Maryland-based Heron Systems designed the AI system using both conventional and unconventional moves, including aggressive gunfire and flying toward their adversary at close range without letting up until the last possible moment. The experiments were part of Phase 1 of DARPA's Air Combat Evolution program, focused on exploring how AI technology can help automate various aspects of air-to-air combat. Future phases will test live-fly sub-scale aircraft scenarios, commercial UAVs, and finally combat aircraft. See Oriana Pawlyk, "Rise of the Machines: AI Algorithm Beats F-16 Pilot in Dogfight," *Military.com*, August 24, 2020, https://www.military.com/daily-news/2020/08/24/f-16-pilot-just-lost-algorithm-dogfight.html; and Thomas Newdick, "AI-Controlled F-16s Are Now Working as a Team in DARPA's Alpha Dogfights," *The Drive*, March 22, 2021, https://www.thedrive.com/the-war-zone/39899/darpa-now-has-ai-controlled-f-16s-working-as-a-team-in-virtual-dogfights.
137. Patrick Tucker, "It's Time to Wargame Against an AI-Enabled China," *Defense One*, June 23, 2021, https://www.defenseone.com/technology/2021/06/its-time-wargame-against-ai-enabled-china/174922/.
138. Wong, *Deterrence in the Age of Thinking Machines.*
139. Ibid., pp. 39–59.
140. Alex Wilner and Casey Babb, "New Technologies and Deterrence: Artificial Intelligence," in *Netherlands Annual Review of Military Studies 2020. Deterrence in the 21st Century—Insights from Theory and Practice*, edited by Frans Osinga and Tim Sweijs (New York: Springer, 2021), pp. 401–416, p. 405.
141. Sanjay Bisht et al., "Command Agent Belief Architecture to Support Commander Decision Making in Military Simulation," *Defense Science Journal* 68, no. 1 (2018): pp. 46–53.
142. Game and decision theory is an integral element of AI-ML classifiers; that is, one player is challenging the other in finding the optimal hyperplane by giving them the most difficult points to classify. For example, multi-agent reinforcement learning agents learn by trial-and-error interaction with the environment—and with other agents that consist of multiple autonomous interacting units—each with their own sensory systems and goals (e.g., surveillance, navigation, target tracking for nuclear or non-nuclear missiles, or cyber defense).
143. See Nathan Berg, and Gerd Gigerenzer, "As-If Behavioral Economics: Neoclassical Economics in Disguise?" *History of Economic Ideas* 18, no. 1 (2010): pp. 133–166; and Cass R. Sunstein, "The Good, the Bad, and the Ugly: A Unified Approach to Behavioural Welfare Economics," *Journal of Benefit-Cost Analysis* 1 (2016): pp. 12–68.
144. The importance of understanding the dynamics, judgments, and rationale of game theoretic algorithms in complex and competitive environments is an active field of open research. See Russell and Norvig, *Artificial Intelligence*, and Peter Vrancx, Anne Nowe, and Yann- Michael De Hauwere, eds, *Reinforcement Learning: Adaptation, Learning, and Optimization* (Berlin: Springer, 2012), pp. 441–470.
145. Studies using initial game and decision theoretic models have shown some progress in understanding nuclear deterrence to alleviate the burden of human cognitive biases. See Geoffrey P. Morgan et al., "Sociocultural Models of Nuclear Deterrence," *IEEE Transactions on Computational Social Systems* 4, no. 3 (2017): pp. 121–134.
146. "Escalation dominance" is a situation in which one side can escalate a conflict in ways that will be disadvantageous (or costly) to the adversary. The other side cannot respond

in kind, either because it has no escalation options or the options it does have do not improve its predicament. Herman Kahn, *On Escalation: Metaphors and Scenarios.*

147. Defense experts have warned it is critical that detection and early-warning systems of the C2 are independent of other parts of the nuclear decision-making chain. System accidents caused by hidden interactions between these functions can lead to unexpected and potentially catastrophic outcomes in NC3 systems. See Paul Bracken, "Instabilities in the Control of Nuclear Forces," in *Breakthrough: Emerging New Thinking—Soviet and Western Scholars Issue a Challenge to Build a World Beyond War*, edited by M. Hellman (Walker and Co: New York, 1988), pp. 21–30; Perrow, *Normal Accidents: Living with High-Risk Technologies*; and Sagan, *The Limits of Safety.*
148. See Ingvild Bode and Tom Watts, "Meaning-Less Human Control: Lessons from Air Defence Systems on Meaningful Human Control for the Debate on AWS," *University of Southern Denmark*, February 2021, https://dronewars.net/wpcontent/uploads/2021/02/DW-Control-WEB.pdf; and Michael C. Horowitz and Paul Scharre, "Meaningful Human Control in Weapon Systems: A Primer," *Center for New American Security*, March 2015, https://www.cnas.org/publications/reports/meaningful-human-control-in-weapon-systems-a-primer.
149. Stuart Russell, *Human Compatible* (New York: Viking Press, 2019).
150. "Theory of Mind" is the innate ability of most humans to perceive others' intentions and beliefs, allowing them to make predictions about others' behavior. See Brittany N. Thompson, "Theory of Mind: Understanding Others in a Social World," *Psychology Today*, July 3, 2017.
151. For example, Iraq's troop movements towards the Kuwaiti border in 1990 resulting in US warships in the Persian Gulf being put on alert.
152. V.M. Burenok, R.A. Durnev, and K.U. Kryukov, "Intelligent Armament: The Future of Artificial Intelligence in Military Affairs," *Weapons and Economics* 1, no. 43 (2018): pp. 4–13, http://www.viek.ru/43/4-13.pdf.
153. For example, an ongoing scientific study (EU's iBoaderCtrl program) is testing the use of AI applications (e.g., AI avatars with facial recognition technology) to monitor EU border and national security. Wilner and Babb, "New Technologies and Deterrence."
154. John D. Williams, *The Compleat Strategyst: Being a Primer on the Theory of Games of Strategy* (New York: Dover Publications, 1986).
155. Wong et al., *Deterrence in the Age of Thinking Machines*, pp. 81–82.
156. Morgan et al., *Dangerous Thresholds: Managing Escalation in the 21st century*, pp. 18–19.
157. For example, this explains why people who are opposed to the use of torture on moral grounds are generally resistant to arguments that its use may save lives. See Alan Fiske and Philip Tetlock, "Taboo Trade-Offs: Reactions to Transactions that Transgress the Spheres of Justice," *Political Psychology* 18 (June 1997): pp. 255–297.
158. The US, for example, is known to have deployed AI technology ISR remote-sensing to support its autonomous Perdix UAV swarm operations and its Aegis ballistic missile defense systems' radar seekers. Boulanin, ed., *Artificial Intelligence, Strategic Stability and Nuclear*, pp. 42–43.
159. Matthijs M. Maas, "How Viable is International Arms Control for Artificial Military Intelligence? Three Lessons from Nuclear Weapons," *Contemporary Security Policy* 40, no. 3 (2019): pp. 285–311.

160. Heather Williams, "Asymmetric Arms Control and Strategic Stability: Scenarios for Limiting Hypersonic Glide Vehicles," *Journal of Strategic Studies* 42, no. 6 (2019): pp. 789–813.
161. Joshua New, "Why the United States Needs a National Artificial Intelligence Strategy and What it Should Look Like," *ITIF* December 4, 2018.
162. For example, Matt Turek, "Semantic Forensics (SemaFor) Proposers Day," *Defense Advanced Research Projects Agency*, August 28, 2019.
163. National Security Commission on Artificial Intelligence Interim Report to Congress, November 2019.

3
Inadvertent escalation

A new model for nuclear risk

How might AI-enabled capabilities increase inadvertent escalation risk? This chapter revisits Cold War-era thinking about inadvertent escalation to consider how AI technology—especially AI augmentation of advanced conventional counterforce weapons—might influence inadvertent escalation risk between nuclear-armed adversaries during a conventional crisis or conflict through different mechanisms and pathways. We are now in an era of rapid disruptive technological change, especially in AI technology. This technology is already being infused into military machines, and global armed forces are well advanced in their planning, research and development, and, in some cases, deployment of AI-enabled capabilities.[1] Therefore, the embryonic journey to reorient military forces to prepare for the future digitized battlefield is no longer merely speculation or science fiction.

While much of the recent discussion has focused on specific technical issues and uncertainties involved as militaries develop and diffuse AI applications at the tactical and operational level of war, the strategic and theoretical treatment of these developments has received far less attention.[2] This chapter addresses this gap.[3] It examines the psychological and cognitive features of escalation theorizing to consider whether and how AI technology characteristics, contextualized within the broader political dynamics associated with today's digital information ecosystem, may increase inadvertent escalation risk. It explains how AI technology could be incorporated into nuclear and conventional operations in ways that affect inadvertent escalation risks during a crisis or sub-nuclear conflict in strategically competitive dyads—US-China, India-Pakistan, and US-Russia. How might AI be incorporated into nuclear and conventional operations in ways that affect escalation risk?

The chapter speaks to the broader scholarship in international relations (IR)—notably "bargaining theories of war" (shifts in the balance of power,

AI and the Bomb. James Johnson, Oxford University Press. © James Johnson (2023).
DOI: 10.1093/oso/9780192858184.003.0004

uncertainty, asymmetric information, and commitment problems), which argues that the impact of technology on the cause of war occurs through its political and psychological effects, rather than tactical or operational battlefield shifts.[4] Specifically, the political consequences that flow from changes to the balance of power, and the impact on the redistribution of resources and perceived (by beneficiaries and their rivals) strategic advantage of utilizing a particular capability.[5] A new asymmetric capability, doctrine, or strategy that decreases (or increases) the perceived cost, risk, and lethality of warfare, *ceteris paribus*, should affect the observable mechanisms of conflict such as escalation, only to the extent that it changes actors' *perceptions* about how adversaries might perform in battle.[6] Because of the centrality of information (i.e., capabilities, interests, and intentions), the critical factor is to what degree a particular technology disproportionately affects states' perception of the balance of power.[7] States' perceptions of a shift in the balance of power can undermine effective bargaining and amplify misperception (i.e., which state has the upper hand and for how long), thus making settlement through peaceful bargaining less likely. New technologies like AI threaten to create potentially destabilizing shifts in the balance of power.[8]

The remainder of this chapter is organized as follows. First, it provides a conceptual overview of escalation theorizing. It defines the various terms, analogies, mechanisms, and metaphors associated with escalation, which at its core is fundamentally psychological and perceptual one. Second, it applies Barry Posen's inadvertent escalation analytical framework to examine the effects of AI technology on the causes of inadvertent escalation—that is, the "security dilemma," the Clausewitzian notion of the "fog of war," and military strategy and doctrine.[9] This section revisits this model to conceptualize the psychological underpinnings of the novel ways AI technology and the emerging digital information ecosystem may increase inadvertent escalation risk. Next, the chapter contextualizes how the AI technology and the digitized information ecosystem using "mis/disinformation" (i.e., information that is inaccurate or misleading) asymmetric tactics by state or non-state actors during a crisis between nuclear-armed adversaries might increase inadvertent risk. The chapter then considers how state or non-state actors might use AI technology to conduct mis/disinformation asymmetric operations in ways that might increase inadvertent risk.[10] Finally, the chapter highlights the critical features of inadvertent escalation risk in the emerging AI-nuclear strategic nexus, concludes with policy implications of the AI-nuclear strategic nexus, and suggests possible ways to mitigate inadvertent escalation risk and improve strategic stability.

Conceptualizing inadvertent escalation: Escalation ladders, dominance, and other metaphors

The concept of escalation is, at its core, a fundamentally psychological and perceptual one. Like other related concepts, such as strategic stability (Chapter 1) and deterrence (Chapter 2), escalation relies upon the actor's unique understanding of context, motives, and intentions—especially in the use of capabilities.[11] Understanding how actors resolve these complex psychological variables associated with the cause, means, and effects of a military attack (both kinetic and non-kinetic) remains a perplexing and invariably elusive endeavor.[12] Furthermore, deterring escalation is generally achieved as a result of the fear and uncertainty (or "fear of eruption") of how an adversary might assess capabilities and threats, and respond (or overreact) to a situation, rather than the perceived costs or military advantages of escalation per se.[13] How might uncertainty about digital vulnerabilities affect inadvertent escalation dynamics?

Escalation theorizing came into prominence during the Cold War era with the development of nuclear weapons, particularly the need to conceptualize and control conflict below the level of all-out total war. Nuclear theories of escalation continue to provide the theoretical basis for escalatory strategies and to undergird debates about nuclear deterrence and strategic planning, particularly how a conventional skirmish could become a nuclear war. On escalation, Herman Kahn's work conceptualizes a forty-four-rung escalation ladder metaphor, which moves from low-scale violence to localized nuclear war (or counter-value warfare) to conventional and nuclear attacks against civilian populations (or strategic counter-value warfare).[14] Kahn's "escalation ladder" metaphor hinges on the idea of psychological obstacles, thresholds, or stages of the escalation process that would impose a threshold (or firebreak) to the next rung or step up the ladder in ascending order of intensity.[15] Escalation theory's emphasis on the importance of firebreaks between the rungs underscores the qualitative, psychological (both rational and irrational), and normative difference (or "normative stigma" and taboo) between nuclear and non-nuclear domains.[16]

A study by the RAND Corporation defined escalation as "an increase in the intensity or scope of conflict that crosses a threshold(s) *considered significant by one or more of the participants*."[17] An "unintentional" increased intensity or scope of a situation can be inadvertent, "catalytic" (see Chapter 5), or accidental—encompassing incorrect or unauthorized usage (see Figure 3.1).[18] Intentional escalators knowingly take actions that cross thresholds (or firebreaks) for strategic gain, to send a signal of intent, obtain

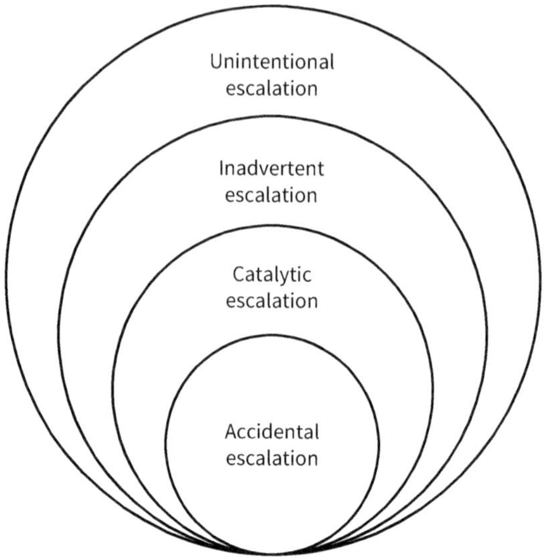

Figure 3.1 "Unintentional escalation"

information about an adversary (i.e., resolve, credibility commitment, or risk acceptance), or avert military defeat (i.e., through pre-emption, a "bolt-from-the-blue" attack, or gray-zone tactics).[19] By contrast, inadvertent escalation occurs when an actor crosses a threshold that it considers benign, but the other side considers significant.[20] The escalator may, for example, send a signal to an adversary that it does not *intend* to cross a threshold but is perceived to do so by the other.[21]

These distinctions are not, however, binary, or mutually exclusive. An escalation mechanism that leads from a crisis or conflict to its outcome can involve more than one of these categories. For example, a "false-flag cyber operation" by a third-party actor targeting a state's nuclear Command, Control, and Communication (NC3) systems,[22] which accidentally sets in train escalatory mechanisms, because the victim perceives the attack as a precursor to a pre-emptive strike by an adversary.[23] In this example, an actor accidentally breaches another's psychological (real or illusory) threshold,[24] triggering counter-escalation dynamics that may be a response to misinformation, fear, or misperception—also known as "catalytic escalation" (see Chapter 5).[25] Within the broader digital information ecosystem associated with the emergent "third nuclear age,"[26] the deliberate use of nuclear weapons that originates from a false, manipulated, or distorted assessment of a situation (e.g., in response to an early-warning system false alarm) can quickly muddy intentionality lines (see Chapter 5).[27]

Besides, in the digital age, empirical assessments and hypothesis testing about availabilities like AI and cyber and stability are inherently problematic due to, *inter alia*: the virtual (non-kinetic) nature of these domains; the technical challenge of understanding the balance of capabilities; the covert features of AI-cyberspace; and the lack of historical case samples (i.e., cyber exploits and vulnerabilities of NC3 systems) or technical data points (i.e., the probability or scope of cyber-attacks against NC3 systems) to corroborate and verify findings. Cyber-attacks, cyber espionage (i.e., for gathering intelligence), and other forms of cyber exploitation use almost identical techniques, and thus differ only in their ends. In a recent survey of international military experts and decision-makers supported by the Dutch Ministry of Foreign Affairs, seventy to eighty percent of the participants agreed that the inclusion of AI (especially AI-enhanced cyber intrusions) in nuclear command and control erodes strategic stability and increases the risk of inadvertent escalation.[28]

Used during a crisis, therefore, these operations are inherently escalatory in nature. A cyber intrusion conducted by State A detected in State B's NC3 system could be either a benign espionage operation, or the precursor of an offensive cyber-attack. Because B would be hard pressed to determine A's intentions *before* the operation was carried out, a worst-case scenario assessment would likely result.[29] While some behavior is interpreted as escalatory by almost all rational actors (e.g., the use of nuclear weapons to respond to a low-level conventional conflict), others (e.g., cyber espionage and intelligence gathering) are far more ambiguous. A cyber espionage operation against a state's dual-use C2 systems, for instance, may be interpreted either as a defensive countermeasure, espionage, or an attempt to create a window of opportunity for a larger strategic strike.[30] An additional risk factor arises from the fact that cyber weapons are usually very well concealed—easily detected cyber capabilities will likely be swiftly countered. Thus, if State A penetrates State B's NC3 systems clandestinely, A will have a tactical advantage over B, of which B remains ignorant. Aware of its upper hand vis-à-vis B, A might feel no compulsion to refrain from further escalation. B, unaware of the intrusion—operating under the assumption that its power remains intact—may inadvertently escalate a situation it would be wise to avoid.[31]

Much like the paucity of empirical research in the nuclear domain, therefore, how actors *perceive* their (and an adversary's) capabilities, and particularly how cognitive, structural, and material norms of behavior, organizational, and other strategic independent variables—rather than the capabilities balance alone—can influence nuclear decision-making and effect escalation dynamics, has become a fruitful avenue of investigation.[32] Experimental war-gaming, and other simulation exercises, have also offered insights into others' motivations and perceptions of strategic decision-making and escalation dynamics.[33] In short, the binary distinction between deliberate and unintentional use of nuclear weapons is inherently problematic.[34] Ultimately, whether the impact of unintended escalation risk is stabilizing or destabilizing depends on the actor's relative strength, and the fear it instills in its adversary.[35] Escalation can, therefore, be a strategic bargaining tool (i.e., for deterrence, damage limitation, and coercion) and a risk to be controlled and potentially mitigated.[36]

Escalation situations typically involve "competition in risk-taking" and resolve,[37] thus either side can intensify a situation, providing the other side does not match that escalation in kind. If escalation is not matched and victory achieved, the "cost of the increased effort would be low in relation to the benefits of victory."[38] In asymmetric situations (in both directions), however, the demonstration of cyber capabilities can make it difficult for others to respond, short of escalating with conventional options to signal resolve—progressively escalating a situation and increasing the risk that things spiral out of control. Technological advances in AI technology and cyber capabilities, coupled with the increasingly commingled nature of the state's nuclear and conventional C2 systems, have enabled possible solutions to overcome the robustness of permissive action links (PALs), thus increasing the vulnerability of these systems.[39]

As Bernard Brodie noted in 1965, "since the beginning of the nuclear era there has been *in the minds of men* a strong tendency to distinguish between nuclear and non-nuclear weapons combined with a *fear of and aversion to the former*" and this distinction has tended "to *grow stronger with time* rather than weaken" (emphasis added).[40] To be sure, recent scholarship questioning the validity of the public aversion (or "taboo") on the use of nuclear weapons (especially as tools to fight terrorism and nuclear proliferation), and policy debates about the use of tactical nuclear weapons—which surrounded the Trump Administration's Nuclear Posture Review—demonstrates the prophetic nature of Brodie's "firebreak theory."[41] Besides, the "fear" and "aversion" associated with nuclear weapons have been used in studies to explain the role that emotion and cognition play in the misperceptions about weapons' effectiveness. Misperceptions can create an "emotional springboard" for the process of reasoning that can influence nuclear non-use, deterrence, and taboos, which in turn informs policymaking.[42]

The escalation ladder metaphor, like any theoretical framework, has limitations. Actors would not necessarily move sequentially and inexorably from the lower rungs ("sub-crisis maneuvering") to the higher rungs ("spasm or insensate war")—that is, rungs can be skipped and go up as well as down. Instead, there are many pathways and mechanisms—for instance, conventional vs. nuclear, tactical vs. strategic, or counterforce vs. counter-value—between low-intensity and all-out nuclear confrontation. Besides, states can be at different rungs or thresholds along the "relatively continuous" pathways to war.[43] Despite its limitations, however, Kahn's "escalation ladder" is a useful metaphorical framework with which to reflect on the possible available options (e.g., a show of force, reciprocal reprisals, costly signaling, and pre-emptive attacks), progression of escalation intensity, and scenarios in a competitive nuclear-armed dyad. This chapter argues that the introduction of AI is creating new ways (or "rungs") and potential shortcuts up (and down) the ladder, which might create new mechanisms for a state to perceive (or misperceive) others to be on a different rung, thus making some "rungs" more (or less) fluid or malleable.

Strategic rivals require a balanced understanding of how the other views the escalation hierarchy. During a crisis or conflict, continuous feedback about an adversary's intentions, where it views itself on the escalation ladder, and how shifts in the scope or intensity of a situation (i.e., kinetic, non-kinetic, or rhetorical) may be perceived.[44] Because of the inherently subjective nature of escalation, actions perceived as escalatory by one state can be misunderstood as thus by others.[45] What characteristics of AI technology may create new rungs on the escalation ladder that increase the inadvertent risk of escalating a conventional conflict to nuclear war?

An important feature of escalation theory is the notion of "escalation dominance": when a nuclear power forces an adversary to back down because the cost and risk associated with further escalation outweigh the perceived benefits.[46] The state that possesses a position of escalation dominance, *ceteris paribus*, has unique tactical advantages on a particular rung of the escalation ladder. Moreover, a state that has most to lose in a situation escalating, or fears escalation the least (i.e., it is not the one doing the escalation), will axiomatically achieve an element of escalation dominance. Escalation dominance is thus a function of several variables: (1) where two states are positioned on the escalation ladder; (2) an assessment of their respective capabilities (both offensive and defensive) on a particular rung; (3) each side's perception of the probability and likely outcome of moving to a different rung; and related (4) the perceived ability of one side of the other to effectuate this shift. Furthermore, both sides' fear and aversion of intensifying a situation is a crucial psychological variable in electing between a nuclear or conventional attack to access escalation risk and secure escalation dominance. State A's decision to launch a nuclear attack on State B to control escalation risk (i.e., dissuade B from countering with additional nuclear attacks) might equally risk further escalation—thus failing to realize its goal of preventing further ascension up the rungs of the escalation ladder.

A new model of inadvertent escalation in the digital age

In his work on inadvertent escalation, Barry Posen identifies the major causes of inadvertent escalation as the "security dilemma," the Clausewitzian notion of the "fog of war,"[47] and offensively oriented military strategy and doctrine.[48] This section applies Posen's framework to examine the effects of AI technology on the causes of inadvertent escalation. In the light of recent technological change, this section revisits Posen's analytical framework to examine the psychological underpinnings of security dilemma theorizing (misperception, cognitive bias, and heuristics) to consider how and why the novel characteristics of AI technology and the emerging digital information ecosystem may destabilize "crisis stability" and increase inadvertent escalation risk.[49] According to psychologist Daniel Kahneman, the vast majority of cognitive biases evinced by decision-makers (e.g., cognitive dissonance, attribution error, prospect theory, and groupthink) tend to favor "hawkishness"—that is, a predisposition toward overconfidence, assertiveness, and aggression rather than conciliation or empathy with others.[50]

The security dilemma concept

According to "security dilemma"[51] theorizing, the ease of carrying out and defending against military attacks (the "offense-defense balance"), and the ambiguity about whether a weapon is offensive or defensive (the "offense-defense distinguishability"), can cause crisis instability and deterrence failure because these characteristics create fear and uncertainty about an adversary's intentions. That is, where they harbor "benign" (i.e., non-aggressive/defensive) or "malign" (i.e., aggressive/offensive) intent toward the other side.[52] Robert Jervis defines the security dilemma as the "unintended and undesired consequences of actions *meant to be defensive*," which can, nonetheless, decrease the security of others—that is, one state's gain in security can inadvertently undermine the security of others.[53] Actors tend to perceive the accumulation of offensive capabilities (and attendant offensive strategies) by others as threatening, and assuming the worst respond with countermeasures—for example, arms racing, crisis mobilization, raising the alert status (a concern for inadvertent escalation), and pre-emptive war.[54]

According to Jervis, it is "the fear of being exploited" that "most strongly drives the security dilemma."[55] As noted, the fear (both rational and irrational) of conventional skirmishes crossing the nuclear Rubicon can be found in nuclear policymakers' statements as far back as the 1950s.[56] Therefore, security dilemma logic can be used to consider both peacetime spirals of political mistrust and shifts in the military balance, crisis stability dynamics, and escalation mechanisms once military conflict begins. The security dilemma concept is an inherently inadvertent phenomenon; that is, weapons indistinguishability (i.e., offense vs. defense) allows states to inadvertently (for rational political or military reasons) threaten others, which can spark spirals of mutual distrust, strategic competition, and military action.[57]

There are several ways in which the security dilemma can act as a catalyst for inadvertent escalation that may be compounded in the digital age.[58] First, while nuclear-armed states have a shared interest in avoiding nuclear war, they also place a high value on their nuclear forces—both vital security assets and as symbols of national prestige and status.[59] As a corollary, conventional weapons—devised by a nuclear power for defensive purposes or counterforce missions—may nonetheless be viewed by an adversary as an offensive threat to their nuclear survivability (discussed below).[60] Second, escalatory rhetoric and other aggressive responses by a threatened state (especially in situations of military asymmetry) may be misperceived as unprovoked malign intent, not as a response to the initiator's behavior, thus prompting action and reaction spirals of escalation. Third, and related, the state of heightened tension and the compressed decision-making pressures during a conventional conflict would radically increase the speed by which action and reaction spirals of escalation unravel. In the digital age these dynamics would be amplified, reducing the options and time for de-escalation and increasing the risks of horizontal (the scope of war) and vertical (the intensity of war) inadvertent escalation.[61]

Security dilemma theorizing also ties in with the concept of the "capability/vulnerability paradox" in IR.[62] That is, one state's pursuit of a new resource to compete against other states introduces a vulnerability that is perceived as threatening. This paradox suggests that when a state's military capabilities are dependent on a particular resource (i.e., that may be exploited or dominated by an adversary), both sides have incentives for first strike and pre-emptive action—the resource vulnerability and not the new capability per se generate these destabilizing outcomes.[63] Much like the security dilemma, therefore, the potential impact of the "capability/vulnerability paradox" has increased in the context of technological advancements associated with the "information revolution" in military affairs (i.e., the centralization of information and dependencies on digital information to conduct modern warfare), which now includes AI.[64] As Richard Danzig notes, "digital technologies ... are a security paradox: even as they grant unprecedented powers, they also make users less secure."[65] In this sense, as a cause of inadvertent escalation, the security dilemma connects to the broader strategic and security literature on misperception, emotions, and cognition, evincing a strong qualitative connection between human psychology and war.[66] How might the confusion and uncertainty of war increase inadvertent risk?

Clausewitzian "fog of war"

Inadvertent escalation risk can also be caused by the confusion and uncertainty (or "fog of war") associated with gathering, analyzing, and disseminating relevant information about a crisis or conflict—which has important implications for the management, control, and termination of war.[67] As Carl von Clausewitz observed in his text on strategic military theory, "a great part of the information obtained in war is contradictory, a still greater part is false, and by far the greatest part is of a doubtful character."[68] Traditional nuclear deterrence, crisis stability, and strategic stability theorizing works from the questionable premise that actors are rational, and presumes that they are therefore able to understand their environment and coordinate policy instruments in ways that cut through the "fog of war."[69] In reality, through misperception, miscalculation, accidents, and failures of communication "events can readily escape control," and although these escalation dynamics are often unintended, potentially catastrophic events can result.[70]

The confusion and uncertainty associated with the "fog of war" can increase inadvertent risk in three ways: (1) it can complicate the task of managing and controlling military campaigns at a tactical level (or situational battlefield awareness); (2) it can further compound the problem of offense-defense distinguishability; and (3) it can increase the fear of a surprise or pre-emptive attack (or a "bolt from the blue").[71] Taken together, these mechanisms can result in unintentional (and possibly irrevocable) outcomes and thus obfuscate the meaning and the intended goals of an adversary's military actions.[72]

The dense "fog of war" increases the risk of inadvertent escalation because "misperceptions, misunderstandings, poor communications, and unauthorized or unrestrained offensive operations could reduce the ability of civilian authorities to influence the course of war."[73] A recent reassessment of lessons learned from the Cuban Missile Crisis reveals, for example, that both President Kennedy and British Prime Minister Harold MacMillan were apprehensive about the risk of inadvertent escalation—rather than a deliberate nuclear exchange between the US and the Soviets—caused by fear, misperception, heightened tension, and human frailties, such as exhaustion. MacMillan complained about crisis decision-making amidst anxiety, fear, and being in a state of exhaustion in his diaries.[74]

With the advent of the speed, uncertainty, complexity, and cognitive strains (see below) associated with conducting military operations on the digitized battlefield, the prospect that decision-making by "dead reckoning"[75]—dependent on the experience and sound judgment of defense planners when information is scarce—can prevent C2 failures caused by the "fog" seems fanciful. In short, while improved and faster communications and information flow could, in theory, mitigate some of the negative consequences of cognitive bias during crisis and conflict—reducing escalation risk caused by misunderstanding and misperception—the confusion and anxiety caused by increased speed of battlefield interactions, and the spread of mis/disinformation in the digitized information ecosystem through various modalities (e.g., memes, deepfake videos, tweets, post, rumors, conspiracies, and fake news), will likely decrease the time for decision-making (for "cooling off"), and increase inadvertent escalation risk.

While Cold War decision-makers also faced extreme domestic and foreign pressures to decipher the adversary's intentions correctly, these pressures are being amplified in the current digitized information ecosystem—reducing the timeframe for decision-makers during a crisis, and increasing the cognitive and heuristic burdens caused by information overload and complexity.[76] Cognitive pressures caused by the sheer volume of information flow (both classified and open sources) are being compounded by the degraded reliably and politicization (or "weaponization") of information. These pressures can, in turn, make decision-makers more susceptible to cognitive bias, misperceptions, and heuristics (or mental shortcuts)[77] to approach complex problem-solving—either directly or indirectly informing decision-making.

While disinformation and psychological operations in deception and subversion are not a new phenomenon,[78] the introduction of new AI-enhanced tools in the emerging digital information ecosystem enables a broader range of actors (state and non-state) with asymmetric techniques to manipulate, confuse, and deceive—undermining the quality of intelligence and increasing the likelihood of a "disinformation cascade."[79] Disinformation operations might erode the credibility of, and undermine public confidence in, a state's retaliatory capabilities by targeting specific systems (i.e., command and control (C2)) or personnel (primarily via social media) who perform critical functions in maintaining these capabilities.[80] For example, cyberweapons (e.g., "left of launch" operations) were allegedly used by the US to undermine Iranian and North Korean confidence in their nuclear forces and technological preparedness.[81]

The potential utility of social media to amplify the effects of a disinformation campaign was demonstrated during the Ukrainian Crisis in 2016 when military personnel's cell phones were the victims of Russian information operations.[82] However, the use of these novel techniques during a nuclear crisis and how they might impact the "fog of war" is less well understood, and empirically untested.[83] During a nuclear crisis, a state might attempt, for instance, to influence and shape the domestic debate of an adversary (e.g., shift preferences, exacerbate domestic-political polarization, or coerce/coopt groups and individuals on social media) in order to improve its bargaining hand by delegitimizing (or legitimizing) the use of nuclear weapons during an escalating situation, or bring pressure to bear on an adversary's leadership to sue for peace (i.e., terminate) or de-escalate (or not further escalate) a situation—a tactic which may, of course, dangerously backfire.[84]

Moreover, a third-party (state or non-state) actor, in order to achieve its nefarious goals, could employ active information techniques (e.g., spreading false information of a nuclear detonation, troop movement, or missiles leaving their garrison) during a crisis between nuclear rivals (e.g., India-Pakistan, US-China, Russia-NATO) to incite crisis instability.[85] Public pressures on decision-makers from these crisis dynamics, and evolving at a pace that may outpace events on the ground, might impel (especially thin-skinned and inexperienced) leaders, operating under the shadow of the deluge of twenty-four-hour social media feedback and public scrutiny, to take actions that they might not otherwise have.[86]

In sum, the emerging AI-enhanced digitized information environment, while not fundamentally altering the premise upon which actors (de)escalate a situation, based on imperfect information, uncertainty, and risk associated with the "fog of war," introduces novel tools of mis/disinformation and an abundance of information, radically altering the cognitive pressures placed on decision-makers during crisis and conflict.[87] Besides, decision-makers' subjection to an abundance of disparate and often unverified information will indubitably influence actors' (collective or individual) policy preferences and perceptions, thus complicating the security dilemma challenge of understanding an adversary's capabilities, intentions, doctrine, and strategic thinking, with potentially profound repercussions for escalation dynamics. To be sure, how policy preferences are formulated (and the perceptions and assumptions undergirding them), and the learning that occurs, may be more important than the actual policy choice that emerges.[88] How might these pressures affect states' regional conflicts with different military doctrine, objectives, and attitudes to risk?

Offensive military doctrine and strategy

With its multiple conventional-nuclear, offense-defense, and regional-transregional dimensions, modern strategic warfare is inherently complex—and that complexity is only made more onerous with the emerging technology such as AI and new war-faring domains in outer space, cyberspace, and the information ecosystem (see below). Because of a lack of understanding between (nuclear-armed and non-nuclear armed) adversaries about where the new tactical possibilities offered by these capabilities figure on the Cold War-era escalation ladder, the pursuit of new "strategic non-nuclear weapons" (e.g., cyber weapons, drones, missile defense, precision munitions, counterspace weapons) increases the risk of misperception.[89] The fusion of AI technology into conventional weapon systems (i.e., to enhance autonomous weapons, remote sensing for reconnaissance, improving missile guidance and situational awareness) is creating new possibilities for a range of destabilizing counterforce options targeting states' nuclear-weapon delivery and support systems; for example, cyber NC3 "kill switch" attacks or tracking adversaries' nuclear-armed submarines and mobile missile launchers.[90]

Russia, the US, and China are currently pursuing a range of dual-capable (conventional and nuclear-capable) delivery systems (e.g., hypersonic guide vehicles, stealth bombers, and a range of precision munitions) and advanced conventional weapons (drones, space-based, and cyber weapons[91]) that are capable of achieving strategic effects—that is, provide credible options to respond to or pre-empt deterrence failure without the need to use nuclear weapons.[92] The potential threat posed by conventional counterforce weapons is compounded by the blurring between dual-use C2 systems (i.e., early-warning, situational awareness, and surveillance) to manage conventional and nuclear missions.[93] Chinese analysts, for example, while concerned about the vulnerabilities of their command and control systems to cyber-attacks, are optimistic about the deployment of AI augmentation (e.g., intelligence surveillance reconnaissance (ISR), intelligent munitions, and unmanned aerial vehicles) to track and target an adversary's forces, and believe these technological upgrades will lower the cost of (economic and political) signaling and deploying military forces.[94]

Advances in AI technology, in conjunction with the technologies it can enable (e.g., remote sensing, hypersonic technology, and robotics and autonomy), increase the speed, precision, lethality, and survivability of strategic non-nuclear weapons, exacerbating the destabilizing effects of these capabilities used by nuclear-armed rivals. These dynamics—by upending the perceived ability of weak states to inflict unacceptable damage on attackers and shifting the perceived costs of an attack by a superior nuclear state—will likely degrade strategic stability, opening new pathways for both horizontal and vertical inadvertent escalation.[95] Moreover, these technological advances have been accompanied by destabilizing doctrinal shifts by certain regional nuclear powers (Pakistan, India, North Korea, and possibly China), which indicates a countenance of the limited use of nuclear weapons to deter an attack (or "escalate to de-escalate"), in situations where they face a superior conventional adversary and the risk of large-scale conventional aggression.[96]

Recent war-gaming experiments have demonstrated that while vulnerabilities caused by cyber operations targeting NC3 systems did not create a strong impetus for pre-emptive nuclear use,[97] cyber exploits in situations when they offer asymmetrical advantages in an otherwise symmetrical contest can incentivize coercive nuclear and conventional postures and destabilizing nuclear strategies (e.g., a launch-on-warning posture).[98] Further, vulnerabilities caused by cyber exploits may prompt pre-delegation or automation that increases the risk of accidental and inadvertent nuclear use.[99] These findings are borne out in recent scholarly literature that suggests damage limitation counterforce incentive mechanisms are primarily a function of the perceived asymmetric advantages offered by cyber-attacks against NC3 systems.[100]

Furthermore, volatility in nuclear-armed states' civil-military relations can create internal pressures to pursue a more aggressive nuclear force posture or doctrine.[101] The assumption that new (and latent) regional nuclear states will act in ways that reflect their interests as rational actors to avoid nuclear war—and thus enhance deterrence (or "rational deterrence theory"[102]) and crisis stability—understates the role of influential military organizations (i.e., offensive-doctrinal bias, parochialism, and behavioral rigidities) in shaping nuclear doctrine that can lead to deterrence failure and nuclear escalation—and often despite national security interests to the contrary.[103] For instance, in a nuclear-armed state where the military officers influence the nuclear strategy, the adoption of offensive doctrines (or other sub-optimal outcomes) may emerge, which reflect volatile civil-military relations rather than strategic realities.[104]

In short, technological advancements to support states' nuclear deterrence capabilities will develop in, and be influenced by, existing military organizations imbued with their norms, cultures, structures, risk acceptance (or avoidance), and mutually exclusive strategic interests.[105] China's "military-civil fusion" concept (a dual-use integration strategy) designed to coopt (or arguably coerce) Chinese commercial entities to support the technological development of the People's Liberation Army to meet the needs of "intelligentized warfare in the future"—is an important test-case in a civilian-led initiative designed to drive military innovation in the pursuit of broader geostrategic objectives.[106] The impact of China's "military-civil fusion" on civil-military relations remains unclear, however.[107]

US counterforce capabilities—to disarm an enemy without resort to nuclear weapons—used in conjunction with air and missile defenses (together known as "damage limitation"[108]) to mop up any residual capabilities after an initial attack will likely increase crisis instability and create "use-it-or-lose-it" situations.[109] The US Defense Department's 2019 missile defense policy, for instance, integrates these counterforce concepts with traditional missile defenses in a policy it labels "comprehensive missile defense."[110] Chinese analysts, for example, have expressed concern that US advances in AI could overwhelm Chinese air defenses, reducing the time available to commanders to respond to an imminent attack—for example, from the US autonomous AI-enabled Long Range Anti-Ship Missile (AGM-158C) designed to target "high-priority targets."[111] The increased optimism in states' ability to use AI-enhancements to find, track, and destroy others' nuclear forces enabled by AI technology—particularly when military-capability imbalances exist—will also be an inherently destabilizing phenomenon.[112] What one side views as conventional operations might be viewed by the other side as a precursor to a disabling counterforce attack (e.g., targeting dual-use C2 centers and air defenses), thus increasing inadvertent escalation risk.[113]

Because of the asymmetry of interest at stake in a regional crisis involving the US, the stakes will likely favor the defending nuclear-armed power.[114] According to prospect theory, a regional power would perceive the relative significance of a potential loss more highly than a gain.[115] That is, when prospect theory is applied to deterrence dynamics, leaders are inclined to take more risks (i.e., risk-acceptant) to protect their positions, status, and reputations, than they are to enhance their position.[116] Thus, having suffered a loss leaders are generally predisposed to engage in excessive risk-taking behavior to recover lost territory—or other positions or reputational damage.[117] If, for instance, Chinese or North Korean leaders are faced with the prospect of an imminent attack on Taiwan, or Pyongyang view their regime survival to be at stake, they would likely countenance greater risks to avoid this potential (existential) loss.[118] From China's vantage point, while it is not clear how US actions in a regional crisis or conflict (deliberate or inadvertent) might affect Beijing's perception of regime security, given the Chinese Communist Party's persistent sense of regime insecurity, it is possible that any dispute would involve a heightened sense of regime vulnerability regardless of US actions or intentions.[119]

This net assessment crisis instability dynamic might also be exacerbated by irrational behavior derived from misperception, cognitive biases, or other emotional impulses (i.e., indignation, vengeance, humiliation, and fear), which make nuclear escalation more likely.[120] Chinese analysts, for example, tend to overestimate the US's military AI capabilities relative to open-source reports—often citing outdated or inaccurate projections of US AI "warfighting" budgets, development, and force posture.[121] The framing of Chinese discussion on US military AI projects is analogous to Soviet concerns about the missile gap with the US during the Cold War, and risks compounding Beijing's fear that AI technology could be strategically destabilizing.[122] In a world with imperfect (and asymmetric) information about the balance of power and resolve, and incentives to misrepresent and manipulate perceptions (exploit psychological dispositions and vulnerabilities) and emotions (strategic framing and fear appeals) of the information ecosystem, bargaining failure (caused by perceived shifts in the balance of power) and war are more likely.[123]

Given the confluence of secrecy, complexity, and erroneous or ambiguous intelligence data (notably from open-source intelligence and social media outlets)[124]—akin to the Cold War-era use of tactical atomic weapons by NATO to counter a Warsaw Pact in the inner German broader—AI-augmentation will likely exacerbate compressed decision-making and the inherent asymmetric nature of cyberspace information.[125] For example, using AI-enhanced cyber capabilities to degrade or destroy a nuclear state's NC3 systems—whether as part of a deliberate coercive counterforce attack or in error as part of a limited conventional strike—may create windows of vulnerability that lead to incentives for pre-emptive "use-it-or-lose-it" (i.e., deliberate escalation) situations or inadvertent escalation caused by misperception or miscalculation.[126] Recent experimental war-gaming demonstrated that when a crisis is triggered by a cyber-attack, and the targeted state lacks the ability to conduct proportional cyber countermeasures (i.e., asymmetric in nature), the situation tends to escalate more rapidly—as a demonstration of resolve.[127] In a US-China conflict scenario, for instance, inadvertent escalation risk is increased by a US penchant for counterforce operations targeting an adversary's command and control, the commingled nature of China's (nuclear and conventional) missile forces, US and Chinese preference for the pre-emptive use of cyberweapons, and domestic-political pressures on *both* sides to retaliate for costly losses (either physical/kinetic or nonphysical/political).[128]

These risks should give defense planners pause for thought in using advanced conventional capabilities to project military power in conflicts with regional nuclear powers. In short, conventional doctrines and operational concepts could *exacerbate old* (e.g., third-party interference, civil-military overconfidence, regime type, accidental or unauthorized use, theft of nuclear weapons, nuclear terrorism, or an overzealous commander with predelegation authority) and *create new pathways* (e.g., AI-enhanced ISR and precision missile targeting and guidance, drone swarms, AI-cyber-attacks, and mis/disinformation subversion) to potentially uncontrollable inadvertent escalation. Missile defenses and advanced conventional weapons are unlikely to prevent these escalatory mechanisms once battlefield perception shifts and the nuclear threshold is crossed.[129]

The extent to which a state may succeed in limiting the damage from a nuclear exchange in using technologically enhanced (including AI) counterforce operations continues to be an issue of considerable debate.[130] From the perspective of inadvertent escalation, however, the efficacy of "damage-limitation" counterforce tactics is less significant than whether an adversary views them as such.[131] The fear of China and Russia, for instance, that the US is developing and deploying conventional counterforce capabilities (e.g., cyber-attacks targeting NC3 systems) to blunt their nuclear deterrence risks generating "crisis instability" caused by "use-it-or-lose-it" pressures—that is, pressures to use nuclear weapons before losing the capability to do so.[132]

Nuclear powers maintain different attitudes and perceptions on the escalation risk posed by cyber-attacks and information warfare more generally, however. Chinese analysts, in particular, have expressed an acute awareness of the potential vulnerabilities of their respective NC3 systems to cyber-attacks.[133] The US has begun to take this threat more seriously, whereas Russian strategists, despite bolstering their cyber defenses, appear more sanguine, viewing information warfare as a continuation of peacetime politics by other means.[134] Consequently, the risk of inadvertent escalation caused by misperception and miscalculation will likely increase.[135] In what ways might the new tools and techniques emerging in digitized information exacerbate these dynamics?

Digitized information ecosystem, human psychology, and inadvertent risk

Human cognitive psychology—particularly the study of individual and group relationships with social media—provides important empirical insights into the ability of information operations to impact perceptions, and, in turn, escalation risk. Misperceptions, cognitive bias, and the human psychological features of security dilemma theorizing can also be used to elucidate the escalatory dynamics that can follow from inflammatory, emotionally charged, and other offensive public rhetoric (e.g., fake news, disinformation, rumors, and propaganda) used by adversaries during crisis—or saber-rattling behavior.[136] During, in anticipation of, or to incite a crisis or conflict, a state or non-state actor (e.g., clandestine digital "sleeper cells") could employ sub-conventional (or "gray zone") information warfare campaigns to amplify its impact by sowing division, eroding public confidence, and delaying an effective official response.[137]

The public confusion and disorder that followed a mistaken cell phone false alert, warning residents in Hawaii of an imminent ballistic missile threat, in 2018 serve as a warning of the vulnerabilities of US civil defenses to state or non-state actors seeking asymmetric advantages vis-à-vis a superior adversary.[138] That is, compensating for its inferior conventional (and nuclear) capabilities. North Korea's nuclear strategy, for instance, is believed to be one of "asymmetric escalation"—that is, threatening nuclear first use (e.g., targeting US Andersen Air Force Base in Guam) to degrade a conventional attack while retaining longer-range nuclear capabilities to deter nuclear retaliation by the US.[139] Therefore, Pyongyang's conventional inferiority may motivate it to replicate incidents like the Hawaii false alarm to promulgate a disinformation campaign (i.e., issuing false evacuation orders, or false nuclear alerts, and subverting real ones via social media) to sow mass confusion.[140]

During a crisis in the South China Seas or South Asia, for example, when tensions are running high, state or non-state disinformation campaigns could have an outsized impact on influencing crisis stability—dependent on the interpretation and processing of reliable intelligence—with potentially severe escalatory consequences (see Chapter 5).[141] This impact is compounded when populist leaders (Trump, Orban, or Bolsonaro) rely heavily on social media for information-gathering and open-source intelligence and are thus more susceptible to social media manipulation. *In extremis*, a populist leader may come to view social media as an accurate barometer of public sentiment, eschewing official (classified and non-classified) evidence-based intelligence sources, and regardless of the origins of this virtual voice—that is, from genuine users or fake accounts as part of a malevolent disinformation campaign.[142]

As a consequence, the agenda-setting framing narrative of decision-makers during a crisis would be informed by a fragmented and politicized social media information ecosystem, lacking context and authentication, and thus be more likely to stimulate emotionally visceral reactions than "rational" thought—assumed by classical deterrence theorists.[143] These dynamics, in turn, could amplify rumors, conspiracy theories, and radical polarization, reducing the possibility of achieving a public consensus to inform and legitimatize decisions during a crisis.[144] One possible outcome of these dynamics is the inculcation of an "anything-goes" mentality among social media users, who might become less vigilant or concerned about the quality of information they share on social media, thus eroding social norms and making it increasingly more challenging to establish the truth and debunk falsehoods. For example, during the COVID-19 pandemic, between half and a quarter of the British public, depending on the period, reported coming across false or misleading news about the virus, and an additional quarter were not sure.[145]

The use of social media to intensify domestic polarization, radicalize political in-groups, and increase national popularism are not, however, confined to authoritarian regimes. Social media-fueled "filter bubbles" and "echo chambers,"[146] rapidly diffused by social bots (automated accounts) and cyborgs (hybrid/automated accounts),[147] have amplified inflammatory polarizing falsehoods to support anti-establishment candidates in highly popularist and partisan environments, such as the US 2016 and 2020 elections and the Brexit referendum.[148] WhatsApp groups were used in both the 2018 Brazilian presidential election and the 2019 Indian general election, for example, to distribute misleading and divisive political content.[149] These dynamics can expose decision-makers to increased "rhetorical entrapment" pressure, whereby alternative policy options (viable or otherwise) may be overlooked or dismissed out of hand.[150]

Furthermore, increased public scrutiny levels—especially when combined with disinformation and public panic—could further increase political pressures on leaders whose electoral success determines their political survival.[151] Under crisis conditions, these dynamics may compromise diplomatic de-escalation efforts and complicate other issues that can influence crisis stability.[152] Escalation and de-escalation dynamics are fundamentally about signaling, and such signals may well be lost as the fog of war grows even more dense at the strategic level—for instance, maintaining public confidence in a state's retaliatory capability while simultaneously signaling resolve to adversaries and assurance to allies.[153] State or non-state disinformation campaigns might also be deployed in conjunction with other AI-augmented non-kinetic/political (e.g., cyber-attacks, deepfake technology, or disinformation campaigns via social media amplified by automated bots) or kinetic/military (e.g., drone swarms, missile defense, anti-satellite weapons, or hypersonic weapons) actions to distract decision-makers—thus reducing their response time during a crisis and conferring a tactical or operational advantage to an adversary.[154]

Fake news spread on social media is noxious in a divided electorate with clear in-groups and out-groups, where people are more inclined to accept a statement, hate speech, or news story, providing that it aligns with their pre-existing beliefs.[155] That is, people operating within individual group identities tend to use heuristics, or stereotypes, that are congruent with their worldview to inform their decisions—while eschewing dissonant information that can cause cognitive stress even when it is irrational to do so.[156] In this way, the real-time, spontaneous, and self-reinforcing nature of social media facilitates, and thus exacerbates, the susceptibility of individuals to mis/disinformation that conforms with their world views—requiring very little cognitive consideration.

Because people often do not accept fact-checking corrections that run contrary to their group identities or worldview, debunking false information becomes less effective and, paradoxically, often counterproductive—strengthening pre-existing beliefs, biases, and flawed rhetorical arguments and tropes.[157] In other words, the stronger these beliefs are, the more resistant they are to correction. Thus, misperceptions are more likely to persist when the misinformation is from a credible source (i.e., populist leaders and elite groups) and when the information is persistent and repeated (i.e., by bots or sharing).[158] Moreover, using micro-targeting by social media platforms (i.e., curating users' identities, preferences, and prejudices) is compounding the challenge for users to recognize and dismiss mis/disinformation that conforms with their beliefs.[159]

During times of uncertainty, people's sense of in-group belonging and out-group enmity increases and is manifested through "shared sensemaking" (e.g., sharing conspiracies, rumors, and fake news), and so amplifying the perceived differences between groups.[160] This polarization can be intensified if intergroup competition is seen as a zero-sum game.[161] For example, in the aftermath of a terrorist attack in India's Jammu and Kashmir in 2019, a disinformation campaign (e.g., fake news and false and doctored images), which spread via social media amid a heated national election,[162] inflamed emotions and domestic-political escalatory rhetoric, prompting calls for retaliation against Pakistan, and bringing two nuclear-armed adversaries close to conflict.[163] The crisis provides a sobering glimpse of how information and influence campaigns between two nuclear-armed adversaries can affect crisis stability and the concomitant risks of inadvertent escalation. In short, the catalyzing effect of costly signaling and testing the limits of an adversary's resolve—which did not previously exist—to enhance security can increase inadvertent escalation risks, leaving both sides less secure.

The effect of escalatory imbued rhetoric in the information ecosystem may be a double-edged sword for inadvertent escalation risk. On the one hand, public rhetorical escalation can mobilize domestic support and signal deterrence and resolve to an adversary—making war less likely. On the other hand, "costly talk" can sow public fear, distrust (e.g., confidence in the legitimacy and reliability of NC3 systems), and threaten a leader's reputation and image (e.g., the credibility of strategic decision-makers and robustness of nuclear launch protocols) domestically, which, in turn, generate pressures to make political commitments, inadvertently making adversaries of unresolved actors. Public escalatory rhetoric can, therefore, create commitments that did not previously exist.[164] For example, following the Hague's Permanent Court of Arbitration ruling against China over the territorial disputes in the South China Seas in 2016, the Chinese government had to resort to social media censorship to stem the flood of nationalism, calling for war with US ally the Philippines.[165]

Furthermore, domestic public disorder and confusion—caused, for example, by a disinformation campaign or cyber attacks—can act as an escalatory force, putting decision-makers under pressure to respond forcefully to foreign or domestic threats, to protect a states' legitimacy, self-image, and credibility. Leaders tend to be influenced more by relevant in-group attributes (i.e., elites, public opinion, or electoral imperatives), while more susceptible to the perceived threats from out-groups (i.e., the "enemy image")—for example, political expediency, strategic necessity, or alliance considerations.[166] Moreover, elites have a natural advantage in enlisting digital media to disseminate polarizing and persuasive mis/disinformation, which can exacerbate informational cascades and encourage polarization and mistrust.[167] During crisis and warfare, this psychological phenomenon is generally intensified when in-groups are often glorified (i.e., high levels of solidarity) and others are demonized (i.e., high levels of xenophobia or jingoism).[168]

While groupthink cognitive bias may exist in democracies (i.e., where decisions disproportionally reflect the personal influence of individual leaders run contrary to independent or collective opinion), authoritarian regimes with tightly controlled bureaucracies and institutions with strong leaders—who surround themselves with "yes men" for example—are more generally more prone to these biases.[169] Russia, for example, is reportedly considering integrating AI technology into its C2 information gathering, analysis, and decision-making support system (Russia's National Defense Management Center), to defend it against US conventional counterforce capabilities and to monitor in real-time the geopolitical and military-political situation.[170]

These rhetorical escalation dynamics can simultaneously reduce the possibility for face-saving de-escalation efforts by either side—analogous to Thomas Schelling's "tying-hands mechanism."[171] During heightened tensions between the US and North Korea in 2017, for instance, the Trump administration's heated war of words with Kim Jong Un, whether a madman's bluff or in earnest (or "rattle the pots and pans") raised the costs of Kim Jong Un backing down—that is, with regime survival at stake—thus increasing inadvertent escalation risk, and, simultaneously, complicating de-escalation.[172] Because of the fear that its nuclear (and conventional) forces are vulnerable to a decapacitating first-strike, rhetorical escalation between a conventionally inferior and superior state is especially hazardous.[173] Research would be beneficial on how today's information ecosystem might affect decision-making in different political systems.[174]

Ultimately, states' willingness to engage in nuclear brinkmanship—the ultimate contest of resolve and ability to tolerate high risk and uncertainty—will depend upon information (and mis/disinformation), cognitive bias (the perception of it and the value attached to it), and what is at stake, as well as the balance of uncertainty in each other's damage limitation effectiveness.[175] If one side considers the potential consequences of not going to war as intolerable (i.e., regime survival, the "tying-hands," or "use-it-or-lose-it" pressures), then de-escalation measures, such as off-ramps and firebreaks, will be unable to prevent crisis instability from intensifying. This leaves ample room for first-strike and crisis instability, and the risk of nuclear confrontation if a crisis occurs.[176] Finally, the extent to which public pressure emanating from the contemporary information environment affects whether nuclear war remains "special" or "taboo" will be critical for reducing the risk of inadvertent escalation by achieving crisis stability during a conventional war between nuclear-armed states. Thus, any response to a violation of the nuclear taboo would raise fundamental questions of reputation and justice, requiring a close calibration of military and political objectives—including both sides' civil-military relations and attitude to leveraging the information landscape to exert psychological pressure as a tool of escalation management.

Conclusion

To what extent might AI-enabled capabilities increase inadvertent escalation risk? In a global security environment characterized by great-power strategic competition and regional strategic asymmetry (capabilities, domains, and stakes), new rungs, firebreaks, and thresholds on the escalation ladder are already challenging conventional assumptions of deterrence, strategic stability, and escalation. This chapter underscores the need for greater clarity and discussion on the specific characteristics of AI technology that may create new (or disrupt old) rungs on the metaphorical escalation ladder, which could, in turn, increase the risk of inadvertently transitioning crises between nuclear-armed (and especially regional) states from conventional to nuclear confrontation. The chapter builds on and adapts the foundational work on inadvertent escalation conducted at the end of the Cold War—on the cusp of the "second nuclear age." Specifically, it examines the psychological underpinnings of escalation theorizing to elucidate whether and how characteristics of AI technology, contextualized with the broader digital information ecosystem, might destabilize crisis stability, and increase inadvertent escalation risk.

The chapter highlights three critical features of inadvertent escalation risk in the emerging AI-nuclear strategic nexus. First, while nuclear-armed states have a shared interest in avoiding nuclear war (expressed by the "rational deterrence theory"), they also place a high value on their nuclear forces, which advanced conventional weapons enhanced by AI technology inadvertently threaten—especially in asymmetric military situations. The synthesis of AI technology into conventional weapon systems to enhance autonomy, remote sensing, missile guidance, and situational awareness creates new possibilities for a range of novel destabilizing conventional counterforce possibilities (e.g., cyber-attacks on NC3 and locating survivable nuclear retaliatory capabilities). In a regional asymmetric crisis or conflict, AI-powered tools to find, track, and destroy a state's nuclear forces may be viewed by the other side as a precursor to a disabling counterforce attack, thus increasing incentives to escalate a conventional situation inadvertently. Further, these technological developments have been accompanied by destabilizing doctrinal shifts by several regional nuclear powers (Russia, North Korea, and possibly China); thereby compounding the problem of commingled dual-use conventional and nuclear weapons missions.

Second, escalatory rhetoric and other aggressive behavior, amplified by the digital information ecosystem, might be misperceived as unprovoked malign intent—not as a response to the initiator's behavior—leading to action and reaction spirals of potentially irrevocable inadvertent escalation. Cognitive and heuristic burdens caused by information overload and complexity will likely make decision-makers more susceptible to cognitive bias, misperceptions, and heuristics to approach complex problem-solving. Also, new AI-enhanced tools are enabling a more comprehensive range of actors (state and non-state) with asymmetric techniques (e.g., facial recognition, cyber penetration, new kinds of radar, drone video, and disinformation) to improve their bargaining hand by delegitimizing (or legitimizing) the use of nuclear weapons during an escalating situation—or suing for peace to de-escalate a situation.[177] The study also demonstrates that the impact of escalatory rhetoric in the information ecosystem could be a double-edged sword for inadvertent escalation risk at the political decision-making level. On the one hand, public rhetorical escalation can mobilize domestic support and signal deterrence and resolve to an adversary, thus making war less likely. On the other hand, sowing public fear, distrust, and threatening a leader's reputation and image domestically can inadvertently make enemies of unresolved actors.

Third, and related, the state of heightened tension, uncertainty, complexity, and compressed decision-making (or "the fog of war") of modern digitized warfare will likely be dramatically increased with the infusion of AI technology. That is, restricting the options and time available for de-escalation, compounding the problem of offense-defense distinguishability, and increasing the risks of both horizontal and vertical inadvertent escalation. Moreover, the increasing dependencies and concomitant vulnerabilities on digital technologies (especially AI) and information to conduct modern warfare create a new security paradox. This "paradox" could create resource vulnerabilities that generate first-strike and pre-emption incentives, predicated upon "use-it-or-lose-it" pressures, whether real or illusory.

How can decision-makers mitigate the inadvertent escalation risks associated with AI and nuclear systems? Possible ways forward include, *inter alia*, arms control and verification, changes to norms and behavior, unilateral measures and restraint, and bilateral and multilateral stability dialogue (see the conclusion below). AI technology is already raising a multitude of questions about warfare and shifts in the balance of power, which are challenging traditional arms control thinking.[178] Traditional arms control and non-proliferation frameworks of nuclear governance are not necessarily obsolete, however.[179] Instead, we will need to depart from conventional siloed, rigid, and stove-piped approaches and search for innovative frameworks and novel approaches to meet the challenges of the rapidly evolving dual-use technology (see Chapter 4), the linkages between conventional and nuclear weapons, and the informational changes in the new nuclear age. In designing such agreements, a sensible starting point would be an asymmetric arms control framework that emphasizes the importance of dynamism, allowing for mutual adjustment in force posture in ways that differ from the traditional "like-for-like" approach to arms control.[180]

In today's nuclear multipolar world, with great-power techno-military competition (US-China and US-Russia) and regional asymmetric dynamics, there is a political imperative to address the challenges for inadvertent escalation by engaging in broader strategic stability talks about the development of new and innovative normative frameworks. Against the unpropitious backdrop of increasing populism, great-power competition, and nuclear multipolarity—amplified by and manipulated in the digital information ecosystem—long-term regional and global stability and nuclear security efforts will likely continue to be jettisoned by short-termism, political fragmentation, and the pursuit of fleeting first-mover advantages.[181]

Technological solutions are not a panacea, when the problem (inadvertent escalation risk) manifests because of technological creativity—and compliance and verification require sacrificing potential future breakthroughs and exploiting the benefits of a particular innovation. Moreover, the solution is not reducible to structural processes and assumptions of "rational" actors—agency still matters.[182] A viable solution to the emerging AI-nuclear strategic challenge must, as a corollary, be as much a political as a technological one. Solutions will have to help actors "escape an *irrational* situation where it is precisely *rational* [decision-making] behavior that may be most dangerous"—that is, reducing perceived vulnerability in the short-term at the expense of future inadvertent risk.[183] The next chapter examines how and why dual-use AI technology may affect the security dilemma between nuclear-armed adversaries. What (if anything) makes AI different from other emerging technologies?

Notes

1. Vincent Boulanin et al., *Artificial Intelligence, Strategic Stability and Nuclear Risk* (Stockholm: SIPRI Publications, June 2020).
2. Notable exceptions include: Kenneth Payne, "Artificial Intelligence: A Revolution in Strategic Affairs?" *Survival* 60, no. 5 (2018): pp. 7–32; Michael C. Horowitz, "When Speed Kills: Lethal Autonomous Weapon Systems, Deterrence and Stability," *Journal of Strategic Studies* 42, no. 6 (2019): pp. 764–788; Mark Fitzpatrick, "Artificial Intelligence and Nuclear Command and Control," *Survival* 61, no. 3 (2019): pp. 81–92; Michael

C. Horowitz et al., "Strategic Competition in an Era of Artificial Intelligence," in *Artificial Intelligence and International Security* (Center for New American Security), July 2018; and James Johnson, "'Catalytic Nuclear War' in the Age of Artificial Intelligence; Autonomy: Emerging Military Technology and Escalation Risk between Nuclear-Armed States," *Journal of Strategic Studies* (2021), doi: 10.1080/01402390.2020.1867541.

3. Sections of this chapter are derived in part from James Johnson, "Inadvertent Escalation in the Age of Intelligence Machines: A New Model for Nuclear Risk in the Digital Age," *European Journal of International Security* 7, no. 3 (2022): pp. 337–359. Copyright Taylor & Francis, available online: doi:10.1017/eis.2021.23.
4. See, for example, James D. Fearon, "Rationalist Explanations for War," *International Organization* 49, no. 3 (1995): pp. 379–414; Robert Powell, "War as a Commitment Problem," *International Organization* 60, no. 1 (2006): pp. 169–203; Daniel R. Headrick, *The Tools of Empire: Technology and European Imperialism in the Nineteenth Century* (Oxford: Oxford University Press, 1981); and Robert Powell, *In the Shadow of Power: States and Strategies in International Politics* (Princeton, NJ: Princeton University Press, 1999).
5. For a critique of strategic theorists' preoccupation with technical details (or "technocentric approaches") at the expense of political considerations, see Lawrence Freedman, "Has Strategy Reached a Dead-End?" *Futures* 11, no. 2 (1979): pp. 122–131; Michael Howard, "The Forgotten Dimensions of Strategy," *Foreign Affairs* (1979), https://www.foreignaffairs.com/articles/1979-06-01/forgotten-dimensions-strategy; Colin S. Gray, *Theory of Strategy* (Oxford: Oxford University Press 2018); and Bill Owens, *Lifting the Fog of War* (New York: Farrar, Straus and Giroux, 2000).
6. See William H. McNeill, *The Pursuit of Power: Technology, Armed Force, and Society since AD 1000* (Chicago, IL: University of Chicago Press, 1984); and Kier A. Lieber, *War and the Engineers: The Primacy of Politics over Technology* (Ithaca, NY: Cornell University Press, 1995).
7. Bryan R. Early and Erik Gartzke, "Spying from Space: Reconnaissance Satellites and Interstate Disputes," *Journal of Conflict Resolution* (March 2021), doi: 10.1177/0022002721995894, p. 4.
8. Michael C. Horowitz, "Artificial intelligence, International Competition, and the Balance of Power," *Texas National Security Review* 1, no. 3 (2018): pp. 37–57.
9. Barry R. Posen, *Inadvertent Escalation* (Ithaca, NY: Cornell University Press, 1991).
10. "Misinformation" refers to false or out-of-context information that is presented as fact regardless of an intent to deceive (e.g., fake news, deepfakes, false rumors, and insults). "Disinformation" is a sub-set of misinformation that is intentionally false and intended to deceive or mislead (e.g., malicious propaganda, spear phishing, and spoofing). The principal effect of misinformation is to elicit fear and suspicion among a population. Both misinformation and disinformation involve the sharing of bad or otherwise debunked information, with varying intents and purposes. For a recent exploration of the effects of misinformation and disinformation on attributes of the digital ecosystem, see Rachel Armitage and Cristian Vaccari, "Misinformation and Disinformation," in *The Routledge Companion to Media Disinformation and Populism*, edited by Howard Tumber and Silvio Waisbord (London: Routledge, 2021), pp. 38–49; and Cristian Vaccari and Andrew Chadwick, "Deepfakes and Disinformation: Exploring the Impact of Synthetic Political Video on Deception, Uncertainty, and Trust in News," *Social Media + Society* (February 19, 2020), https://journals.sagepub.com/doi/full/10.1177/2056305120903408.

11. See Lawrence Freedman, *Deterrence* (New York: Polity, 2004); Lawrence Freedman, *The Evolution of Nuclear Strategy* (Basingstoke: Palgrave Macmillan 2003); Thomas Schelling, *Arms and Influence* (New Haven and London: Yale University Press, 1966); Glen H. Snyder, *Deterrence and Defense: Towards a Theory of National Security* (Princeton, NJ: Princeton University Press, 1961); Patrick M. Morgan, *Deterrence Now* (Cambridge: Cambridge University Press, 2003); Michael Quinlan, *Thinking About Nuclear Weapons: Principles, Problems, Prospects* (Oxford: Oxford, University Press 2009); Colby Elbridge and Michael Gerson, eds, *Strategic Stability: Contending Interpretations* (Carlisle, PA: Army War College, 2013); and Barry Buzan and Eric Herring, *The Arms Dynamic in World Politics* (London, Boulder, CO: Lynne Reinner, 1998).
12. See Robert Powell, *In the Shadow of Power: States and Strategies in International Politics* (Princeton, NJ: Princeton University Press, 1999).
13. On uncertainty as a cause of war, see Geoffrey Blainey, *The Causes of War* (New York: Free Press, 1988); Powell, *In the Shadow of Power: States and Strategies in International Politics*; and Rose McDermott, "Decision-Making Under Uncertainty," in *Proceedings of a Workshop on Deterring Cyberattacks: Informing Strategies and Developing Options for US policy*, edited by National Research Council (Washington DC: National Academies Press, 2010), pp. 227–242.
14. Kahn sub-divides these forty-four rungs into seven units with seven thresholds (or firebreaks), which denote important inflection points along the escalation continuum. Herman Kahn, *On Escalation: Metaphors and Scenarios* (New York: Prager, 1965), p. 40.
15. Ibid., pp. 39–40.
16. See Nina Tannenwald, *The Nuclear Taboo: The United States and the Normative Basis of Nuclear Non-Use* (New York: Cambridge University Press, 2007).
17. Escalation, in a broader sense, can be both violent (kinetic) and non-violent (non-kinetic) in nature and occur during a military exchange ("vertical escalation") or represent an expansion in the scope and range of a conflict ("horizontal escalation"). In contrast, "political escalation" refers to non-military changes in the scope or intensity of a situation—i.e., rhetorical, an articulation of expansive objectives, or changes to the accepted rules of engagement). See Forrest E. Morgan, and Karl P. Mueller et al., *Dangerous Thresholds: Managing Escalation in the 21st century* (Santa Monica, CA: RAND Corporation, 2008), Ch. 2.
18. See Bruce G. Blair, *The Logic of Accidental Nuclear War* (Washington, DC: Brookings Institution, 1993); and Scott D. Sagan, *The Limits of Safety: Organizations, Accidents, and Nuclear Weapons* (Princeton, NJ: Princeton University Press, 1993).
19. For discussion on why a state might contemplate intentional escalation, see Kelly M. Greenhill and Peter Krause, eds, *Coercion: The Power to Hurt in International Politics* (New York: Oxford University Press, 2018), pp. 11–12.
20. See Posen, *Inadvertent Escalation*; and Bradley A. Thayer, "The Risk of Nuclear Inadvertence," *Security Studies* 3, no. 3 (Spring 1994): pp. 428–493.
21. For example, amidst heightened tensions in the Baltics, NATO's actions to bolster its deterrence posture in Eastern Europe might be perceived by Moscow as preparation for a pre-emptive military offensive, thus risking inadvertent counter-escalation by Russia. Ulrich Kühn, *Preventing Escalation in the Baltics: A NATO Playbook* (New York: Carnegie Endowment for International Peace, 2018).

22. A false-flag cyber operation is designed to deflect attribution to a neutral party, and the actor behind the attack takes steps to impersonate or use the distinctive infrastructure, tactics, techniques, or procedures to appear as if it had been the work of another. For example, the Olympic Destroyer cyber-attack against the 2018 Pyeongchang Winter Olympic Games is regarded as having been a false-flag operation in which Russia's military intelligence service designed its attack to appear as if it had been the work of North Korea. Andy Greenberg, "The Untold Story of the 2018 Olympics Cyberattack, the Most Deceptive Hack in History," *Wired*, October 17, 2019, https://www.wired.com/story/untold-story-2018-olympics-destroyer-cyberattack/.
23. Herbert Lin, "Escalation Dynamics and Conflict Termination in Cyberspace," *Strategic Studies Quarterly* 6, no. 3 (2012): pp. 46–70.
24. However, not all threats in the use of force are escalatory; escalation occurs only when at least one actor views this action (i.e., rhetoric or signaling) as shifting the scope or intensity of a situation. See Michael Brecher, "Crisis Escalation: Model and Findings," *International Political Science* 17, no. 2 (1996): pp. 215–230.
25. Johnson, "'Catalytic Nuclear War' in the Age of Artificial Intelligence & Autonomy."
26. Rebecca Hersman, "Wormhole Escalation: The New Nuclear Age," *Texas National Security Review* 2, no. 3 (2020): pp. 91–109; Nicholas Miller and Vipin Narang, "Is a New Nuclear Age upon Us? Why We May Look Back on 2019 as the Point of No Return," Foreign Affairs (December 2019), https://www.foreignaffairs.com/articles/2019-12-30/new-nuclear-age-upon-us; and Andrew Futter and Benjamin Zala, "Strategic Non-Nuclear Weapons and the Onset of a Third Nuclear Age," *European Journal of International Security* 6, no. 3 (2021): pp. 1–21.
27. Kahn, *On Escalation: Metaphors and Scenarios*, p. 285.
28. Michal Onderco and Madeline Zutt, "Emerging Technology and Nuclear Security: What Does the Wisdom of the Crowd Tell Us?" *Contemporary Security Policy* (2021), doi: 10.1080/13523260.2021.1928963.
29. See Ben Buchanan and Fiona S. Cunningham, "Preparing the Cyber Battlefield: Assessing a Novel Escalation Risk in a Sino-American Crisis," *Texas National Security Review* 3, no. 4 (Fall 2020): pp. 55–81.
30. Technological advances in AI technology and cyber capabilities, coupled with the increasingly commingled nature of the state's nuclear and conventional command and control systems, have enabled solutions to overcome the robustness of permissive action links (PALs) and increase these vulnerabilities systems. Bruce Blair, *The Logic of Accidental Nuclear War* (Washington, DC: Brookings Institution, 1993).
31. Erik Gartzke and Jon R. Lindsay, "The Cyber Commitment Problem and the Destabilization of Nuclear Deterrence," in *Bytes, Bombs, and Spies: The Strategic Dimensions of Offensive Cyber Operations*, edited by Herbert Lin and Amy Zegart (Washington, DC: Brookings Institution, 2018), pp. 195–235.
32. See Jervis, "Deterrence and Perception," pp. 3–30; Jervis, *Perception and Misperception in International Politics*; Jervis, Lebow, and Stein, *Psychology and Deterrence*; and Graham Allison and Phillip Zelikow, *Essence of Decision* (New York: Longman, 1999).
33. See Robert Mandel, "Political Gaming and Foreign Policy Making During Crises," *World Politics* 29, no. 4 (1977): pp. 610–625; Robert Mandel et al., "Policy-Making Perspectives on War Simulations," 24, no. 2 (1980): pp. 359–375; and Schelling, *Arms and Influence*.

34. Sico van der Meer, "Reducing Nuclear Weapons Risks: A Menu of 11 Policy Options," *Policy Brief*, Clingendael Netherlands Institute of International Relations, June 2018.
35. Miles, "The Dynamics of Strategic Stability and Instability," p. 429 and 437.
36. Deterrence and coercion require the actor that is deterring to share information about the military balance with the actor that is being deterred. Early and Gartzke, "Spying from Space: Reconnaissance Satellites and Interstate Disputes," p. 5.
37. Kahn, *On Escalation: Metaphors and Scenarios*, p. 289.
38. Ibid.
39. See Bruce Bueno de Mesquita, James Morrow, and Ethan Zorick, "Capabilities, Perception, and Escalation," *American Political Science Review* 91, no. 1 (1997): pp. 15–27.
40. Bernard Brodie, *Escalation and the Nuclear Option* (Santa Monica, CA: RAND Corporation, 1965), p. 64.
41. See Darryl Press, Scott Sagan, and Benjamin Valentino, "Atomic Aversion: Experimental Evidence on Taboos, Traditions, and the Non-Use of Nuclear Weapons," *American Political Science Review* 107, no. 1 (2013): pp. 188–206.
42. Frank Sauer, *Atomic Anxiety: Deterrence, Taboo and the Non-Use of US Nuclear Weapons* (New York: Springer, 2015), p. 176.
43. Ibid. p. 38.
44. Brecher, "Crisis Escalation: Model and Findings."
45. Greenhill and Krause, eds, *Coercion: The Power to Hurt in International Politics*, pp. 3–33.
46. On the core logic of escalation dominance, see Kahn, *On Escalation: Metaphors and Scenarios*; Kahn, *Thinking about the Unthinkable in the 1980s* (New York: Touchstone, 1985); and Matthew Kroenig, "Nuclear Superiority and the Balance of Resolve: Explaining Nuclear Crisis Outcomes," *International Organization* 67, no. 1 (2013): pp. 141–171.
47. For critiques of scholars who question the relevance of Clausewitzian military theory in the context of the revolution in military affairs associated with modern warfare such as hybrid and gray zone war, sub-conventional war, and more recently "cyberwar," see Chiara Libiseller and Lukas Milevski, "War and Peace: Reaffirming the Distinction," *Survival* 63, no. 1 (2021): pp. 101–112; Donald Stoker and Craig Whiteside, "Blurred Lines: Gray-Zone Conflict and Hybrid War—Two Failures of American Strategic Thinking," *Naval War College Review* 73, no. 1 (2020): pp. 13–48; Hew Strachan, *The Direction of War: Contemporary Strategy in Historical Perspective* (Cambridge: Cambridge University Press 2013); and Colin M. Fleming, *Clausewitz's Timeless Trinity: A Framework for Modern War* (Farnham: Taylor and Francis 2013).
48. Posen, *Inadvertent Escalation*, pp. 12–27.
49. "Crisis stability" refers to the presumption that control can be maintained during crisis or conflict to ensure that nuclear weapons are not used—that is, a situation where neither side has an incentive to strike first or pre-emptively. Cold War-era debates on the concept centered on how specific capabilities, postures, and military doctrine could escalate (either intentionally or inadvertently) to nuclear crisis or war. See Schelling, *Arms and Influence*; Richard Lebow, "Clausewitz and Nuclear Crisis Stability," *Political Science Quarterly* 103, no. 1 (1988): pp. 81–110; and Forrest Morgan, *Crisis Stability and Long-Range Strike: A Comparative Analysis of Fighters, Bombers, and Missiles* (Santa Monica, CA: RAND Corporation, 2013).
50. Daniel Kahneman and Jonathan Renshon, "Why Hawks Win," *Foreign Policy* no. 158 (2007): pp. 34–38. For an opposing view, see Dominic Johnson, *Strategic Instincts:*

The Advantages of Cognitive Biases in International Politics (Princeton, NJ: Princeton University Press, 2020).

51. For seminal works on the "security dilemma" see Herbert Butterfield, *History and Human Relations* (London: Collins, 1951); John Herz, *Political Realism and Political Idealism: A Study in Theories and Realities* (Chicago, IL: University of Chicago Press, 1951); and Robert Jervis, "Cooperation under the Security Dilemma," *World Politics* 30, no. 2 (1978): pp. 169–214.
52. The "offense-defense balance" refers to the relative ease of carrying out and defending against military attacks—primarily by military technology and geographical factors. For seminal works on the concept, see Sean M. Lynn-Jones, "Offense-Defense Theory and its Critics," *Security Studies* 4, no. 4 (1995): pp. 660–691; Glaser and Kaufmann, "What is the Offense-Defense Balance and How Can We Measure It?"; and Jervis, "Cooperation under the Security Dilemma." For studies on the "cyber offense-defense balance," see Gartzke and Lindsay, "Weaving Tangled Webs: Offense, Defense, and Deception in Cyberspace"; Ilai Saltzman, "Cyber Posturing and the Offense-Defense Balance," *Contemporary Security Policy*, 34, no.1 (2013): pp. 40–63; Rebecca Slayton, "What is the Cyber Offense-Defense Balance? Conceptions, Causes, and Assessment," *International Security*, 41, no. 3 (2017): pp. 72–109; and Ben Garfinkel, and Allen Dafoe, "How Does the Offense-Defense Balance Scale?" *Journal of Strategic Studies*, 42, no. 6 (2019): pp. 736–763.
53. Jervis, "Cooperation under the Security Dilemma," pp. 169–170.
54. Schelling, *Arms and Influence.*
55. Jervis, "Cooperation under the Security Dilemma," p. 172.
56. Brodie, *Escalation and the Nuclear Option.*
57. Jervis, "Cooperation under the Security Dilemma," pp. 170 and 193.
58. Posen, *Inadvertent Escalation*, pp. 12–15.
59. See Scott Sagan, "Why Do States Build Nuclear Weapons? Three Models in Search of a Bomb," *International Security* 21 (1996): pp. 54–86; and Tannenwald *The Nuclear Taboo.*
60. See Futter and Zala, "Strategic Non-Nuclear Weapons and the Onset of a Third Nuclear Age," pp. 1–21.
61. It is also plausible that improvements in battlefield awareness and decision-making AI-enabled tools afford commanders more time to make decisions during combat. See Paul Scharre, "Autonomous Weapons and Stability," *PhD Thesis* (King's College London, 2020), https://ethos.bl.uk/OrderDetails.do?uin=uk.bl.ethos.806777; and James Johnson, "Artificial Intelligence in Nuclear Warfare: A Perfect Storm of Instability?" *The Washington Quarterly* 43, no. 2 (2020): pp. 197–211.
62. Jacquelyn Schneider, "The Capability/Vulnerability Paradox and Military Revolutions: Implications for Computing, Cyber, and the Onset of War," *Journal of Strategic Studies* 42, no. 6 (2019): pp. 841–863.
63. As a helpful counterpoint, while the proliferation and dependency on the digital information ecosystem increase, the ubiquity of information and the intrinsic inability to control all information might create disincentives for first strike. Schneider, "The Capability/Vulnerability Paradox and Military Revolutions," p. 842.
64. For example, see Eliot A. Cohen, "A Revolution in Warfare," *Foreign Affairs* (1996): pp. 37–54; Cohen, "Change and Transformation in Military Affairs," *Journal of Strategic Studies* 27, no. 3 (2004): pp. 395–407; Erik J. Dahl, "Network Centric Warfare

and Operational Art," *Defence Studies* 2, no. 1 (Spring 2002): pp. 17–34; Arthur Cebrowski and John Garstka, "Network-Centric Warfare: Its Origin and Future," *US Naval Institute Proceedings* 124, no. 1 (1998): pp. 28–35; and Michael Raska, "The Sixth RMA Wave: Disruption in Military Affairs?"; *Journal of Strategic Studies* (2020) doi: 10.1080/01402390.2020.1848818

65. Richard Danzig, "Surviving on a Diet of Poisoned Fruit: Reducing the National Security Risks of America's Cyber Dependencies," *Center for a New American Security*, July 21, 2014, https://www.cnas.org/publications/reports/surviving-on-a-diet-of-poisonedfruit-reducing-the-national-security-risks-of-americas-cyber-dependencies.
66. See Jervis *Perception and Misperception in International Politics*; Jacques Hymans, *The Psychology of Nuclear Proliferation: Identity, Emotions and Foreign Policy* (Cambridge: Cambridge University Press, 2006); Janice G. Stein, "Building Politics into Psychology: The Misperception of Threat," *Political Psychology* 9, no. 2 (1988): pp. 45–71; Barbara Farnham, *Avoiding Losses/Taking Risks: Prospect Theory and International Conflict* (Ann Arbor, MI: University of Michigan Press, 1994); Rose McDermott, James Fowler, and Oleg Smirnov, "On the Evolutionary Origin of Prospect Theory Preferences," *Journal of Politics* 70, no. 2 (2008): pp. 335–350; Philip Tetlock and Richard Boettger, "Accountability: A Social Magnifier of the Dilution Effect," *Journal of Personality & Social Psychology* 57, no. 3 (1989): pp. 388–398; and Jonathan Mercer, "Human Nature and the First Image: Emotion in International Politics," *Journal of International Relations & Development* 9 (2006): pp. 288–303.
67. The strategic value of "information" has long been recognized in military affairs: (1) as a force multiplier to alter the balance of power during war and peace; and (2) providing knowledge about an adversary's capabilities, which can make political leaders more effective at recognizing acceptable bargains and channels for cooperation, thus ensuring peace. Early and Gartzke, "Spying from Space: Reconnaissance Satellites and Interstate Disputes," p. 5.
68. Carl von Clausewitz, *On War* (North Charleston, SC: Jazzybee Verlag, 2017), p. 46.
69. See Bill Owens, *Lifting the Fog of War* (New York: Farrar, Straus and Giroux 2000); and Jervis, Lebow, and Stein, *Psychology and Deterrence*.
70. Robert Jervis, "Arms Control, Stability, and Causes of War," *Political Science Quarterly* 108, no. 2 (1988): pp. 81–110.
71. Ibid. p. 20.
72. The historical record is replete with examples of how misunderstandings or incomplete information about ongoing military operations can contribute to escalation—for example, China's entry into the Korean War in 1950; Germany's Blitz campaign on the British Isles in 1940; and the Cuban crisis in 1962. See Rosemary Foot, *The Wrong War: American Policy and the Dimensions of the Korean Conflict 1950–1953* (Ithaca, NY: Cornell University Press, 1985); F.M. Sallagar, *The Road to Total War: Escalation in World War Two* (Santa Monica, CA: RAND Corporation, 1969); and Richard K. Betts, *Soldiers, Statesmen, & Cold War Crisis* (Cambridge: Harvard University Press, 1977).
73. Posen, *Inadvertent Escalation*, p. 22.
74. Harold Macmillan, *The Macmillan Diaries, Volume II: Prime Minister and After 1957–1966*, edited and introduced by Peter Catterall (Basingstoke: Macmillan, 2003), pp. 508–513.

75. The term "dead reckoning" refers to analytical predictions, intuitions, predictions, or judgments of captains or pilots to navigate a ship, derived from the environment's internal instruments and knowledge. Kahn, *On Escalation: Metaphors and Scenarios*, pp. 211–212.
76. Rebecca Hersman et al., "Under the Nuclear Shadow: Situational Awareness Technology and Crisis Decision-making," *Center for Strategic and International Studies*, March 18, 2020, https://ontheradar.csis.org/analysis/final-report/.
77. Heuristics are rules of thumb that facilitate fast, intuitive problem solving, which influence estimates of outcomes and probabilities over time. Heuristics are critical to decision-making because without them, problems would be subject to a large amount of mental "deliberation cost"—which AI's might be able to reduce. See Cass R. Sunstein, "Algorithms, Correcting Biases," *Social Research* 86, no. 2 (2019): pp. 499–511; Daniel Kahneman, *Thinking, Fast and Slow* (New York: Farrar, Straus and Giroux, 2011); Martin Kaplan, Tatiana Wanshula, and Mark Zanna, "Time Pressure and Information Integration in Social Judgment," in *Time Pressure and Stress in Human Judgment and Decision Making*, edited by Ola Svenson and John Maule (Springer, Boston, MA, 1993), pp. 255–267; and Carsten De Dreu, "Time Pressure and Closing of the Mind in Negotiation," *Organizational Behavior and Human Decision Processes* 91, no. 2 (2003): pp. 280–295.
78. See Peter J Phillips and Gabriela Pohl, "Disinformation Cascades, Espionage & Counter-Intelligence," *International Journal of Intelligence, Security and Public Affairs* (October 2020); and Thomas Rid, *Active Measures: The Secret History of Disinformation and Political Warfare* (New York: Farrar, Straus and Giroux, 2020).
79. For example, see David Sanger, *The Perfect Weapon: War, Sabotage, and Fear in the Cyber Age* (New York: Broadway Books, 2019); and Audrey K. Cronin, *Power to the People: How Open Technological Innovation is Arming Tomorrow's Terrorists* (New York: Oxford University Press, 2019).
80. Kristiina Müür, Holger Mölder, and Vladimir Sazonov, "A Comparative Overview of Online News," in *Russian Information Warfare against the Ukrainian State and Defence Forces*, edited by Vladimir Sazonov et al. (NATO Strategic Communications, 2016), pp. 70–99.
81. In 2017, the US Defense Department officially adopted a new declaratory policy that seeks to deny its adversaries the ability to deliver missiles to target, in part by striking these missiles before they launch. US Department of Defense, *Left of Launch Declaratory Policy* (Washington DC: Joint Chiefs of Staff, 2017) Section 1684; and Riki Ellison, "Left of Launch," *Missile Defense Advocacy Alliance* (2015) https://missiledefenseadvocacy.org/alert/3132/.
82. Marie Baezner and Patrice Robin, *Cyber and Information Warfare in the Ukrainian Conflict* (Zurich: Center for Security Studies, ETH Zurich, 2018).
83. Harold Trinkunas, Herbert Lin, and Benjamin Loehrke, *Three Tweets to Midnight: Effects of the Global Information Ecosystem on the Risk of Nuclear Conflict* (Stanford, CA: Hoover Institution Press, 2020), pp. 150–152; and Johnson, "'Catalytic Nuclear War' in the Age of Artificial Intelligence & Autonomy."
84. Russia, for example, reportedly conducted an active information campaign against several NATO states to influence the public discourse and policymakers to destabilize NATO's missile defense operations in Europe. See Hege Eilersten, "Russia's Ambassador Warns: Missile Shield Will Endanger Norway's Borders," *High North News* February 22,

2017, https://www.highnorthnews.com/en/russias-ambassador-warns-missile-shield-will-endanger-norways-borders.

85. See Robert Ayson, "After a Terrorist Nuclear Attack: Envisaging Catalytic Effects," *Studies in Conflict & Terrorism*," 33, no. 7 (2010): pp. 571–593; and Hayes, Peter, *Non-State Terrorism and Inadvertent Nuclear War*, Nautilus Institute for Security and Sustainability Special Reports, January 18, 2018.
86. Trinkunas, Lin, and Loehrke, *Three Tweets to Midnight: Effects of the Global Information Ecosystem on the Risk of Nuclear Conflict*, p. 152.
87. AI technology may also be used to enhance cyber defenses against heavy information data streams (or "disinformation cascades"), which is currently a burgeoning field of research. For example, see Liang Xiao et al., "IoT Security Techniques Based on Machine Learning: How Do IoT Devices Use AI to Enhance Security?" *IEEE Signal Processing Magazine* 35, no. 5 (2018): pp. 41–49.
88. Joseph Nye, "Neorealism and Neoliberalism," *World Politics* 40, no. 2 (1988): pp. 235–251.
89. "Strategic non-nuclear weapons" are also referred to as strategic conventional weapons or advanced conventional weapons. See James Acton, "Russia and Strategic Conventional Weapons: Concerns and Response," *Nonproliferation Review* 22, no. 2 (2015): pp. 141–154; and Futter and Zala, "Strategic Non-Nuclear Weapons and the Onset of a Third Nuclear Age," pp. 1–21.
90. See Andrew Futter, *Hacking the Bomb: Cyber Threats and Nuclear Weapons* (Washington, DC: Georgetown University Press, 2018), pp. 117–125.
91. See US *Annual Report to Congress: Military and Security Developments Involving the People's Republic of China 2020* (Washington, DC: Office of the Secretary of Defense, 2020); and US National Air and Space Intelligence Center, *Ballistic and Cruise Missile Threat* (Washington, DC: Office of the Secretary of Defense, 2020).
92. For example, Russia deploys dual-use ground-launched cruise missiles; India and Pakistan possess dual-payload ground-launched missiles. China's DF-26 intermediate-range ballistic missile is dual-capable, and the US, China, Russia, India, and Pakistan all have nuclear-capable aircraft capable of supporting conventional systems. See James M. Acton, "Is it a Nuke? Pre-Launch Ambiguity and Inadvertent Escalation," Carnegie Endowment for International Peace, April 9, 2020, https://carnegieendowment.org/2020/04/09/is-it-nuke-pre-launch-ambiguity-and-inadvertent-escalation-pub-81446; Hans M. Kristensen, Robert S. Norris, and Julia Diamond, "Pakistani Nuclear Forces, 2018," *Bulletin of the Atomic Scientists* 74, no. 5 (2018): pp. 348–358, p. 355; and Hans M. Kristensen and Matt Korda, "Indian Nuclear Forces, 2018," *Bulletin of the Atomic Scientists* 74, no. 6 (2018): pp. 361–366, p. 363, doi: 10.1080/00963402.2018.1533162.
93. See James M. Acton et al., eds, *Entanglement: Russian and Chinese Perspectives on Non-Nuclear Weapons and Nuclear Risks* (Washington, DC: Carnegie Endowment for International Peace, 2017); and Hersman et al., "Under the Nuclear Shadow: Situational Awareness Technology and Crisis Decision-making."
94. See John Schaus and Kaitlyn Johnson, "Unmanned Aerial Systems' Influences on Conflict Escalation Dynamics," *Center for Strategic and International Studies*, August 7, 2018, https://aerospace.csis.org/unmanned-aerial-systems-influences-on-conflict-escalationdynamics/.
95. See Keir A. Lieber and Darryl G. Press, "The New Era of Counterforce: Technological Change and the Future of Nuclear Deterrence," *International Security* 41, no. 4

(Spring 2017): pp. 9–49"; Paul Bracken, "The Cyber Threat to Nuclear Stability," *Orbis* 60, no. 2 (2016): pp. 188–203; Dean Wilkening, "Hypersonic Weapons and Strategic Stability," *Survival* 61, no. 5 (2019): pp. 129–148; and Paul Scharre Horowitz, and Alexander Velez-Green, "A Stable Nuclear Future? The Impact of Autonomous Systems and Artificial Intelligence," arXiv (December 2019) https://arxiv.org/abs/1912.05291.

96. For example, Russia's military doctrine explicitly states it would consider nuclear weapons to respond to a large-scale conventional attack. While China maintains an official no-first-use policy, defense analysts continue to debate the veracity of this pledge, particularly in the event of conventional military aggression against a superior adversary in the Taiwan Straits or targeting China's dual-use military platforms. On Russia, see Alexey Arbatov, Petr Topychkanov, and Vladimir Dvorkin, *Entanglement as a New Security Threat: A Russian Perspective* (Washington, DC: Carnegie Endowment for International Peace, 2017), pp. 25–26; and Paul K. Davis et al., *Exploring the Role Nuclear Weapons Could Play in Deterring Russian Threats to the Baltic States* (Santa Monica, CA; RAND Corporation, 2019) pp. 27–29. On China, see Caitlin Talmadge, "Would China Go Nuclear? Assessing the Risk of Chinese Nuclear Escalation in a Conventional War with the United States," *International Security* 41, no. 4 (2017): pp. 50–92; and Eric Heginbotham et al., *China's Evolving Nuclear Deterrent: Major Drivers and Issues for the United States* (Santa Monica, CA; RAND Corporation, 2019).
97. By contrast, a recent game theoretic study found that damage limitation conventional counterforce capabilities could be a force for stability rather than instability as they induce challengers to drop out of crises because of their weak bargaining position. Benjamin W. Bahney and Braden C. Soper, "The Delicate Balance of Nuclear Survivability: Using Deterrence Theory to Explain How Imperfect Damage Limitation Capabilities Impact Crisis Bargaining" (draft paper).
98. Jacquelyn Schneider, Ben Schechter, and Rachael Shaffer, "Cyber Operations and Nuclear Use: A Wargaming Exploration," 2021 International Studies Association Meeting (draft paper).
99. Several scholars have argued that decisions to put forces on nuclear alert or to automate nuclear decision-making can increase the risk of accidental or inadvertent escalation to nuclear use. See Blair, *The Logic of Accidental Nuclear War*; Scott D. Sagan, "Proliferation Pessimism and Emerging Nuclear Powers," *International Security* 22, no. 2 (1997): pp. 193–202; and Scott D. Sagan and Jeremi Suri, "The Madman Nuclear Alert: Secrecy, Signaling, and Safety in October 1969," *International Security* 27, no. 4 (2003): pp. 150–183.
100. See Lieber and Press, "The New Era of Counterforce: Technological Change and the Future of Nuclear Deterrence," pp. 9–49; Ian Bowers and Henrik Stålhane Hiim. "Conventional Counterforce Dilemmas: South Korea's Deterrence Strategy and Stability on the Korean Peninsula," *International Security* 45, no. 3 (2021): pp. 7–39; Austin Long and Brendan Rittenhouse Green, "Stalking the Secure Second Strike: Intelligence, Counterforce, and Nuclear Strategy," *Journal of Strategic Studies* 38, no. 1–2 (2015): pp. 38–73; and Christopher Clary and Vipin Narang, "India's Counterforce Temptations: Strategic Dilemmas, Doctrine, and Capabilities," *International Security* 43, no. 3 (2019): pp. 7–52.
101. See Scott Sagan, "The Perils of Proliferation: Organization Theory, Deterrence Theory, and the Spread of Nuclear Weapons," *International Security* 18, no. 4 (1994): pp. 66–107.

102. Several defensive-realist international relations scholars have argued that self-interested rational unitary actors will likely behave in ways that reduce the dangers associated with the proliferation of nuclear weapons. Kenneth N. Waltz, "Nuclear Myths and Political Realities," *American Political Science Review* 84, no. 3 (1990): pp. 731–745.
103. Some models of rationality assume that individual rationality (defined in classical economic and behavioral science terms) leads to collective rationality, while others acknowledge that collective irrationality or sub-optimal outcomes may, paradoxically, arise because of individual pursuit of their *rational self-interest*. See Russell Hardin, "Collective Action as an Agreeable N-Prisoners' Dilemma," *Behavioral Science* 16, no. 5 (September 1971): pp. 472–481; Mancur Olson, Jr., ed., *The Logic of Collective Action: Public Goods and the Theory of Groups* (New York: Schocken, 1971); and Schelling, *The Strategy of Conflict*.
104. States that have volatile civil-military relations are also more vulnerable to accidents involving nuclear operations. Sagan, "The Perils of Proliferation," pp. 98–99.
105. Psychologists and sociologists have long recognized that risk acceptance (or avoidance) has an important cultural aspect. See Ortwin Renn and Bernd Rohrmann, "Cross-Cultural Risk Perception: State and Challenges," in *Cross-Cultural Risk Perception: Technology, Risk, and Society*, edited by Ortwin Renn and Bernd Rohrmann (Boston: Springer, 2000), pp. 211–233.
106. Suppose the Chinese government wants a technology that a particular commercial entity controls; extra-legal influence or coercion can compel the company to turn it over. In that case, forced technology transfers are not thought to occur routinely or considered effective. See Elsa B. Kania and Lorand Laskai, *Myths and Realities of China's Military-Civil Fusion Strategy* (Washington, DC: Center for a New American Security, 2021).
107. Ibid.
108. For an early conceptualization of "damage limitation" that includes the notion of a cyber-like "perfect damage limitation weapon … which disarms without harming adversary troops," see William A Stewart, "Counterforce, Damage Limiting and Deterrence," P-3385 (Santa Monica, CA: RAND Corporation, 1967).
109. Both the US and Russian nuclear doctrines maintain the option for counterforce operations to limit the damage it would suffer from a nuclear exchange or belief that the other side might launch a counterforce attack. Observers have also debated whether India and China are moving in the same direction. See US Department of Defense, *Nuclear Posture Review* (Washington, DC: Office of the Secretary, 2018), p. 23; Christopher Clary and Vipin Narang, "India's Counterforce Temptations: Strategic Dilemmas, Doctrine, and Capabilities," *International Security* 43, no. 3 (2018/2019): pp. 7–52; and Talmadge, "Would China Go Nuclear?"
110. US Department of Defense, *2019 Missile Defense Review* (Washington DC: US Department of Defense, 2019), pp. 32–33 and p. 60.
111. Zhang Yao et al., "Performance Analysis and Research of LRASM, the Next Generation of US Anti-ship Missile," *Aerodynamic Missile Journal*, July 15, 2018, https://kns.cnki.net/KCMS/detail/detail.aspx?dbcode=CJFQ&dbname=CJFDLAST2018&filename=FHDD201807006&v=MjI3NzZyV00xRnJDVVI3cWZZT1pzRmlybVdyN0FJeVhQYXJHNEg5bk1xSTlGWW9SOGVYMUx1eFlTN0RoMVQzcVQ=; and John Keller, "Air Force asks Lockheed Martin to Build Three More LRASM Anti-Ship Missile Systems for High-Priority targets," *Military & Aerospace Electronics*, December 8, 2018, https://

www.militaryaerospace.com/computers/article/16726716/air-force-askslockheed-martin-to-build-three-more-lrasm-anti-ship-missile-systems-for-highpriority-targets.

112. See Johnson, "Artificial Intelligence in Nuclear Warfare."
113. Posen, *Inadvertent Escalation*, pp. 1–27.
114. David Ochmanek, and Lowell H. Schwartz, *The Challenge of Nuclear-Armed Regional Adversaries* (Santa Monica, CA: RAND Corporation, 2008); and Avery Goldstein, "First Things First: The Pressing Danger of Crisis Instability in US-China Relations," *International Security* 37 (2013): pp. 49–89.
115. See Jeffrey Berejikian, "A Cognitive Theory of Deterrence," *Journal of Peace Research* 39, no. 2 (2002): pp. 165–183; and Daniel Kahneman and Ames Tversky, "Prospect Theory: An Analysis if Decision Making under Risk," *Econometric* 47 (March 1979): pp. 263–91.
116. See Jack S. Levy, "Loss Aversion, Framing, and Bargaining: The Implications of Prospect Theory for International Conflict," *International Political Science Review* 17, no. 2 (1996): pp. 179–195.
117. For a recent study that finds elite actors' decision to go nuclear are heavily influenced by how they affect their personal status and reputation, see Pauly B.C. Reid, "Would U.S. Leaders Push the Button? Wargames and the Sources of Nuclear Restraint," *International Security* 43, no. 2 (November 2018): pp. 151–192.
118. For example, several analysts warned about the inadvertent escalation risks associated with the US's Air-Sea Battle operational concept to counter China's anti-access, area-denial capabilities in the Western Pacific. See Joshua Rovner, "AirSea Battle & Escalation Risks," *Policy Brief*, no. 12, University of California Institute on Global Conflict and Cooperation (January 2012).
119. Avery Goldstein, "China's Grand Strategy under Xi Jinping: Reassurance, Reform, and Resistance," *International Security* 45, no. 1 (2020): pp. 164–201.
120. In addition to these psychological factors, there are of course other explanations for the irrational behavior of states, including, *inter alia*, imperfect information (i.e., a lack of inaccurate, or manipulated information), organizational pathologies and inertia, conflicting policy goals, or situations where the speed of change of information flow outpaces the ability of policies to respond—which is a key concern in the digital information age. For example, see Jack Snyder, *Myths of Empire: Domestic Politics and Political Ambitions* (Ithaca, NY: Cornell University Press, 1991); and Allison and Zelikow, *Essence of Decision: Explaining the Cuban Missile* (New York: Longman, 1999).
121. For example, see Zhang Shen, Ji Zili, and Wang Wenhua, "Overview of the Development of US Military Intelligent Weapons and Equipment," *Military Digest*, September 1, 2019, https://kns.cnki.net/KCMS/detail/detail.aspx?dbcode=CJFQ&dbname=CJFDLAST2019&filename=JSWN201917015&v=MjQ4NTk4ZVgxTHV4WVM3RGgxVDNxVHJXTTFGckNVUjdxZllPWnNGaXJtVkwzQUx6N2NZTEc0SDlqTnFJOUVZWVI=; and Zhang Shen, Ji Zili, and Wang Wenhua, "US Military to Speed Up the Development of Smart Weapons," *Defense Science & Technology Industry*, August 15, 2019, https://kns.cnki.net/KCMS/detail/detail.aspx?dbcode=CJFQ&dbname=CJFDLAST2019&filename=ZGBG201908016&v=MTg5OTBoMVQzcVRyV00xRnJDVVI3cWZZT1pzRmlybVZydkxQeXJKYWJHNEg5ak1wNDlFWW9SOGVYMUx1eFlTN0Q=.
122. Currently, China lags behind the US in the development of military AI. Chinese AI engineers face significant technical obstacles in developing and deploying AI applications,

including constraints on service members' technical literacy and the availability of data and computing power. Ryan Fedasiuk, *Chinese Perspective on AI & Future Military Capabilities* (Washington DC: CSET Policy Brief, 2021).

123. See Fearon, "Rationalist Explanations for War"; Powell, "War as a Commitment Problem," and Kelly Greenhill and Ben Oppenheim, "Rumor Has It: The Adoption of Unverified Information in Conflict Zones," *International Studies Quarterly* 61, no. 3 (2017): pp. 660–676.
124. For example, reports suggest that during the 2017 US-North Korean tensions, both Kim Jong Un and Donald Trump got much of their information from social media and other media outlets. Tom O'Connor, "Like Trump, North Korea's Kim Jong Un Gets His News from TV and Twitter," *Newsweek*, November 2, 2017, https://www.newsweek.com/north-korea-kim-jong-un-relies-social-media-us-news-ignores-trump-701672.
125. See James Johnson, "The AI-Cyber Nexus: Implications for Military Escalation, Deterrence, and Strategic Stability," *Journal of Cyber Policy* 4, no. 3 (2019): pp. 442–460; and James M. Acton, "Cyber Warfare & Inadvertent Escalation," *Daedalus* 149, no. 2 (2020): pp. 133–149.
126. Posen, *Inadvertent Escalation*, pp. 13–14. For an alternative view that asymmetric capabilities used to attack an adversary's NC3 might mitigate the perceptions of a window of vulnerability, thus creating incentives for stability—akin to traditional mutually assured destruction—see Ashton B. Carter, "Assessing Command System Vulnerability," in *Managing Nuclear Operations*, edited by Ashton B. Carter, John D. Steinbruner, and Charles A. Zraket (Washington, DC: Brookings Institution, 1987), pp. 560–573.
127. Benjamin Jenson and Brandon Valeriano, "Cyber Escalation Dynamics: Results from War Game Experiments International Studies Association," Annual Meeting Panel: War Gaming and Simulations in International Conflict, March 27, 2019, http://web.isanet.org/Web/Conferences/Toronto%202019-s/Archive/71e7820c-e61c-4187-ab8c-28de83dfd660.pdf.
128. Goldstein, "First Things First: The Pressing Danger of Crisis Instability in US China Relations," pp. 49–89; and David Gompers and Martin Libicki, "Cyber Warfare and Sino-American Crisis Instability," *Survival* 56 (2014): pp. 7–22.
129. See Horowitz, Scharre, and Velez-Green, "A Stable Nuclear Future?"
130. For example, see Lieber and Press, "The New Era of Counterforce," pp. 9–49; Charles L. Glaser and Steve Fetter, "Should the United States Reject MAD? Damage Limitation and US Nuclear Strategy toward China," *International Security* 41, no. 1 (2016): pp. 63–70; Austin Long and Brendan Rittenhouse Green, "Stalking the Secure Second Strike: Intelligence, Counterforce, and Nuclear Strategy," *Journal of Strategic Studies* 38, nos 1–2 (2015): pp. 38–73; Rafael Loss and Joseph Johnson, "Will Artificial Intelligence Imperil Nuclear Deterrence?" *War on the Rocks*, September 19, 2019, https://warontherocks.com/2019/09/will-artificial-intelligenceimperil-nuclear-deterrence/; and Wilkening, "Hypersonic Weapons and Strategic Stability."
131. Scholars of the "nuclear revolution school" argue, by contrast, that damage limitation capabilities need to be perfect to affect deterrence—as the costs of deterrence failure are too high to bear irrespective of the force balance or posture. See Robert Jervis, *The Illogic of American Nuclear Strategy* (Ithaca, NY: Cornell University Press, 1985) and Robert Jervis, *The Meaning of the Nuclear Revolution: Statecraft and the Prospect of Armageddon* (Ithaca, NY: Cornell University Press, 1989).

132. For example, a cyber-attack could be used to "blind" or spoof an adversary's early-warning systems in advance of launching a broader kinetic counterforce strike on its nuclear forces. See Acton, "Cyber Warfare & Inadvertent Escalation."
133. See Fiona S. Cunningham and M. Taylor Fravel, "Assuring Assured Retaliation: China's Nuclear Posture and US-China Strategic Stability," *International Security* 40, no. 2 (2015): pp. 15–23; Bracken, "The Cyber Threat to Nuclear Stability," pp. 197–200; and Alexei Arbatov, Vladimir Dvorkin, and Sergey Oznobishchev, *Non-Nuclear Factors of Nuclear Disarmament: Ballistic Missile Defense, High-Precision Conventional Weapons, Space Arms* (Moscow: IMEMO RAN, 2010), https://www.files.ethz.ch/isn/144178/10002.pdf.
134. By contrast, NATO states tend to view information warfare as limited and tactical.
135. See Vincent Boulanin, ed., *The Impact of Artificial Intelligence on Strategic Stability and Nuclear Risk, Volume I: Euro-Atlantic Perspectives* (Stockholm: Stockholm International Peace Research Institute, May 2019); Edward Geist and Andrew J. Lohn, *How Might Artificial Intelligence Affect the Risk of Nuclear War?* (Santa Monica, CA: RAND Corporation, 2018); and Michael Horowitz, Paul Scharre, and Alexander Valez-Green, "A Stable Nuclear Future? The Impact of Autonomous Systems and Artificial Intelligence," (2019), https://arxiv.org/abs/1912.05291.
136. See Jervis, *Perception and Misperception in International Politics.*
137. Hersman, "Wormhole Escalation: The New Nuclear Age," pp. 91–109.
138. On January 13, 2018, the Hawaii Emergency Management Agency (HI-EMA) sent a text using the Federal Emergency Alert System to people across Hawaii that read "BALLISTIC MISSILE THREAT INBOUND TO HAWAII. SEEK IMMEDIATE SHELTER. THIS IS NOT A DRILL." A HI-EMA employee had sent out the alert when he failed to hear the word "exercise" repeated at the beginning of a drill and mistook the drill for an actual missile warning. Observers noted that it would have been surprisingly easy to accidentally select an actual alert instead of a system test from the user's interface. See Amy Wang, "Hawaii Missile Alert: How One Employee 'Pushed the Wrong Button' and Caused a Wave of Panic," *Washington Post*, January 14, 2018, https://www.washingtonpost.com/news/post-nation/wp/2018/01/14/hawaii-missile-alert-how-one-employee-pushed-the-wrong-button-and-caused-a-wave-of-panic/.
139. Van Jackson, *On the Brink: Trump, Kim, and the Threat of Nuclear War* (New York: Cambridge University Press, 2019); and Vipin Narang, "Why Kim Jong Un Wouldn't be Irrational to Use a Nuclear Bomb First," *Washington Post*, September 8, 2017, https://www.washingtonpost.com/outlook/why-kim-jong-un-wouldnt-be-irrational-to-use-a-nuclear-bomb-first/2017/09/08/a9d36ca4-934f-11e7-aace-04b862b2b3f3_story.html?utm_term=.09e6a478b6d3.
140. For example, in 2017, US forces in Korea received false SMS and Facebook messages ordering an evacuation. Kim Gamel, "US Forces in Korea Warns of Fake Evacuation Messages," *Stars & Stripes*, September 21, 2017, https://www.stripes.com/news/pacific/us-forces-korea-warns-of-fake-evacuation-messages-1.488792.
141. Greg Sargent, "Could Trump Help Unleash Nuclear Catastrophe with a Single Tweet?" *Washington Post*, December 26, 2016, https://www.washingtonpost.com/could-trump-help-unleash-nuclear-catastrophe-with-a-single-tweet.
142. For an alternative view, which posits that people incorrectly assume that humans are "blank slates" that quickly change their minds when presented with new pieces of information, and thus overstate the persuasive power of psychological disinformation

operations, see Alexander Lanoszka, "Disinformation in International Politics," *European Journal of International Security* 4, no. 2 (2019): pp. 227–248; and Claes Wallenius and Sofia Nilsson, "A Lack of Effect Studies and of Effects: The Use of Strategic Communication in the Military Domain," *International Journal of Strategic Communication* 13, no. 5 (2019): pp. 404–417.

143. See, for example, Dan Ariely, *Predictably Irrational: The Hidden Forces that Shape Our Decisions* (New York: Harper Perennial, 2010); Daniel Kahneman, Paul Slovic, and Amos Tversky, eds, *Judgment Under Uncertainty: Heuristics and Biases* (Cambridge: Cambridge University Press, 1982); Jonathan Baron, *Thinking and Deciding*, 4th ed. (Cambridge: Cambridge University Press, 2008); and Robert B. Cialdini, *Influence: The Psychology of Persuasion*, rev. ed. (New York: Harper Business, 2006).

144. The connection between polarization, misinformation, and misperception is an active area of research. For example, see Nicholas DiFonzo, Jason Beckstead, Noah Stupak, and Kate Walders, "Validity Judgments of Rumors Heard Multiple Times: The Shape of the Truth Effect," *Social Influence* 11, no. 1 (2016): pp. 22–39; and D.J. Flynn, Braden Nyhan, and Jason Reifler, "The Nature and Origins of Misperceptions: Understanding False and Unsupported Beliefs about Politics," *Political Psychology* 38 (2017): pp. 127–150.

145. See Andrew Chadwick, Cristian Vaccari, and Benjamin O'Loughlin, "Do Tabloids Poison the Well of Social Media? Explaining Democratically Dysfunctional News Sharing," *New Media & Society*, 20, no. 11 (2018): pp. 4255–4274.

146. "Filter bubbles" and "echo chambers" are often cited as central drivers of political polarization and societal fragmentation. Both terms are based on the notion that people are excluded from information that is different from what they already believe; they also can refer to a partisan divide in ideological camps. See Judith Moller, "Filter Bubbles and Digital Echo Chambers," in *The Routledge Companion to Media Disinformation and Populism*, edited by Howard Tumber and Silvio Waisbord (London: Routledge, 2021), pp. 92–101.

147. Recent studies demonstrate that social bots play a disproportionate role in spreading disinformation from low-credibility sources; bots amplify these sources in the early dissemination stages *before* the information goes viral. See C. Shao et al., "The Spread of Low-Credibility Content by Social Bots," *Nature Communications* 9, no. 4787 (2018): pp. 1–8.

148. Camille Francois and Herb Lin, "The Strategic Surprise of Russian Information Operations on Social Media in 2016 in the United States: Mapping a Blind Spot," *Journal of Cyber Policy* 6, no. 1 (2021): pp. 9–30; Glenda Cooper, "Populist Rhetoric and Media Misinformation in the 2016 UK Brexit Referendum," in *The Routledge Companion to Media Disinformation and Populism*, edited by Tumber and Waisbord, pp. 397–411.

149. Vidya Narayanan et al., "News and Information over Facebook and WhatsApp During the Indian Election Campaign," *COMPROP Data Memo*, May 2019, http://comprop.oii.ox.ac.uk/wp-content/uploads/sites/93/2019/05/India-memo.pdf; and Guy Schleffer and Benjamin Miller, "The Political Effects of Social Media Platforms on Different Regime Types," *Texas National Security Review* (Summer 2021), https://tnsr.org/2021/07/the-political-effects-of-social-media-platforms-on-different-regime-types/#_ftnref149.

150. See Frank Schimmelfennig, "The Community Trap: Liberal Norms, Rhetorical Action, and the Eastern Enlargement of the European Union," *International Organization* 55, no. 1 (2001): pp. 47–80; John Postill, "Populism and Social Media: A Global Perspective,"

Media, Culture & Society 40, no. 5 (July 2018): pp. 754–765; and Sven Engesser et al., "Populism and Social Media: How Politicians Spread a Fragmented Ideology," *Information, Communication and Society* 20, no. 8 (August 2017): pp. 1109–1126.

151. Christopher Gelpi, "Democracies in Conflict: The Role of Public Opinion, Political Parties, and the Press in Shaping Security Policy," *Journal of Conflict Resolution* 61, no. 9 (2017): pp. 1925–1949.

152. Recent scholarship, for instance, suggests that the US public is more likely to be a force for spurring rather than constraining the use of nuclear weapons—especially if tactical advantages are clear. See Scott D. Sagan and Benjamin A. Valentino, "Revisiting Hiroshima in Iran: What Americans Really Think about Using Nuclear Weapons and Killing Non-Combatants," *International Security* 42, no. 1 (2017): pp. 41–79; and Daryl G. Press et al., "Atomic Aversion: Experimental Evidence on Taboos, Traditions, and the Non-Use of Nuclear Weapons," *American Political Science Review* 107, no. 1 (2013): pp. 188–206.

153. Rose McDermott, Anthony Lopez, and Peter Hatemi, "'Blunt Not the Heart, Enrage It': The Psychology of Revenge and Deterrence," *Texas National Security Review* 1, no. 1 (2017): pp. 68–89.

154. Today, given the incipient nature of these technologies, there are active debates about the impact of social media on the dissemination and diffusion of (true and false) information on social media. For example, contrary to conventional wisdom, while bots tend to accelerate the spread of true and false news, false news (primarily political) spreads more than the truth because humans, not bots, are more predisposed to spread it. Soroush Vosoughi, Deb Roy, and Sinan Aral, "The Spread of True and False News Online," *Science* 359, no. 6380 (2018): pp. 1146–1151.

155. Several recent studies have investigated the effect of algorithmic content-based filter systems on singular platforms, for example, Facebook, Google, and Google news. These studies suggest the potential of algorithms to exclude users from information incongruent with their belief systems, though the automatic selection can influence selective exposure. Moller, "Filter Bubbles and Digital Echo Chambers," p. 95; and Anna Sophie Kümpel, "The Matthew Effect in Social Media News Use: Assessing Inequalities in News Exposure and News Engagement on Social Network Sites (SNS)," *Journalism* (2020), doi: 10.1177/1464884920915374. For example, see Libby Hogan and Michael Safi, "Revealed: Facebook Hate Speech Exploded in Myanmar During Rohingya Crisis," *Guardian*, April 2, 2018, https://www.theguardian.com/world/2018/apr/03/revealed-facebook-hate-speech-exploded-in-myanmar-during-rohingya-crisis; and Michael Safi, "Sri Lanka Accuses Facebook Over Hate Speech After Deadly Riots," *Guardian*, March 14, 2018, https://www.theguardian.com/world/2018/mar/14/facebook-accused-by-sri-lanka-of-failing-to-control-hate-speech.

156. See Jamie Bartlett, *The People vs. Tech: How the Internet is Killing Democracy and How We Save It* (London: Ebury Press, 2018); and Leon Festinger, *A Theory of Cognitive Dissonance* (Stanford, CA: Stanford University Press, 1957).

157. Stephen Lewandowsky, Ullrich K.H. Ecker, and John Cook, "Beyond Misinformation: Understanding and Coping with the 'Post-Truth' Era," *Journal of Applied Research in Memory and Cognition* 6, no. 4 (2017): pp. 353–339.

158. Nathan Walter and Riva Tukachinsky, "A Meta-Analytic Examination of the Continued Influence of Misinformation in the Face of Correction: How Powerful Is It, Why

Does it Happen, and How to Stop it," *Communication Research*, 47, no. 2 (2020): pp. 155–177.

159. Recent research indicates that even if deepfakes may not mislead individuals, many people respond to them with uncertainty, thus reducing trust in all information received on social media. See Chadwick, Vaccari, and O'Loughlin, "Do Tabloids Poison the Well of Social Media?"

160. Social psychologists have found that conspiracy theories can act as a coping mechanism to manage uncertainty, confusion, and complexity. From that perspective, conspiracy theories—half soothing, half unsettling—are a cultural coping mechanism to deal with a complex and uncertain world. Karen M. Douglas et al., "Understanding Conspiracy Theories," *Political Psychology* 40, no. 1 (2019); pp. 3–35.

161. Angela Sherman, "Perceived Polarization: Reconciling In-Group and Intergroup Perceptions Under Uncertainty," *Group Processes & Intergroup Relations* 12, no. 1 (2009): pp. 95–109.

162. Most notably, disinformation spread via Facebook's WhatsApp, which falsely claimed that the Indian National Congress Party leader had offered a bribe to the suicide bomber's family. Despite Facebook's efforts to contain the nefarious campaign, the misinformation was disseminated to more than 2.8 million Facebook users. Neha Thirani Bagri, "Back Story: When India and Pakistan Clashed, Fake News Won," *Los Angeles Times*, March 15, 2019, https://www.latimes.com/world/la-fg-india-pakistan-fake-news-20190315-story.html.

163. Social media disinformation campaigns can impact information flow in a crisis in two ways: they undermine the transmission of information from the decision-makers to the public and from the public to decision-makers. In combination, these dynamics can impair the timeliness, reliability, and accuracy of publicly disseminated information. Trinkunas, Lin, and Loehrke, *Three Tweets to Midnight: Effects of the Global Information Ecosystem on the Risk of Nuclear Conflict*, pp. 69–73.

164. Scott D. Sagan, "The Commitment Trap: Why the United States Should Not Use Nuclear Threats to Deter Biological and Chemical Weapons Attacks," *International Security* 24, no. 4 (2000): pp. 85–115.

165. Kenneth Tan, "Chinese Censors Harmonize Online Posts Calling for War Following South China Sea Ruling," *Shanghaiist* (blog), July 13, 2016, http://shanghaiist.com/2016/07/13/hague_ruling_censored/amp/.

166. People tend to favor their own in-group and its constituents while disparaging out-groups and their constituents—which can fan the flames of ethnic conflict, the repression of minority groups, and genocide (e.g., the Balkans War and the Rwandan genocide in the 1990s, and the ongoing Israeli-Pakistan tensions). See Michael A. Hogg, "A Social Identity Theory of Leadership," *Personality and Social Psychology Review* 5, no. 3 (2001): pp. 184–200; and Henri Tajfel, "Social Identity and the Intergroup Behavior," *Social Science Information* 13, no. 2 (1974): pp. 65–93.

167. See Sven Engesser et al., "Populism and Social Media: How Politicians Spread a Fragmented Ideology," *Information, Communication & Society* 20, no. 8 (2017): pp. 1109–1126; and Jessica T. Feezell, "Agenda Setting through Social Media: The Importance of Incidental News Exposure and Social Filtering in the Digital Era," *Political Research Quarterly* 71, no. 2 (2018): pp. 482–494.

168. For example, during the US-led invasion of Iraq in 2003, studies demonstrated extreme levels of ethnic-based solidarity and xenophobia within Iraqi public opinion. See Ronald Inglehart, Mansoor Moaddel, and Mark Tessler, "Xenophobia and In-Group Solidarity in Iraq: A Natural Experiment on the Impact of Insecurity," *Perspectives on Politics* 4, no. 3 (2006): pp. 495–505; and Lawrence LeShan, *The Psychology of War: Comprehending its Mystique and its Madness* (New York: Helios, 2002).
169. See Gautam Mukunda, *Indispensable: When Leaders Really Matter* (Boston: Harvard Business School Press, 2012); and Michael Horowitz, Allan Stam, and Cali Ellis, *Why Leaders Fight* (Cambridge: Cambridge University Press, 2015).
170. See Jeffrey Edmonds et al., "Artificial Intelligence and Autonomy in Russia," *Center for Naval Analysis* (May 2021), https://www.cna.org/CNA_files/centers/CNA/sppp/rsp/russia-ai/Russia-Artificial-Intelligence-Autonomy-Putin-Military.pdf, p. 88.
171. The "tying-hands mechanism" refers to the idea that states seek to increase the credibility of their threats and resolve by taking costly actions (i.e., "go public" with their threats and demands) that increase the costs of backing down were the other side to counter-escalate but which would otherwise incur few costs. See Schelling, *Arms and Influence*.
172. Trinkunas, Lin, and Loehrke, *Three Tweets to Midnight: Effects of the Global Information Ecosystem on the Risk of Nuclear Conflict*, pp. 117–121.
173. Morgan et al., *Dangerous Thresholds: Managing Escalation in the 21st Century*, p. 105.
174. For a recent study on the different effects of social media platforms in democracies and non-democracies, see Sarah Kreps, *Social Media and International Relations* (New York: Cambridge University Press, 2020), pp. 32–43; and Schleffer and Miller, "The Political Effects of Social Media Platforms on Different Regime Types."
175. Because there may be substantial uncertainty around the effectiveness of advanced technologies like AI, the reliability of the technology—and both states' perception of this effectiveness—can impact brinkmanship outcomes.
176. Richard Ned Lebow, "The Deterrence Deadlock: Is There a Way Out?" *Political Psychology* 4, no. 2 (1983): pp. 333–354.
177. In 2020, for example, Colombia and Russia applied AI-augmented facial recognition technology to combat the coronavirus by detecting whether someone was not wearing a mask or had a high temperature using thermal cameras and sounding an alarm. Megan Janetsky, "How a Colombian Market is Using AI to Combat Covid-19 Outbreaks," *BBC News*, July 18, 2020, https://www.bbc.co.uk/news/world-latin-america53388656?intlink_from_url=https://www.bbc.co.uk/news/topics/ce1qrvleleqt/artificialintelligence&link_location=live-reporting-story; and Sarah Rainsford, "Coronavirus: Russia Uses Facial Recognition to Tackle Covid-19," *BBC News*, April 3, 2020, https://www.bbc.com/news/av/world-europe-52157131.
178. Amandeep Singh Gill, "Artificial Intelligence and International Security: The Long View," *Ethics and International Affairs* 33, no. 2 (2019): pp. 169–179.
179. Alexey Arbatov, *An Unnoticed Crisis: The End of History for Nuclear Arms Control?* (Moscow: Carnegie Moscow Center, June 2015), https://carnegieendowment.org/files/CP_Arbatov2015_n_web_Eng.pdf.
180. Heather Williams, "Asymmetric Arms Control and Strategic Stability: Scenarios for Limiting Hypersonic Glide Vehicles," *Journal of Strategic Studies* 42, no. 6 (2019): pp. 789–813.

181. James Johnson, "The End of Military-Techno Pax Americana? Washington's Strategic Responses to Chinese AI-Enabled Military Technology," *The Pacific Review* 34, no. 3 (2021): pp. 351–378.
182. Futter and Zala, "Strategic Non-Nuclear Weapons and the Onset of a Third Nuclear Age," pp. 18–19.
183. Hakan Wilberg, Ib Damgaard Petersen, and Paul Smoker, eds, *Inadvertent Nuclear War* (New York: Pergamon Press, 1993), p. 18.

4

AI-security dilemma

Insecurity, mistrust, and misperception under the nuclear shadow

This chapter applies the classical "security dilemma" international relations (IR) theory to consider the impact of dual-use AI technology on the uncertainties (estimates of adversaries' "intentions" and "motives")[1] and insecurities within competitive strategic dyads—especially US-China.[2] How might dual-use disruptive AI technology influence security dilemma dynamics? The chapter argues that specific structural (dual-use technology, offense-defense balance, asymmetric distribution of power, and regime type) and non-structural (socially constructed dynamics and cognitive-psychological) traits of dual-use AI technology intersect with the security dilemma theory in ways that might intensify these dynamics in competitive dyads—or "AI-security dilemma dynamics." Specifically, security dilemma dynamics will shape and inform how AI influences relations between strategic dyads.

The "security dilemma concept"—and the broader spirals and offense-defense models[3]—have been used to illuminate many events and challenges in international relations, including, *inter alia*: explaining the origins of World War I and the Cold War;[4] the cause of several ethnic-related conflicts;[5] managing arms races and designing arms reduction, non-proliferation, and confidence-building measures;[6] and to understand great-power transitions—namely a rising China and a declining US hegemon.[7] The chapter revisits the classical security dilemma theory to consider the implications of advances in dual-use AI technology for the US-China competitive strategic dyad. The research does not expound on AI's nature or speculate how it might alter warfare's character or nature—which is discussed elsewhere.[8] Instead, it contextualizes AI with other emerging dual-use technologies to elucidate the potential dangers it poses to the security dilemma in competitive dyads where conflict of interest exists, and empathy is scarce. What (if anything) makes AI different from other emerging technologies? And do those characteristics matter for the mechanisms described in the security dilemma?

AI and the Bomb. James Johnson, Oxford University Press. © James Johnson (2023).
DOI: 10.1093/oso/9780192858184.003.0005

The security dilemma can mean different things to different people. Defensive realists, for instance, have used the concept to demonstrate the potential for cooperation and peace between rival states. While offensive realists stress the inevitability of power accumulation, great-power competition, and warfare as the rational outcome,[9] neoliberal scholars emphasize the positive effect democratic institutions can have on alleviating the dilemma, thus producing peaceful outcomes.[10] While constructivists argue that because the security dilemma is a social condition, the condition can be mitigated through the reconstitution of identity in anarchical international politics.[11] The chapter applies Shiping Tang's "the BHJ formulation" definition of the security dilemma, derived from the original theorizing and subsequent development of the concept by Herbert Butterfield, John Herz, and Robert Jervis.[12]

Whereas scholars have noted the centrality of cognitive-psychological factors influencing the severity of the security dilemma, the literature on the intersection between human psychology, emerging technology, and international security is limited.[13] Through illuminating the security dilemma-technology-psychology nexus in AI technology and great-power strategic rivalry, this chapter addresses the gap. In responding to this omission, the research not only establishes a theoretical framework for exploring dual-use AI technology and the security dilemma (or the "AI-security dilemma"), but also explains the potential dynamics and dangers of this rapidly evolving international security phenomena.

Drawing on multidisciplinary literature, the research highlights the importance of human cognitive-psychology and security dilemma theorizing to understand better the nature and potential dangers posed by the "AI security dilemma" within competitive dyads—for example, new pathways for misperception that increase the risk of inadvertent escalation between nuclear-armed rivals.[14] While the security dilemma's psychological features are not prerequisites in its manifestation, they nonetheless can play an indispensable role in how the dilemma evolves—i.e., intensifies or is alleviated.[15] These psychological features are interwoven into the chapter's empirical core to describe how psychology has influenced the US-China security dilemma, and will likely continue to do so.

The remainder of the chapter is organized as follows. First, it outlines a conceptual framework of the "security dilemma" concept derived from the definitions, elaborations, and reconceptualization of the term—describing the security dilemma's theoretical parameters, preconditions, limitations, and associated IR concepts and approaches; notably, the offense-defense balance, deterrence theory, and action-reaction and arms racing spiral dynamics. Are the common dichotomous categories often used in strategic studies—such as offensive-defensive, dual-use, civil-military, war-peace—helpful in the context of intelligent machines in the digital age?

Next, the chapter considers how particular *structural* (dual-use technology, offense-defense balance, asymmetric distribution of power, and regime type) and *non-structural* factors (socially constructed dynamics and cognitive-psychological) associated with AI technology might intensify (or deepen) the security dilemma—or "security dilemma dynamics."[16] That is, the outcome of intensifying a security dilemma associated with pursuing dual-use AI technology in ways that risk spirals of insecurity and competitive military preparations—*in extremis*, inadvertent (or unintentional) armed conflict.

Third, the chapter uses human and political psychology (misperception, images, and reputation) to consider the impact of dual-use AI technology on the US-China security dilemma. The chapter concludes by describing some policy implications of this research, including arms control and confidence-building, and other measures that may mitigate (or at least ameliorate) AI-security dilemma dynamics in competitive security dyads.

Security dilemma conceptual framework

The "security dilemma" concept founding fathers are Herbert Butterfield, John Herz (who coined the term "security dilemma"), and Robert Jervis.[17] From this established theoretical canon, three features (or preconditions) must be present for a "genuine" security dilemma to exist.[18] This specificity ensures that the security dilemma concept is not reduced to "a meaningless and ambiguous term associated with any deterioration in relations."[19]

First, and perhaps most central, structural anarchy in international politics, causing uncertainty,[20] and fear.[21] Moreover, the desire to pursue self-help (i.e., means to escape the security dilemma by accumulating power) ensures security and survival.[22] According to Glenn Snyder, "uncertainty about the aims of others is inherent in structural anarchy."[23] As long as uncertainty exists about others' intentions, therefore, the effects of the security dilemma loom large. Robert Jervis developed the security dilemma as a fundamentally structural condition in international relations. According to Jervis, "the heart of the security dilemma argument is that an increase in one state's security can make others less secure, not because of misperceptions or imaged hostility, but because of the anarchic context of international relations."[24] Jervis defines the security dilemma as the "unintended and undesired consequences of actions *meant to be defensive* … many of how a state tries to increase its security decrease the security of others"—that is, one state's gain in security can inadvertently undermine the security of others.[25]

Under the condition of anarchy, because states cannot be sure about others' present and, especially, future intentions, actors are predisposed to fear the possibility that the other side might turn out to be an aggressor. Thus, leading to self-fulfilling cycles (or spiral-like situations) of insecurity, competition, and action-reaction dynamics that create and perpetuate the security dilemma—through self-reinforcing positive feedback mechanisms.[26] For defensive realists, the structural features of international politics are at the heart of the security dilemma—and thus drive the spirals and conflict that can arise from this condition. They posit that—unlike offensive realists—structure alone does not drive conflict and war.[27] Rather, human psychology and perception are immutable features of conflict and war.[28]

Second, an accumulation of military power includes *both* offensive and defensive capabilities—that is, the offense-defense balance concept.[29] Robert Jervis argues that the security dilemma is more intense when the offense-defense balance in the international political system inclines toward the offense (or offense dominant).[30] That is, all else being equal,[31] the relative ease (i.e., relative efficacy of investments and the first-mover advantages in a conflict) by which offensive military campaigns can be launched compared to defensive ones (or defense dominant), will affect whether actors feel more threatened and thus more predisposed to preventative or pre-emptive attacks to mitigate this perceived threat.[32] In this scenario, status-quo State A views security-seeking State B as threatening, thereby prompting deliberate *offensive* action—threatening to launch pre-emptive or preventive attacks—by A to signal resolve and the cost of war to deter B responding.[33] This scenario depicts a possible outcome of a profound (or exacerbated) security dilemma. In sum, an actor's fear of an uncertain geostrategic environment coupled with a preference for offensive capabilities and doctrine will increase the likelihood that conflicts of interest—particularly mutually exclusive political goals—and situations more intractable, thus deepening the security dilemma.

Because states consider that military power accumulation will enhance their security, even a predominately defensive configurated military will inevitably contain some offensive capabilities—resulting in the offense-defense distinguishability problem described below. In short, the amassing of additional (and often superfluous) military power does not ensure increased security; the opposite invariably occurs. In sum, structural anarchy leads to uncertainty, uncertainty to fear, and fear of military power accumulation. These features can then trigger a latent (or dormant) security dilemma between states, leading to spirals of competition, arms racing, and ultimately war.[34]

Third, a lack of malign (or deliberately threatening) intentions on both sides.[35] That is, two states who desire security (or security seekers/status-quo vs. power-seeking/revisionist states),[36] without the *intention* of threatening the other.[37] According to Jervis, "when determining … a *true security dilemma*, we have to consider whether those actions were *purely defensive*."[38] Defensive realist scholars have used various theories and approaches (e.g., rational-choice deterrence theory, cooperation theory, and Charles Osgood's "graduated reciprocation in tension" formula[39]) to assert that security-seeking (or benign/status-quo) states can demonstrate their intentions with "costly signals"[40] (e.g., arms control agreements, unilateral arms reductions, changes to force structure and doctrine, and a defensive military posture), that malign/revisionist states would be unwilling to take—thereby reducing the intensity of the security dilemma and the risk of inadvertent escalation.[41]

Given that structural anarchy and some (offensive and/or defensive) accumulation of power can in most cases be taken as a given, then an absence of malign intents is the crucial precondition for depicting a situation as a "genuine" security dilemma. This is not to say that a conflict of interest does not exist, however. The core thesis of the security dilemma theory is, to be sure, predicated on the assumption that a conflict of interest (i.e., a divergence between strategic ends and means) between two (or more) states exists—especially when the pursuit of those vital interests (or spheres of interest) meaningfully diverges with those of a strategic rival, or a potential one. Whether a particular conflict is an objective (i.e., genuine/real) or subjective (i.e., perceived/illusory) one, or reconcilable or non-reconcilable, are complex and contested issues in the scholarship.[42]

In sum, a security dilemma is compatible with a real-world conflict of interests between states. For example, in today's complex multipolar nuclear order, the threshold for the use of nuclear weapons, the types, numbers, and utility (i.e., mutual vulnerability, a second-strike capacity, or strategic stability) of nuclear weapons, and doctrines of use (i.e., how they define the "escalation ladder")[43] will vary considerably.[44] Because actors cannot accurately predict others' behavior, they will likely be inferred through the lens of pre-existing preconditions (or "confirmation bias"), assumptions of rationality, and cognitive biases (see below).[45]

The "AI security dilemma": A dilemma aggravator *primus inter pares*?

This section highlights several structural/material (dual-use technology; offense-defense balance and distinguishability; domestic-politics and regime type[46]) and non-structural/non-material (human psychology; perceptions; images; reputation[47]) factors that can aggravate AI-security

Table 4.1 Security dilemma regulators and aggravators

Security dilemma regulators/aggravators	AI technology
Uncertainty over others' intentions/motives	To a high degree
Psychological-cognitive features	To a high degree
Offense-defense distinguishability	To a high degree
Offense-dominant perception	To a moderate degree—not explicit/inconsistent
New pathways for inadvertent escalation	To a moderate degree—not explicit
Technological surprise	To a moderate degree—not explicit/contested
Military-use latency/scalability	To a high degree—explicit
Great-power competition	To a high degree—explicit
Asymmetric and non-state utility	To a high degree—inconsistent
Challenges for arms control/non-proliferation	To a moderate degree
First-mover advantages	To a high degree—inconsistent/contested

dilemma dynamics—the outcome of exacerbating or intensifying a security dilemma—in ways that increase the dangers of spirals of insecurity and competitive military preparations that may risk strategic rivals stumbling into armed conflict (see Table 4.1).[48]

Dual-use AI technology and offense-defense distinguishability

The notion that dual-use technology (i.e., technologies with both civilian and military applications[49]) creates uncertainties that can heighten states' insecurities with the deleterious spiraling effects associated with an intense security dilemma (or the "dual-use security dilemma") has received renewed attention amongst security and strategic scholars.[50] Scholarship that considers dual-use distinguishability and the security dilemma tends to focus on state-to-state interactions,[51] which in turn influence security dilemma dynamics. Others have highlighted the connection between the latency potential of dual-use of commercial technologies (e.g., nuclear, 3D printing, and space technology) and the uncertainties that can influence security dilemma dynamics—for example, driving the proliferation of nuclear weapons and counter-space capabilities.[52] Less research, however, has focused on the unique characteristics of these technologies that give rise to uncertainty and spirals of insecurity.[53]

The uncertainties and insecurities that dual-use technologies (AI, cyberspace, space technology, nuclear technologies, GPS technology, and 3D printing) can exacerbate are influenced by both their technical characteristics as well as social constructs associated with constructivist IR traditions that are concerned with how actors perceive particular norms and knowledge.[54] Social constructs and mechanism can interact with the security dilemma and dual-use technologies in two ways:[55] (1) the social production of the characteristics of dual-use technologies; and related (2) actors' construction of a particular dual-use technology of their adversaries as a security threat; set in a broader social context which includes a state's identity, history, strategic and organizational culture, and "securitization" experiences—that is, an extreme version of politicization that enables extraordinary means to be used in the name of security.[56] Through these mechanisms, an actor's behavior (i.e., securitizing dual-use technology) can impact others' security and, in turn, they pursue countermeasures that can lead to spirals of insecurity and conflict escalation.

Driven by these perceived technological imperatives, states accumulate military capabilities and form security alliances and partnerships in order to hedge against the uncertainty associated with the indistinguishability of dual-use technology (i.e., civilian vs. military use and offensive vs. defensive use) that can precipitate spirals of insecurity, arms racing, and ultimately war.[57] According to Amir Lupovici, "the very act of securitizing an opponent's technology without taking any extraordinary measures can alarm an opponent … these may result in a securitizing chain reaction that decreases both actors' security"—that is, igniting a dormant security dilemma.[58] Thus, how an actor constructs a particular dual-use technology will crucially influence whether it is perceived as a military threat or a peaceful civilian application; and if, in the case of the former, whether it will be used for offensive or defensive purposes—the offense-defense balance is discussed below.

The uncertainties surrounding the distinguishability of dual-use technologies like AI—and the difficulty of achieving arms control agreements with these technologies—means that the pursuit of them by strategic competitive dyads (e.g., US-China, Pakistan-India, Israel-Iran, and US-Russia) can lead states to make decisions that exacerbate the security dilemma. For instance, developing similar dual-use technologies, accumulating countermeasures (defensive and offensive) to the perceived threat, or launching a pre-emptive or preventative strike in a use-them-or-lose-them scenario.

First, suppose an actor fears losing the lead or first-mover advantages in a rapidly developing and strategically critical dual-use technology such as AI. In this case, an actor will likely mitigate this perceived security threat (i.e., developing military AI-countermeasures). Because even relatively small and gradual processes in AI technology can have disproportionately strategic effects (i.e., military decision-making, situational awareness, and augmentation of advanced weapon systems),[59] the looming threat of technological breakthroughs in any of these areas will persuade states to assume the worst and drive efforts to close the technological gap with rivals (or sustain the lead).[60]

In its final report to the US congress, the National Security Commission on AI emphasized that because of the "quintessential dual-use" nature of AI technology—transforming all aspects of military activity—proliferation is inevitable, and the US must "lead a coalition of democracies," in partnership with the private sector, to develop AI technology that will "translate into military advantage."[61] In classical security dilemma language, the Commission warned that "*defending against* AI-capable adversaries [notably China] operating at machine speeds without employing AI is an *invitation to disaster*" (emphasis added).[62] In sum, mutual distrust and an atmosphere of uncertainty surrounding dual-use technology between competitors—viewed through the lens of securitization—means that any additional or new advances will deepen the security dilemma.

These mechanisms can lead to action and reaction spiraling dynamics: State A's dual-use technological advances—or even the possibility of advance—are construed as a threat by State B; thus, B responds to this perceived threat with defensively (or security-seeking) motivated capabilities, thus eliciting additional technological countermeasures by A.[63] For instance, China's acquisition of AI dual-use technologies has, amongst other things (e.g., sanctions imposed on China's dominant 5G and information technology giants Huawei and ZTE[64]) fueled pre-existing US perceptions of Chinese revisionist goals, and, irrespective of whether these perceptions are accurate, has exacerbated the US-China security dilemma.[65] Counterfactual thinking might be used to consider the relative impact of AI technology (i.e., as an observable implication) on the US-China security dilemma compared, *ceteris paribus*, to other dual-use technologies. Notwithstanding the limitations of counterfactual hypothesis (subjectivity, circular logic, non-falsifiable, and "epistemological slumming" etc.),[66] the findings of this research indicate that the security dilemma would exist absent AI. AI technology can, therefore, be conceptualized as an aggravating regulator and potentially powerful amplifier of the dilemma—both as an enabler of existing emerging technologies and a disruptive force as an independent variable.

Second, even when employed solely for civilian/commercial purposes, dual-use technologies like AI are emblematic of a state's overall national power and the potentiality that this latent capacity might be leveraged to enhance military power. A civilian space program, for example, symbolizes national power and a latent capacity to weaponize space assets.[67] US defense planners have expressed concern that, in the event of a crisis or conflict, US satellite navigation and communication systems—especially missile-launch and nuclear-detection satellites that reside in higher altitudes—might be vulnerable to Chinese dual-use GPS (known as *BeiDou*) alongside its anti-satellite weapons, satellite and GPS jammers, ground-based missiles and cyber capabilities.[68] In response to claims that China was weaponizing space—and in language demonstrative of a deep US-China security dilemma—Wang Yiwei, an international relations scholar at Renmin University, opined that "as for whether China will try to take out [US] GPS, I don't think China has the ability or the inclination. *That's how America thinks*" (emphasis added).[69]

Third, and related, state control and the severity of the security dilemma. When state control over a particular technology's development, utilization, and distribution is weak or widely dispersed—notably to private sector actors motivated by economic gain—the degree of dual-use distinguishability uncertainty generally increases.[70] This issue creates an interesting point of departure: whether Jervis' traditional offense-defense distinguishability regulator of the security dilemma can be supplanted by (or combined with) the civilian-military distinguishability of these technologies.[71] That is, the ease by which commercially derived dual-use technology like AI can be diffused to the military domain to augment (offensive and defensive) weapon systems.[72] The reformulation of the traditional offense-defense distinguishability concept to view AI technology can help us to elucidate contemporary security dilemma spiraling dynamics that have arisen as levels of investment in AI increase, and states struggle to delineate the intentions of rivals—a relationship of long-term hostility and competition—the pursuit of innovations which possess the latent capacity to enable and enhance military power.[73]

Some observers argue that the application of dual-use AI technology (e.g., natural language, deepfake technology, logistics support, facial/image recognition, and drone swarming) will enable civilian applications to serve economic, productivity, societal, and domestic-political goals, not just warfaring ones. To avoid regime change and legitimize the political status-quo, strong authoritarian regimes like China have, for example, used their economic cleft and growing coercive ("sharp") power to embrace technology including AI and become "digital autocracies"—that is, restricting domestic access use of the internet and social media behind a "Great Firewall" whilst harnessing new tools in the digital domain to repress dissent.[74] However, others emphasize the more adversarial nature of specific AI technology that lends itself to high-performance adversarial military domains (e.g., offensive cyber, command and control), and target recognition automation).[75] According to Kenneth Payne, in these adversarial domains, "marginal quality might prove totally decisive" because "other things being equal, we can expect higher-quality AI to defeat inferior rivals comprehensively."[76] This suggests the need for a nuanced treatment of dual-use technology more generally, and AI—and AI sub-set technology—in particular, to determine the civilian-military distinguishability of these technologies.[77]

Recent scholarship on the offense-defense balance has demonstrated the importance of scaling effects over time on this balance.[78] Scalability is one of the key distinguishing features of AI technology. That is, the marginal cost of increasing the effectiveness of a machine learning (ML) algorithm or replicating this software across multiple domains (e.g., software for coordinated drone swarms) is relatively low. It is likely to decrease further as computing power increases.[79] The researchers found that the "more highly automated and digitally-oriented conflict suggests that scaling phenomena may become much more salient."[80] Because of the first-mover advantages associated with military AI to defeat inferior rivals—and its possession as a means to asymmetrically empower conventionally weaker states (and non-state actors) vis-à-vis superior military rivals—the implications of these findings for the effects of dual-use AI and the security dilemma are significant. Besides, how actors perceive information about an adversary's intentions in this technological symbiosis—and juxtaposed, how it thinks the other side views its civil-military technological transfers—will have important implications for the success of arms control agreements or other normative measures (i.e., confidence-building measures) to limit the severity of the security dilemma.[81]

Another important takeaway from this self-reinforcing dynamic is: because of the uncertainties and trust issues associated with the security dilemma, technological advances (linear, prosaic, or breakthroughs) in dual-use technology can exacerbate the severity of the dilemma, irrespective of whether the source and/or intended use is for primarily civilian or military purposes—the latter will likely be feared, and countermeasures devised to mitigate worst-case scenario expectations.[82] Though the US retains the first-mover advantage in a range of AI technology with direct—and some cases singular—relevance in a military context (e.g., cyber offense, target recognition, IS, ML, and semiconductors), as China approaches parity and likely surpasses the US in several dual-use AI technologies and AI enabling technologies (e.g., facial recognition, drones, 5G),[83] the US will likely interpret any incremental progress—and especially surprises or scientific milestones—through the lens of securitization, rooted in the uncertainties and insecurities of the security dilemma.

These dynamics were explicitly evinced in the 2018 US Nuclear Posture Review (NPR). The NPR emphasized that geopolitical tensions and emerging technology in the nuclear domain could coalesce with unexpected technological breakthroughs in new and existing innovations to "change the nature of the threats" faced by the US and the "capabilities needed to counter them."[84] Like the classical offense-defense distinguishability concept, civilian-military distinguishability is at its core an intrinsically psychological phenomenon—that is, a perceptual and subjective psychological regulator of the security dilemma.[85]

Socially constructed US-China security dilemma dynamics

The wide dispersion of dual-use technology to state and non-state actors (notably AI and cyber) can drive normative, institutional, and social constructed narratives and popular memes, which frame these technologies as fundamentally threatening—that is, they can be weaponized or militarized.[86] The more the beliefs and ideas are repeated and adopted by others, the more their uptakc and endurability increases—known in psychology as a cognitive "bandwagon effect."[87] This socially constructed dynamic—in addition to legal and normative factors—can securitize technology further, thus prompting states to devise (offensive and defensive) measures to counter a perceived (future) threat.[88] Moreover, states' perceptions about the linkage between others' present and future behavior will have a crucial bearing on this socially constructed dynamic. In other words, the security dilemma is not a fixed or constant phenomenon; it can operate at different times and at varying degrees of intensity. For instance, the rapid development of AI technology has been driven primarily by commercial interest and private-sector investment.[89] As a result, state control over the development and use of AI (for civilian and military purposes) has been circumscribed, compared to other dual-use technologies such as space, cyber, and nuclear.[90]

The spiraling dynamics of mistrust and insecurity that accompanied the recent political spate involving China's Huawei and the development of dual-use 5G networks led to the US's securitization of 5G technology and the resultant deterioration of the US-China security dilemma.[91] Some observers traced the roots of recent US-China tensions involving dual-use technology to the early 2010s, particularly Beijing's focus on securing information, strengthening cybersecurity (or China's "cyber sovereignty" concept),[92] and establishing indigenous technological champions, both to defend the regime against US technological hegemony (i.e., technological decoupling or separation),[93] and reduce Chinese dependence on US dual-use technology (e.g., semiconductors, AI, quantum computing, and supercomputers).[94] In 2017, for example, Chinese President Xi Jinping explicitly signaled an acceleration of the military "intelligentization" agenda to better prepare China for future warfare against a near-peer adversary—namely the US.[95]

Official US reports alleging nefarious backdoors between Huawei and the Chinese military—to support intelligence-collection and espionage operations against the US—were used to justify the embargo of Chinese 5G wares to cajole US allies into following suit.[96] Moreover, the fact that others (notably Germany[97]) were reluctant to assuage Washington's draconian policies against Chinese 5G equipment vendors demonstrated that US securitizing countermeasures were in part driven by the socializing and non-deterministic effects (i.e., images, memes, and perceptions) of the security dilemma.[98] As Jervis observed, "when a statesman has developed a certain image of another country [because of the effects of cognitive dissonance], he will maintain that view even in the face of large amounts of discrepant information."[99] This cognitive propensity in part helps explain the entrenched and enduring quality of Washington's image of China; despite the existence of opinions and information to the contrary.

In an escalation of these spiraling dynamics, the Trump administration declared a formal "national emergency" in the development of dual-use communication technology; in particular, China's ability to exploit US perceived vulnerabilities "with potentially catastrophic effects, and thereby constitutes an *unusual and extraordinary threat to the national security*, foreign policy, and economy."[100] Furthermore, the 2017 National Security Strategy promoted the notion of a "national security innovation base," to prevent adversaries "filling in their capability gaps [in critical technologies including AI]" at the expense of US "competitive advantages,"[101] narrowing the traditional delineated duality between US civilian (agile, fast-pace, reliant on open-source competition) and military research (slow, sequential, long-term horizons, user vs. market-driven) and development (R&D).[102]

These countermeasures coalesced into an organizational and investment agenda that included increased funding for the National Science Foundation, additional incentives to spur semiconductor manufacturing and R&D,[103] and investment in open radio access networks to create a cost-effective alternative to Huawei in 5G communications.[104] Driven by similar action and reaction dynamics, the DoD's Defense Innovation Unit in 2017 proposed greater scrutiny and restrictions on Chinese investment in Silicon Valley companies.[105] From Beijing's perspective, these moves were viewed as duplicitous and hostile, reinforcing an atmosphere of ongoing bilateral tension and increasing the risk of escalation during a future regional crisis or conflict. In short, these securitizing reactions typified a broader concern within the US defense community, that synergies created by Beijing's dual-use integration strategy (known as China's "military-civil fusion" concept),[106] will allow shared commercial technology expertise, intellectual property, and coopted commercial entities to be transferred to China's military to support its ambitious techno-nationalistic (i.e., non-benign) geopolitical objectives.[107]

For example, President Xi stated that AI, "big-data," cloud storage, cyberspace, and quantum communications, were amongst the "liveliest and most promising areas for military-civil fusion."[108] Towards this end, in 2017, Beijing established the Military-Civil Fusion Development Commission to expedite the transfer of AI technology from Chinese commercial research centers to the military.[109] US counteractions, in turn, will likely solidify China's endurable self-image (discussed below), heighten its threat perceptions, and reinforce the rationale driving Beijing's pursuit of indigenous dual-use technology in the first place. Whether political leaders (democratic or autocratic) will be able to successfully navigate the inevitable shift in political, economic, and societal power structures—especially control over data and information—as AI technology matures, and the impact of this shift on the security dilemma if this delicate balance is mismanaged, is an important topic for future research.[110]

The status and reputation associated with critical dual-use technological innovation (e.g., AI, space technology, quantum communications, GPS, and 5G networks) can prompt leaders to contemplate countermeasures to obviate the risk that the status and reputation of a state might be challenged or tarnished, which in turn may trigger spirals of insecurity associated with a deteriorating security dilemma.[111] Moreover, an actor will incur reputational costs if it does not follow through on statements it makes (or "reputation signaling") about what it says it will do, increasing its resoluteness and attachment to strategically high-value policies.[112] Because of the anticipated national security value US strategic near-peer competitors (especially China) attach to dual-use AI-related technologies (ML, autonomy and robotics, 5G networks), some observers have characterized the rapid diffusion of dual-use AI as the US's new "Sputnik moment"—and a harbinger for a future AI-enabled arms race.[113]

Scholarship on the diffusion of military technology demonstrates that how states react to and assimilate innovations can have profound implications for strategic stability and the likelihood of war.[114] In particular, the pace at which military organizations diffuse new technologies can influence states' relative advantages from being the first-mover—which tends to be reversely correlated to the adopted speed of innovations.[115] In contrast to previous technological breakthroughs (e.g., stealth technology) whose application is almost purely confined to the military domain, innovations like AI, derived from predominately commercial and university research—rather than classified military research—tend to be diffused faster.[116] In the case of AI, this diffusion will likely be accelerated by the global interest in developing this general-purpose use technology, and the declining costs and availability of computing power to support these efforts.[117] According to Lt. General John Shanahan, former Director of the US Joint Artificial Intelligence Center, unlike other forms of dual-use technology "the barriers to entry" for commercially driven AI innovation are low, and "we are seeing a true democratization of technologies" that will make it "increasingly difficult to prevent the use of AI-enabled capabilities by those who are intent in causing harm."[118]

As the costs and availability of computing power—an essential ingredient for AI-ML systems—decrease, however, technically advanced militaries will likely pull away from those actors who are more (or entirely) reliant on mimicry and espionage.[119] That is, akin to a form of "AI-driven asymmetric warfare" where weaker and less sophisticated states (and non-state actors) coopt commercially developed AI technology to threaten more militarily powerful states with AI-enhanced cyber-attacks and AI-enabled disinformation campaigns—for example, deepfake technology, false-flag cyber operations, and other information manipulation campaigns.[120] Besides, this trend will likely be amplified if either the cost or complexity of AI-ML algorithms increases, thus allowing AI first-movers to maximize their competitive advantages.[121]

Is a digital arms-race inevitable and already unfolding? The former chief acquisition officer for the US Air Force, Will Roper, recently warned of the risks of the US falling behind in a "digital arms race with China," and that, in response, the US "must wake up fast ... the *only thing worse than fearing AI itself is fearing not having it*" (emphasis added).[122] Whereas the reality of an "AI arms race" may be exaggerated or misunderstood, framing US-China status power competition in this way—alongside incendiary (and especially jingoistic) rhetoric on both sides—risks driving the adoption of operational concepts and doctrine that increase the risk of spiral-like situations that neither side desires.[123] While the "AI arms race" trope may be a misrepresentation of the security competition for AI technology (or "technology race"), this widespread mischaracterization will likely deepen the US-China security dilemma nonetheless. Former Director of the DoD's Joint Artificial Intelligence Center Lieutenant General Jack Shanahan recognized this nuance, which is invariably lost in the narrative that surrounds AI. That is, a perpetual state of strategic competition short of armed conflict at best, and action-reaction arms racing dynamics leading to military escalation at worst.[124] According to Shanahan, "it's strategic competition, *not an arms race*. They're [China] going to keep doing what we're doing; we [the US] acknowledge that." Shanahan added "what I don't want to see is a future where our *potential adversaries* [China] *have a fully AI-enabled force, and we* [the US] *do not*" (emphasis added).[125] Deriving the strength of or commitment to an argument on the plausibility of a possible conclusion is known as "belief bias" (see the following section).[126]

Misperception, reputation, and images: US-China military-techno competition

> "We don't see things as they are; we see them as we are."
>
> —**Anais Nin**[127]

This section uses human and political psychology—misperception, images, and reputation—to consider the role that dual-use AI technology has played (as an aggravating regulator) in the trajectory and character of the US-China security dilemma.[128] These cognitive-psychological factors, while not supplanting the core structural and ideational features of the security dilemma, are instructive in illuminating the complex encumbrances—rooted in the objective/genuine and subjective/illusory conflict of interest—that surround the formation of US and Chinese goals, policy decisions, and countermeasures.

Against the backdrop of a competitive power transition—when perceptions of malign intent are accentuated, and empathy is eschewed—psychological dimensions of the US-China dyad can impede cooperation and ensure that issues, even where genuine mutual interest exists (e.g., climate change and global pandemics), remain irreconcilable and intractable.[129] Even without these mutual cognitive-psychological impediments, however, (genuine) conflict of interest would likely still exist—albeit in a potentially less vitriolic and more manageable form. In other words, were Washington and Beijing to perceive each other's words and deeds accurately, strategic competition would likely persist; these psychological factors exacerbate the security dilemma's conditions.[130]

Scholars have identified several cognitive-psychological factors at work in the US-China relationship—and more generally in great-power rivalries—that can influence the severity of the security dilemma. First, the immutable feature of a security dilemma requires great powers to assume that adversaries (or perceived challengers) hold malign intentions.[131] That is, states are predisposed to exaggerate the malevolent impulses or "hostility" of others; both sides emphasizing their benign motivations (i.e., US benevolent hegemony and China's "peaceful rise" meme),[132] exceptionalism (cultural values, policies, and grand strategy), and "reasonableness"[133] (or security-seeking rationality).[134] The denigration or devaluing of an idea or policy because it originates from an adversary is known as "reactive devaluation" bias.[135] This bias is exacerbated by an actor's assumption that the other possesses the same information as they do—or information symmetry. Both sides operate under the assumption that their benign intentions are axiomatic and valid to others and, therefore, will not be threatened by security-seeking policies. After China entered the Korean War in 1950, then US Secretary of State Dean Acheson stated that "no possible shred of evidence *could have existed in the minds of the Chinese* … about the open [peaceful] intention of the forces of the United Nations" (emphasis added).[136]

According to US analyst Derek Mitchell, "the United States is a rather self-centered, proud, nationalistic, and self-reverent society with a sense of its uniqueness and exceptionalism in world history … all attributes that characterize China as well"[137] That is, both sides are trapped in the logic of "mirror-imaging"—associated with "fundamental attribution error"—meaning actors tend to perceive threatening behavior as necessarily threatening (i.e., an actor's actions reveals its intentions).[138] The endurability and hardening of these beliefs can be compounded by people's tendency to ascribe retroactively positive (and thus reinforcing) attributes to an opinion or belief previously held—known as "post-purchase rationalization bias."[139] A capacity for subjective self-deception (or delusion) allows actors to deal with the psychological stresses that arise during heightened tension and crisis—when empathy is likely to be in short supply.[140] Besides, these cognitive predispositions may exacerbate a crisis if an actor assumes that an adversary must know that they are acting reasonably, and thus will be less sympathetic to the notion that their "defensive" policies are perceived as threatening to an adversary—making it harder to convince others to step away from the brink or compromise.[141]

Scholars contend that the security dilemma can be exacerbated by the secrecy (i.e., military capabilities and strategic intentions) and the opacity of non-democratic states.[142] Conversely, the relative openness of democratic states who publicly debate and signal resolve (i.e., a hawkish military policy to enhance deterrence) may be viewed by an adversary as more credible because its decision-making process is more transparent—thereby exacerbating the security dilemma.[143] An abundance of information can lead actors to overestimate the importance it attaches to particular situations and events—a cognitive bias known as "availability heuristic."[144] Moreover, like other non-kinetic capabilities, such as cyber and space, a state's AI technological prowess is less quantifiable than counting missile warheads, numbers, and types of kinetic armaments, or economic power—which has important implications for effective deterrence. That is, if one side is unable to form a reliable assessment of the other side's capabilities, it may feel less compelled to restrain its behavior.

Because China is more likely to heed signals of resolve and deterrence, hardline or bellicose rhetoric from US open policy debates could intensify security dilemma dynamics—regardless of whether it accurately reflects official policy and doctrine. For example, Beijing has long been critical of the US DoD's annual reports to Congress—and especially the open debates that follow—on Chinese military developments and goals. China's Defense Ministry spokesman Colonel Wu Qian labeled the DoD's 2020 report as a "wanton distortion" of China's intentions and rebuked the US as the "fomenter of regional unrest, the violator of the international order, and the destroyer of world peace."[145]

Second, and related, great powers struggle to empathize with others—particularly in the context of a hegemon and a rising power. Despite the efforts of officials, analysts, and scholars to highlight the intensifying US-China security dilemma,[146] there has been a dearth of empathy or mutual understanding in the strategic interactions in this complex, interdependent, and increasingly strained dyad. To be sure, as US-China relations hit their lowest ebb in decades, the US DoD's 2020 report to Congress on the Chinese military developments stated:

> *What is certain is that* (the ruling Communist Party) *has a strategic end* state that it is working towards … and its accompanying military modernization left unaddressed, will have serious implications for US national interests and the security of the international rules-based order (emphasis added).[147]

Third, and paradoxically, great powers are often preoccupied with domestic political issues (particularly image building), and thus neglect or deprioritize external affairs even as their influence in world politics rises—

"great state autism."[148] The Trump administration's preoccupation with the domestically focused "America First" (and related "make America great again") tropes, and President Xi's "China Dream" revitalization initiative are cases in point.[149] When the shortcomings of these ambitious internal preoccupations manifest in competitive dyads—especially when narrowing relative power differentials feature—issues can become intractable, allowing latent security dilemmas to be actualized.[150] Mutual threat perceptions arise as each side believes itself to be the target of the other's malign actions.[151]

Furthermore, these domestic preoccupations can compound the common misperception in great-power relations of viewing an adversary as "more centralized, planned, and coordinated than it is."[152] This tendency is partly attributable to people's tendency to overestimate the ability to make accurate determinations from information that appears coherent—known as the "illusion of validity" bias.[153] Thus, behavior caused by accidents, confusion, poor coordination, or policies arising from stove-piped bureaucrat structures, for example, may be viewed as part of strategically opportunistic, monolithic, coherent (i.e., not subject to preference recalibration), and malevolent agenda.[154] Put another way, while decisions taken during a crisis may not reflect coordinated, cohesive, and centralized planning (e.g., discretion is given to commanders if nuclear weapons are put on alert status), an adversary might infer the existence of a grand strategic plan held by the out-group nonetheless: this threat must be responded to with heightened resolve and military power.[155]

Fourth, as described above, great-power threat perceptions are generally heightened when the offense-defense balance (i.e., the subjective and perceptual nature of the offense-defense distinguishability[156]) is perceived to be rapidly narrowing in near-peer dyads—demonstrated in US-China military-technological competition since the late 1990s and particularly the 2010s.[157] Today, by most metrics, China is the US's closest peer competitor in AI and AI-enabling technology. China's 2017 "Next Generation AI Development Plan" identified AI as a core "strategic technology" and focus of "international competition."[158] Beijing's official goal is to "seize the strategic initiative" and achieve "world-leading levels" of both civilian and military AI by 2030.[159] Towards this end, Beijing has targeted more than US$150 billion in state-led investment, and lowered the entry barriers to collect, process, and disseminate data within China.[160] For instance, Chinese state-directed investment in the US AI market has become increasingly active, and in several instances Chinese investment has competed in direct competition with the US DoD.[161]

In response to the perceived national security challenge posed to the US (the status-quo/security seeking power), technological primacy from China (the revisionist challenger) in the pursuit of dual-use AI and AI-enabling technology may be used to augment China's military power asymmetrically vis-à-vis the US.[162] The Pentagon authored several AI-related programs and initiatives designed to support US military operations on the digitized battlefield (e.g., the Third Offset; Project Maven; Defense Advanced Research Projects Agency (DARPA)'s AI Next Campaign; the establishment of the Joint Common Foundation; the DoD's AI Strategy and Data Strategy reports).[163] Taken together, these initiatives demonstrate the perceived threat that China's pursuit of dual-use technology poses to US national security—in ways that might enhance Chinese power vis-à-vis US hegemony, and regardless of whether these perceptions are accurate.[164]

Political psychologists note that actors tend to underestimate ambiguity and the extent to which their actions threaten others, and simultaneously overestimate others' hostility and their cognitive faculties; for example, viewing oneself as less biased than others when inferring from others' words and behavior their intentions—known as a "blind spot bias."[165] Moreover, assuming that both sides are "risk acceptant" (i.e., believe they are defending the "status quo"[166]) and that an offensive military response is the *only way* to avoid the anticipated loss, these cognitive biases can also exacerbate the effects of loss aversion resulting from a situation—explained in political psychology by "prospect theory."[167]

This view can be partly attributed to the persistence of images that perpetuate a sense of mutual mistrust, uncertainty, and self-reinforcing cycles of action and counter-reactions. Consequently, Beijing has viewed Washington's stance on China as increasingly hawkish; meanwhile, Washington sees strong-man President Xi Jinping (together with Russian President Vladimir Putin) as a challenge to US "power, influence, and interests" to be countered.[168] In language that intimates a deep security dilemma, Chinese foreign minister Wang Yi asserted in an interview in 2020 that "it is the US that threatens China; not vice versa, thus the *problem is to address the US threats*" (emphasis added).[169] These perceived "threats" are viewed through the lens of a self-image of the Chinese national rejuvenation as a great power—imbued with deep historical collective memory, colonial victimization, and economic triumphalism derived from the "hard work and sacrifice of the Chinese people" rather than foreign largesse—which is being stymied by the egregious "hegemonic and power politics" of the US.[170] Thus, magnifying the sense of mutual distrust, recriminations, and the intractability of managing shared concerns confronting the dyad, such as the COVID-19 global pandemic of 2020.[171]

In addition to these perceptual factors, technological advances associated with the information technology revolution (e.g., AI, cyber, remote sensing, robotics, hypersonic variants) will make it increasingly difficult for states to reduce the vulnerability of their military capabilities without simultaneously reducing others' security, either by a more survivable first-strike capacity or an enhanced capacity to undermine others' second-strike capacity—for example, AI-enhanced cyber-attacks or using AI tools to track and monitor an adversary's nuclear deterrent forces.[172]

Qualitative improvements to emerging technology that enables conventional weapons to assume the role of strategic missions (i.e., threaten the survivability of states' nuclear weapons) traditionally reserved for strategic nuclear weapons,[173] will further complicate offense-defense distinguishability and blur the division of roles and strategic balance between nuclear powers and non-nuclear powers (North-South Korea, China-Taiwan, and Japan-China). These concerns may prompt a shift from extended deterrence to regional alliance deterrence regimes (e.g., South Asia, East Asia, and the Middle East security nexus).[174]

Conclusion

The chapter found that specific structural (dual-use technology, offense-defense balance, asymmetric distribution of power, and regime type) and non-structural (socially constructed dynamics and cognitive-psychological) features of dual-use AI technology might interact with the security dilemma in ways that intensify the condition—or "AI-security dilemma dynamics." The diffusion and synthesis of (predominately commercial) dual-use AI increases the danger of spirals of insecurity and distrust that lead to military preparations and doctrinal shifts (i.e., to achieve first-mover advantages and/or hedge against the uncertainty associated with the indistinguishability of dual-use technology), increasing the risk of inadvertent nuclear use.

The chapter revisited the concept of offense-defense distinguishability to view AI technology. It found that, all things being equal, dual-use distinguishability uncertainty generally increases when the development, utilization, and distribution is weak or widely dispersed, particularly to the private sector. The US-China case study demonstrated that AI technology complicates further the challenges posed by dual-use technology as an aggravator of the security dilemma in competitive dyads—in particular in situations where a rising challenger's ambitions are pitted against a declining status-quo hegemon. In sum, mutual distrust and an atmosphere of uncertainty surrounding dual-use technology between competitors—viewed through the lens of securitization—means that any additional or new advances in AI technology will likely deepen the security dilemma.

Several important policy implications flow from the findings in this chapter. First, and for similar reasons to other dual-use emerging security, arms control and disarmament prospects in dual-use AI technology appear remote.[175] The long-standing Nuclear Non-Proliferation Treaty provides a successful case study in global governance that has successfully minimized the threat posed by the securitization of a new dual-use (i.e., nuclear) technology while enabling the mutual benefits of sharing nuclear technology to strengthen deterrence and strategic stability.[176]

However, dual-use AI technology is more widely diffused than nuclear technology, with the commercial sector heavily involved in its research, development, and operation. This issue is complicated by the lack of standard definitions of AI technology (especially military-use AI), standards for information-sharing, notifications, or engineering methodology, which, like cyberspace, stymies the formulation of a new regulatory or an arms control framework to ensure the safety and robustness of AI technology used in a civilian and, in particular, in a military context.[177] Further, because the lines between both dual-use technology and nuclear and non-nuclear capabilities—that AI can augment and enhance—are blurred, the likelihood of misperceptions caused by uncertainty and complexity increases, and the chances of alleviating the security dilemma decrease.[178]

Second, the backdrop of great-power competition that predisposes actors to securitize dual-use technology and frame others' actions (technical breakthroughs and surprises, state-directed initiatives and political goals, economic gains, and rhetoric) in ways that fuel uncertainties and insecurities on both sides, leads to self-reinforcing, countermeasures, and unintended outcomes.[179] In this sense, AI technology adds to the existing challenges posed by dual-use technology as an aggravator of the security dilemma between states locked in competitive dyads, particularly when a rising challenger's ambitions are pitted against a declining status-quo hegemon. As the US responds symmetrically to the perceived dangers posed by China (and to a lesser extent Russia) in the development of dual-use AI technology, these dynamics might, during peacetime, increase mutual mistrust, and cause issues with shared interest to become intractable (especially as the relative power narrows), allowing misallocation of resources, and military and economic overreach. During a crisis or conflict, these dynamics can increase the risk of inadvertent escalation between nuclear-armed states, compounded by others' misperceptions and motives associated with an intense security dilemma.

Third, and related, while the conflict of interest, uncertainty, and thus security dilemmas are immutable features of anarchical international politics, policymakers make choices that determine the severity, endurability, and the possibility of alleviating or mitigating these outcomes. In other words, the security dilemma is not a fixed or constant phenomenon; it operates at different times and with varying degrees of intensity depending on policymakers' decisions, how to frame issues (i.e., whether to securitize them), and anticipating the probable consequences of these choices. According to one scholar, "policymakers are not only responsible for the effects of security claims they anticipate, but also need to accept responsibility for unintended consequences of their actions."[180]

It is a critical task, therefore, to make policymakers aware of the potential dangers posed by their (deliberate and unintentional) choices and the need to develop empathy with others—to determine whether a conflict of interest is genuine or illusory—as a means to align strategic ends and means, and thus reduce the risk of unnecessary arms racing and inadvertent war.[181] For example, success in future confidence-building measures between strategic rivals to control and regulate military-use AI—to anticipate and mitigate the dangers of spiraling security dynamics—will be contingent on the identification of shared interests and recognition of unintended consequences posed by AI related risk to instill mutual trust—or the "rules of the road."[182] Future research would be beneficial on how regime type (i.e., authoritarian vs. democratic) and divergences in states' attitudes to escalation risk, strategic stability, and deterrence in the context of military technology might influence the security dilemma.[183] How might the introduction of intelligent machines into crisis and conflict scenarios between adversaries impact these attitudes and influence the security dilemma? The next chapter revisits the notion of a "catalytic nuclear war" in light of technological change and an improved understanding of human psychology. How might AI technology in the digital age increase the risk of accidental nuclear war caused by actions of third-party actors?

Notes

1. The term "motives" refers to state actors' geopolitical-military goals, distinct from their "intentions" or the strategies/doctrines they choose to achieve those goals. See Jeffrey A. Frieden, "Actors and Preferences in International Relations," in *Strategic Choice and*

International Relations, edited by David. A. Lake and Robert Powell (Princeton, NJ: Princeton University Press, 1999), pp. 42–45.

2. IR scholars distinguish the "security dilemma concept" from the "security dilemma theory." The security dilemma concept is used to label specific situations and behavior in international relations. In contrast, the security dilemma theory is the corpus of knowledge that seeks to understand the underlying causes, regulations, state-to-state interactions, and implications of the security dilemma. See Shiping Tang, "The Security Dilemma: A Conceptual Analysis," *Security Studies* 18, no. 3 (2009): pp. 587–623.
3. See Sean M. Lynn-Jones, "Offense-Defense Theory and its Critics," *Security Studies* 4, no. 4 (1995): pp. 660–691; Charles L. Glaser and Chaim Kaufmann, "What is the Offense-Defense Balance and How Can We Measure It?" *International Security* 22, no. 4 (1998): pp. 44–82.
4. See Robert Jervis, "Was the Cold War a Security Dilemma?" *Journal of Cold War Studies* 3, no. 1 (2001): pp. 36–60; Andrew Kydd, *Trust and Mistrust in International Relations* (Princeton, NJ: Princeton University Press, 2005); and Dale C. Copeland, *The Origins of Major War* (Ithaca, NY: Cornell University Press, 2000).
5. See Barry Posen, "The Security Dilemma and Ethnic Conflict," *Survival* 35, no. 1 (1993): pp. 27–47; and Monica D. Toft, *The Geography of Ethnic Violence: Identity, Interests, and Territory* (Princeton, NJ: Princeton University Press, 2003).
6. See Charles Glaser, "When Are Arms Races Dangerous?" *International Security* 28, no. 4 (2004): pp. 44–84; Chaim D. Kaufmann, "Possible and Impossible Solutions to Ethnic Civil Wars," *International Security* 20, no. 4 (1996): pp. 139–175; and Robert Jervis, *Perception and Misperception in International Politics* (Princeton, NJ: Princeton University Press, 1976), Ch. 3.
7. On the US-China security dilemma, see Thomas J. Christensen, "China, the US-Japan Alliance, and the Security Dilemma in East Asia," *International Security* 23, no. 4 (1999): pp. 49–80; Thomas J. Christensen, "The Contemporary Security Dilemma: Deterring a Taiwan Conflict," *Washington Quarterly* 25, no. 4 (2002): pp. 7–21; James Johnson, *The US-China Military and Defense Relationship during the Obama Presidency* (New York: Palgrave Macmillan, 2018); and Andrew Scobell, "Perceptions and Misperception in US-China Relations," *Political Science Quarterly* 135, no. 4 (2020): pp. 637–664.
8. For example, see Vincent Boulanin, ed., *The Impact of Artificial Intelligence on Strategic Stability and Nuclear Risk Vol. I Euro-Atlantic Perspectives* (Stockholm: SIPRI Publications: May 2019); Edward Geist and Andrew Lohn, *How Might Artificial Intelligence Affect the Risk of Nuclear War?* (Santa Monica, CA: RAND Corporation, 2018); Michael Horowitz, Paul Scharre, and Alex Velez-Green, *A Stable Nuclear Future? The Impact of Automation, Autonomy, and Artificial Intelligence* (Philadelphia, PA: University of Pennsylvania Press, 2017), and James Johnson, "Artificial Intelligence: A Threat to Strategic Stability," *Strategic Studies Quarterly* 14, no. 1 (2020): pp. 16–39.
9. See Robert Jervis, "Cooperation under the Security Dilemma," *World Politics* 30, no. 2 (1978): pp. 169–214; Charles L. Glaser, "The Security Dilemma Revisited," *World Politics* 50, no. 1 (1997): pp. 171–201; and Shiping Tang, "Fear in International Politics: Two Positions," *International Studies Review* 10, no. 3 (2008): pp. 451–470.
10. See Seth Weinberger, "Institutional Signaling and the Origin of the Cold War," *Security Studies* 12, no. 4 (2003): pp. 80–115; and Michael W. Doyle, "Kant, Liberal Legacies, and Foreign Affairs," *Philosophy and Public Affairs* 12, nos 3/4 (1983): pp. 205–235.

11. See Alexander Wendt, "Anarchy is What States Makes of It," *International Organization* 46, no. 2 (1992): pp. 391–425.
12. Shiping Tang, "The Security Dilemma: A Conceptual Analysis," *Security Studies* 1, no. 3 (2009): pp. 587–623.
13. Notable exceptions include: Jervis, *Perception and Misperception in International Politics*; Kaufman, "Spiraling to Ethnic War"; Ken Booth and Nicholas Wheeler, *The Security Dilemma: Fear, Cooperation, and Trust in World Politics* (New York: Palgrave Macmillan, 2008); Van Evera, "The Cult of the Offensive and the Origins of the First World War," *International Security* 9, no. 1 (1984): pp. 58–107, and James Johnson, "Washington's Perceptions and Misperceptions of Beijing's Anti-Access Area-Denial (A2–AD) 'Strategy': Implications for Military Escalation Control and Strategic Stability," *The Pacific Review* 30, no. 3 (2106): pp. 271–288.
14. See James M. Acton et al., *Entanglement: Russian and Chinese Perspectives on Non-Nuclear Weapons and Nuclear Risks* (Washington, DC: Carnegie Endowment for International Peace, 2017).
15. For the seminal text on misperception in international relations, see Jervis, *Perception, and Misperception in International Politics*.
16. Scholars have identified a range of structural and non-structural factors (or "modifiers" or "regulators") that can influence the severity of the security dilemma. A more expansive list includes factors such as geography, mixing and concentration of ethnic groups, external alliances, and a range of psychological regulators (e.g., prospect theory, analogical reasoning, and the cult of reputation). See Jeffrey W. Taliaferro, "Security Seeking under Anarchy: Defensive Realism Revisited," *International Security* 25, no. 3 (Winter 2000–2001): pp. 128–161; Yuen Foong Khong, *Analogies at War* (Princeton, NJ: Princeton University Press, 1992); Shiping Tang, "Reputation, Cult of Reputation, and International Conflict," *Security Studies* 24, no. 4 (October 2005): pp. 34–62; Jonathan Mercer, *Reputation in International Relations* (Ithaca, NY: Cornell University Press, 1996); Daryl Press, *Calculating Credibility: How Leaders Assess Military Threat During Crisis* (Ithaca, NY: Cornell University Press, 2005); and Toft, *The Geography of Ethnic Violence*; Jack S. Levy, "Prospect Theory, Rational Choice, and International Relations," *International Studies Quarterly* 41, no. 1 (1997): pp. 87–112.
17. Herbert Butterfield, *History and Human Relations* (London: Collins, 1951); John Herz, *Political Realism and Political Idealism: A Study in Theories and Realities* (Chicago, IL: University of Chicago Press, 1951); and Robert Jervis, "Cooperation under the Security Dilemma," *World Politics* 30, no. 2 (1978): pp. 169–214.
18. Several IR scholars argue that the security dilemma is a universal and non-conditional feature of world politics. See, for example, Taliaferro, "Security Seeking under Anarchy: Defensive Realism Revisited," *International Security* 25, no. 3 (2000/2001), pp. 131 and 136; Glaser, "The Security Dilemma Revisited"; and Booth and Wheeler, *The Security Dilemma*.
19. Alan R. Collins, *The Security Dilemma of Southeast Asia* (London: Macmillan, 2000), p. 24.
20. The issue of "uncertainty" has proved a contentious one between realism and alternative approaches to viewing international relations and between the various branches of realism themselves. See Robert O. Keohane, *After Hegemony: Cooperation and Discord in the World Political Economy* (Princeton, NJ: Princeton University Press, 1984); and Celeste

A. Wallander, *Mortal Friends, Best Enemies: German-Russian Cooperation after the Cold War* (Ithaca, NY: Cornell University Press, 1999).

21. The connection between anarchy in IR, the security dilemma, and conflict is theoretically contested. See Andrew Kydd, "Sheep in Sheep's Clothing: Why Security Seekers Do Not Fight Each Other," *Security Studies* 7, no. 1 (1997): pp. 91–121; Marc Trachtenberg, "The Question of Realism: A Historian's View," *Security Studies* 13, no. 1 (2002): pp. 156–194; and Wendt, "Anarchy is What States Makes of It."
22. While it appears logical that "uncertainty" about others' intentions should come before fear, IR scholars have tended to use them together, given that they are inextricably interwoven. Tang, "Fear in International Politics: Two Positions," pp. 451–470.
23. Glenn H. Snyder, "The Security Dilemma in Alliance Politics," *World Politics* 36, no. 4 (1984): pp. 461–495.
24. Jervis, *Perception and Misperceptions*, pp. 75 and 80.
25. Jervis, "Cooperation under the Security Dilemma," pp. 169–170.
26. While the security dilemma can lead to types of unintended and self-defeating outcomes (e.g., arms racing, spirals of uncertainty and competition, and inadvertent war), these outcomes (or "security dilemma dynamics") are not necessarily caused by the security dilemma; thus, deterrence policies can lead to unintended and self-defeating consequences, which may or may not be related to a security dilemma. Jervis, *Perception and Misperception*, pp. 81 and 90.
27. The foundational work on structural realism is Kenneth N. Waltz, *Theory of International Politics* (New York: Random House, 1979). On offensive realism, see John J. Mearsheimer, "The False Promise of International Institutions," International Security 19, no. 3 (1994/95): pp. 9–14; and John J. Mearsheimer, *The Tragedy of Great Power Politics* (New York: W.W. Norton, 2001). On defensive realism, see Jervis, "Cooperation under the Security Dilemma"; Charles L. Glaser, "Realists as Optimists: Cooperation as Self-Help," *International Security* 19, no. 3 (1994/95): pp. 50–90.
28. To reconcile the dichotomy between the causes, drivers, and outcomes of the security dilemma, Jack Snyder conceptualized a security dilemma in which only material regulators are present ("structural security dilemma") and another ("perceptual security dilemma") that depicts a security dilemma in which both material and perceptual and psychological regulators are present. Jack Snyder, "Perceptions of the Security Dilemma in 1914," in *Psychology and Deterrence*, edited by Robert Jervis, Richard Ned Lebow, and Janice Stein (Baltimore: Johns Hopkins University Press, 1985), pp. 153–179.
29. The "offense-defense balance" refers to the relative ease of carrying out and defending against military attacks—primarily by military technology and geographical factors. The offense-defense balance is generally defined relative to a particular goal, type of conflict, and actors. Scholars have used the concept to explain the origins of inter-state war, ethnic and civil conflict, arms control, arms racing, alliance behavior, military doctrine, grand strategy, and the international system's nature and shape. Classical and contemporary scholars diverge on several issues, particularly on the applicability and utility of this theory across various military domains—for example, the offense-defense balance in the cyber domain has received attention in recent literature. For classical studies, see Lynn-Jones, "Offense-Defense Theory and its Critics," pp. 660–691; Glaser and Kaufmann, "What is the Offense-Defense Balance and How Can We Measure It?"; Van Evera, "The Cult of the Offensive and the Origins of the First World War"; and

Jervis, "Cooperation under the Security Dilemma." For studies on the "cyber offense-defense balance", see Gartzke and Lindsay, "Weaving Tangled Webs: Offense, Defense, and Deception in Cyberspace"; Ilai Saltzman, "Cyber Posturing and the Offense-Defense Balance," *Contemporary Security Policy* 34, no. 1 (2013): pp. 40–63; and Rebecca Slayton, "What is the Cyber Offense-Defense Balance? Conceptions, Causes, and Assessment," *International Security* 41, no. 3 (2017): pp. 72–109.

30. Jervis, "Cooperation under the Security Dilemma."
31. Some scholars argue that the supposed first-mover advantages that technology can create during a conflict have historically played a small independent role in state decisions to use military force; other factors usually prove decisive—thus technology merely acts as a means when states engage in escalatory behavior for other reasons. See Keir Lieber, "Grasping the Technological Peace: The Offense-Defense Balance and International Security," *International Security* 25, no. 1 (2000): pp. 71–104.
32. Some scholars argue that authors have also suggested that offense-dominance within specific contexts and domains can enhance stability and reduce the security dilemma's severity. See, for example, Erik Gartzke, "The Myth of Cyberwar: Bringing War in Cyberspace Back down to Earth," *International Security* 38, no. 2 (2013): pp. 41–73. Besides factors related to "offensive dominance," which can worsen stability and increase conflict escalation risk, see Stephen Van Evera, "Offense, Defense, and the Causes of War," *International Security* 22, no. 4 (1998): pp. 5–43.
33. This offensive deterrence strategy received much attention in Israel during the late 1970s and 1980s. See Saadia Amiel, "Deterrence by Conventional Forces," *Survival* 20, no. 2 (1978): pp. 58–62.
34. Like John Herz—and contrary to Herbert Butterfield—Robert Jervis also agreed that the security dilemma is not at the heart of all conflict and war. The security dilemma can cause conflict and war but is not the cause of all wars. Herz originally agreed with Butterfield that the security dilemma was the cause of all conflict but later rebutted Butterfield's claim, noting no security dilemma between Hitler's Germany and other states. Jervis, *Perception, and Misperception in International Politics* pp. 75 and 80; Herz, *Political Realism and Political Idealism*, p. 12.
35. Scholars have identified several means to approach the challenges of knowing what a state's intentions are: (1) monitoring an actor's words and actions vis-à-vis others states; (2) signaling one's benign intentions (or "costly signaling"), and then observing the reactions to these reassurances from the other side; and (3) using past behavior as an index to examine a state's characteristics as a guide to how it might behave in the future—which depends on future probabilities rather than signaling. See Kydd, *Trust and Mistrust in International Relations*; and Shiping Tang, "Correspondence: Uncertainty and Reassurance in International Politics," *International Security* 32, no. 1 (2007): pp. 193–197; and Weinberger, "Institutional Signaling and the Origin of the Cold War," pp. 80–115.
36. See Glaser, "Political Consequences of Military Strategies"; Taliaferro, "Security Seeking under Anarchy"; and Paul Roe, "Actors' Responsibility in Tight, Regular, or Loose Security Dilemmas," *Security Dialogue* 32, no. 1 (2001): pp. 103–116.
37. In practice, while states can demonstrate to adversaries their intentions, the conditions whereby benign actors can reveal their underlying motives without simultaneously increasing their vulnerability—a logical feature of this one-sided uncertainty—are

limited. See Evan B. Montgomery, "Breaking Out of the Security Dilemma: Realism, Reassurance, and the Problem of Uncertainty," *International Security* 31, no. 2 (2006): pp. 151–185. On signaling in IR, see Weinberger, "Institutional Signaling and the Origin of the Cold War," pp. 80–115; Kydd, *Trust and Mistrust*. On reassurances and signaling in IR, see Tang, "Correspondence: Uncertainty and Reassurance in International Politics," pp. 193–197; and Robert Jervis, *The Logic of Images in International Relations*, 2nd ed. (New York: Columbia University Press, 1989).

38. Robert Jervis, "Was the Cold War a Security Dilemma?" *Journal of Cold War Studies* 3, no. 1 (Winter 2001): pp. 36–60, p. 57.
39. See James D. Fearon, "Signaling Foreign Policy Interests: Tying Hands versus Sinking Costs," *Journal of Conflict Resolution* 41, no. 1 (1997): pp. 68–90; Robert Axelrod, *The Evolution of Cooperation* (New York: Basic Books, 1984); and Charles E. Osgood, *An Alternative to War or Surrender* (Urbana: University of Illinois Press, 1962).
40. However, success in states' efforts to use these costly signals to reduce uncertainty and reassure adversaries has been few. For example, in the Cold War's final years, bilateral efforts to restrain the competition between the US and Russia generally failed to produce (arms control or reduction) agreements to address the intensifying security dilemma dynamics. Reassurances can be successful when the balance between the offense and defense is precise, and the balance between them is neutral. Barry Buzan and Eric Herring, *The Arms Dynamic in World Politics* (Boulder, CO: Lynne Rienner, 1998); and Montgomery, "Breaking Out of the Security Dilemma," p. 183.
41. In addition to the security dilemma, other causes of inadvertent escalation include: Clausewitz's concept of the "fog of war" and modern technology (e.g., command and control), which can heighten fears that a rival might launch a successful surprise attack; military organizational ("the cult of the offensive"); civil-military relations; and cognitive factors, including the tendency of actors to underestimate the probability of accidents and randomness in international relations. See Barry R. Posen, *Inadvertent Escalation* (Ithaca, NY and London: Cornell University Press, 1991), pp. 12–22; and Robert Jervis, *How Statesmen Think: The Psychology of International Politics* (Princeton, NJ: Princeton University Press, 2017), p. 69.
42. While defensive realists view conflicts of interest as both genuine and the result of misperception, both are reconcilable and irreconcilable. By contrast, offensive realists consider that all conflicts are objective (i.e., real) and irreconcilable, thus underplaying the role of misperception in intensifying a security dilemma. See Arnold Wolfers, "'National Security' as an Ambiguous Symbol," *Political Science Quarterly* 67, no. 4 (1952): pp. 481–502; and Kenneth Boulding, "National Images and International Systems," *Journal of Conflict Resolution* 3, no. 2 (1959): pp. 120–131.
43. On the "escalation ladder" and nuclear risk, see Herman Kahn, *On Escalation: Metaphors and Scenarios* (New York: Praeger, 1965).
44. See Morgan E. Forrest et al., *Dangerous Thresholds: Managing Escalation in the 21st Century* (Santa Monica, CA: RAND Corporation, 2008); and James Johnson, "Deterrence in the Age of Artificial Intelligence & Autonomy: A Paradigm Shift in Nuclear Deterrence Theory and Practice?" *Defense & Security Analysis* 36, no. 4 (2020): pp. 426–429.
45. On Cognitive Bias and Decision-Making, see Amos Tversky and Daniel Kahneman, "Judgment under Uncertainty: Heuristics and Biases," *Science* 185, no. 4157 (September

1974): pp. 1124–1131; Daniel Kahneman, *Thinking, Fast and Slow* (New York: Farrar, Straus and Giroux, 2011); Linda J. Skitka, Kathleen L. Mosier, and Mark Burdick, "Does Automation Bias Decision-Making?" *International Journal of Human-Computer Studies* 51, no. 5 (1999): pp. 991–1006; and Michela Del Vicario et al., "Modeling Confirmation Bias and Polarization," *Sci Rep* 7, no. 40391 (2017): pp. 1–9.

46. See Taliaferro, "Security Seeking under Anarchy"; Yuen Foong Khong, *Analogies at War*; James D. Fearon, "Domestic Political Audiences and the Escalation of International Disputes," *American Political Science Review* 88, no. 3 (September 1994): pp. 577–592; Randall L. Schweller, "Neorealism's Status Quo Bias: What Security Dilemma?" *Security Studies* 5, no. 3 (1996): pp. 90–121; Gideon Rose, "Neoclassical Realism and Theories of Foreign Policy," *World Politics* 51, no. 1 (1998): pp. 144–172; and Kydd, "Sheep in Sheep's Clothing."
47. See Tang, "Reputation, Cult of Reputation, and International Conflict," pp. 34–62; Mercer, *Reputation in International Relations*; Press, *Calculating Credibility: How Leaders Assess Military Threat During Crisis*; Robert McMahon, "Credibility and World Power: Exploring the Psychological Dimension in Post-war American Diplomacy," *Diplomatic History* 15, no. 4 (1991): pp. 455–471; Jervis, *Perception, and Misperception*; Wendt, "Anarchy is What States Makes of It"; and Boulding, "National Images and International Systems."
48. Beyond the security dilemma, there are many other possible pathways to an interstate conflict involving emerging technologies, including AI: creating first-strike instability; undermining shifting the likelihood of conflict escalation; diffusion of military technology; blurring of nuclear and non-nuclear weapons and doctrine; conventional and power transition theory; shifts in the balance of military power (especially asymmetric power); and the distribution of power and alliance politics more broadly defined—or polarity. While some of these pathways are possible regulators of the security dilemma (and discussed in this chapter), they can exist in the absence of a dilemma. See, for example, Stephen Van Evera, *Causes of War: Power and the Roots of Conflict* (Ithaca, NY: Cornell University Press, 1999); Stephen Biddle, *Military Power: Explaining Victory and Defeat in Modern Battle* (Princeton, NJ: Princeton University Press, 2004); and Michael C. Horowitz, *The Diffusion of Military Power: Causes and Consequences for International Politics* (Princeton, NJ: Princeton University Press, 2010); Sechser, Narang, and Talmadge, "Emerging Technologies and Strategic Stability in Peacetime, Crisis, and War"; and James Johnson, "Artificial Intelligence in Nuclear Warfare: A Perfect Storm of Instability?" *The Washington Quarterly* 43, no. 2 (2020): pp. 197–211.
49. The European Commission defines "dual-use" technology as "items, including software and technology, which can be used for both civil and military purposes." Similarly, the US State Department defines it as "items that have both commercial and military or proliferation applications." Council Regulation (EC) No 428/2009 of May 5, 2009, Official Journal of the European Union (May 29, 2009); and *National Research Council, Biotechnology Research in an Age of Terrorism: Confronting the Dual-Use Dilemma* (Washington DC: National Academies Press, 2004).
50. Tristan A. Volpe, "Dual-Use Distinguishability: How 3D-Printing Shapes the Security Dilemma for Nuclear Programs," *Journal of Strategic Studies* 42, no. 6 (2019): pp. 814–840; Amir Lupovici, "The Dual-Use Security Dilemma and the Social Construction of Insecurity," *Contemporary Security Policy*, doi: 10.1080/13523260.2020.1866845; and Nazli

Choucri and David D. Clark, *International Relations in the Cyber Age: The Co-Evolution Dilemma* (Cambridge, MA: MIT Press, 2018).

51. Notable exceptions in the cyber and space domain include Joan Johnson-Freese, *Space as a Strategic Asset* (New York: Columbia University Press, 2007); Benjamin Buchanan, *The Cybersecurity Dilemma: Hacking, Trust and Fear between Nations* (Oxford: Oxford University Press, 2017); and Dunn M. Cavelty, "Breaking the Cyber-Security Dilemma: Aligning Security Needs and Removing Vulnerabilities," *Science and Engineering Ethics* 20, no. 3 (2014): pp. 701–715.
52. See Columba Peoples, "Life in the Nuclear Age: Classical Realism, Critical Theory and the Techno-Politics of the Nuclear Condition," *Journal of International Political Theory* 15, no. 3 (2019): pp. 279–296; Volpe, "Dual-Use Distinguishability: How 3D-Printing Shapes the Security Dilemma for Nuclear Programs"; Matthew Fuhrmann and Benjamin Tkach, "Almost Nuclear: Introducing the Nuclear Latency Dataset," *Conflict Management and Peace Science* 32, no. 4 (2015): pp. 443–446; and Charles L. Glaser, *Rational Theory of International Politics: The Logic of Competition and Cooperation* (Princeton, NJ: Princeton University Press, 2010).
53. Several recent studies consider the characteristics and implications of the cyber domain and the offense-defense balance. See Erik Gartzke and Jon R. Lindsay, "Weaving Tangled Webs: Offense, Defense, and Deception in Cyberspace," *Security Studies* 24, no. 2 (2015): pp. 316–348; and Saltzman, "Cyber Posturing and the Offense-Defense Balance," pp. 40–63; Ben Garfinkel and Allen Dafoe, "How Does the Offense-Defense Balance Scale?" *Journal of Strategic Studies* 42, no. 6 (2019): pp. 736–763; and Slayton, "What is the Cyber Offense-Defense Balance?"
54. See Ronnie Olesker, "The Securitization Dilemma: Legitimacy in Securitization Studies," *Critical Studies on Security* 6, no. 3 (2018): pp. 312–329; Paul Roe, "The 'Value' of Positive Security," *Review of International Studies* 34, no. 4 (2008): pp. 777–794; and Scott D. Watson, "Macrosecuritization and the Securitization Dilemma in the Canadian Arctic," *Critical Studies on Security* 1, no. 3 (2013): pp. 265–279.
55. Ludovici, "The Dual-Use Security Dilemma and the Social Construction of Insecurity," p. 7.
56. Barry Buzan, Ole Wæver, and Jaap de Wilde, *Security: A New Framework for Analysis* (Boulder, CO: Lynne Rienner Publishers, 1998), p. 25. Scholars have sought to integrate the security dilemma concept epistemologically and conceptually with the broader securitization scholarship. See Eric Van Rythoven, "The Securitization Dilemma," *Journal of Global Security Studies* 5, no. 3 (2020): pp. 478–493.
57. See Booth and Wheeler, *The Security Dilemma: Fear, Cooperation, and Trust in World Politics*; Jervis, "Cooperation under the Security Dilemma"; Roe, "Actors' Responsibility in Tight, Regular, or Loose Security Dilemmas"; and Andrew Kennedy and Darren Lim, "The Innovation Imperative: Technology and US-China Rivalry in the Twenty-First Century," *International Affairs* 94, no. 3 (2017): pp. 553–572.
58. Ludovici, "The Dual-Use Security Dilemma and the Social Construction of Insecurity," p. 2.
59. Kenneth Payne, "Artificial Intelligence: A Revolution in Strategic Affairs?" *Survival* 60, no. 5 (2018): pp. 7–32; James Johnson, "Artificial Intelligence & Future Warfare," *Defense & Security Analysis* 35, no. 2 (2019): pp. 147–169; and Michael C. Horowitz,

"Artificial Intelligence, International Competition, and the Balance of Power," *Texas National Security Review* 1, no. 3 (2018): pp. 37–57.

60. For example, see White House, "Executive Order on Maintaining American Leadership in Artificial Intelligence," February 11, 2019; and The State Council Information Office of the People's Republic of China, "State Council Notice on the Issuance of the New Generation AI Development Plan," July 20, 2017.
61. "Final Report," National Security Commission on Artificial Intelligence, March 2021, https://www.nscai.gov/wp-content/uploads/2021/03/Full-Report-Digital-1.pdf, pp. 7, 28, and 61.
62. Ibid. p. 9.
63. See Ashley J. Tellis, "China's Military Space Strategy," *Survival* 49, no. 3 (2007): pp. 41–72; Johnson-Freese, *Space as a Strategic Asset*; and Kai-Fu Lee, *AI Superpowers: China, Silicon Valley and the New World Order* (New York: Houghton Mifflin Harcourt, 2018).
64. Madison Cartwright, "Internationalising State Power Through the Internet: Google, Huawei and Geopolitical Struggle," *Internet Policy Review* 9, no. 3 (2020): pp. 1–18; Dennis Wang, *Reigning the future: AI, 5G, Huawei, and the Next 30 Years of US-China Rivalry* (New York: New Degree Press, 2020); James Johnson, "The End of Military-Techno Pax Americana? Washington's Strategic Responses to Chinese AI-Enabled Military Technology," *The Pacific Review* 34, no. 3 (2021): pp. 351–378; and Nigel Inkster, "The Huawei Affair and China's Technology Ambitions," *Survival* 61, no. 1 (2019): pp. 105–111.
65. Johnson, "The End of Military-Techno Pax Americana?" p. 14.
66. Philip E. Tetlock and Aaron Belkin, eds, *Counterfactual Thought Experiments in World Politics* (Princeton, NJ: Princeton University Press, 1996).
67. Johnson-Freese, *Space as a Strategic Asset*; and Matthew Mowthorpe, *The Militarization and Weaponization of Space* (New York: Lexington Press, 2004).
68. Office of the Secretary of Defense, *Annual Report to Congress: Military and Security Developments Involving the People's Republic of China, 2020* (Washington, DC: US Department of Defense, 2020), p. 65
69. Karina Manson and Christian Shepherd, "US Military Officials' Eye New Generation of Space Weapons," *Financial Times*, September 2, 2020, https://www.ft.com/content/d44aa332-f564-4b4a-89b7-1685e4579e72.
70. Ludovici, "The Dual-Use Security Dilemma and the Social Construction of Insecurity," p. 17.
71. Jervis, "Cooperation under the Security Dilemma," pp. 169–214.
72. Matthijs M. Maas, "How Viable is International Arms Control for Artificial Military Intelligence? Three Lessons from Nuclear Weapons," *Contemporary Security Policy* 40, no. 3 (2019): pp. 285–311, p. 294.
73. In addition to AI, other emerging dual-use technologies, including 3D printing, cyber, autonomous vehicles, and 5G networks, can produce similar effects, albeit in varying degrees of intensity, scale, and lethality. See Ludovici, "The Dual-Use Security Dilemma and the Social Construction of Insecurity"; Volpe, "Dual-Use Distinguishability: How 3D-Printing Shapes the Security Dilemma for Nuclear Programs"; Gartzke and Lindsay, "Weaving Tangled Webs: Offense, Defense, and Deception in Cyberspace," and Jürgen Altmann and Frank Sauer, "Autonomous Weapon Systems and Strategic Stability," *Survival* 59, no. 5 (2017): pp. 117–142.

74. See Paige Leskin, "Here Are all the Major US Tech Companies Blocked Behind China's 'Great Firewall,'" *Business Insider*, October 10, 2019, https://www.businessinsider.com/major-us-tech-companies-blocked-from-operating-in-china-2019-5; and "China's Cyber Watchdog Orders Top Tech Platforms to Increase Self-Censorship," *Reuters*, July 19, 2017, https://www.reuters.com/article/us-china-censorship/chinas-cyber-watchdog-orders-top-tech-platforms-to-increase-self-censorship-idUSKBN1A41CS.
75. See Horowitz, "Artificial Intelligence, International Competition, and the Balance of Power"; Edward Geist and Andrew Lohn, *How Might Artificial Intelligence Affect the Risk of Nuclear War?* (Santa Monica, CA: RAND Corporation, 2018); and Johnson, "Artificial Intelligence in Nuclear Warfare."
76. Payne, "Artificial Intelligence: A Revolution in Strategic Affairs?" p. 24.
77. See Maas "How Viable is International Arms Control for Military Artificial Intelligence?"; Volpe, "Dual-Use Distinguishability: How 3D-Printing Shapes the Security Dilemma for Nuclear Programs"; and Gartzke and Lindsay, "Weaving Tangled Webs: Offense, Defense, and Deception in Cyberspace."
78. Garfinkel, and Dafoe, "How Does the Offense-Defense Balance Scale?"
79. See Miles Brundage et al., "The Malicious Use of Artificial Intelligence: Forecasting, Prevention, and Mitigation," *arXiv preprint* (2018) arXiv:1802.07228.
80. Ibid., p. 759.
81. Charles L. Glaser, *Rational Theory of International Politics: The Logic of Competition and Cooperation* (Princeton, NJ: Princeton University Press, 2010), pp. 72–92; and Booth and Wheeler, *The Security Dilemma: Fear, Cooperation, and Trust in World Politics*, pp. 89–91.
82. See Mercer, *Reputation in International Relations*; Press, *Calculating Credibility: How Leaders Assess Military Threat During Crisis*; and Jervis, *Perception and Misperception*, Ch. 3.
83. Daniel Castro and Michael McLaughlin, "Who is Winning the AI Race: China, the EU, or the United States?—2021 Update," *ITIF* January 25, 2021, https://itif.org/publications/2021/01/25/who-winning-ai-race-china-eu-or-united-states-2021-update.
84. US Department of Defense, *Nuclear Posture Review* (Washington, DC, 2018), p. 14.
85. Tang, "The Security Dilemma: A Conceptual Analysis," pp. 621–622.
86. While beyond the scope of this chapter, it is important to acknowledge that non-state actors (i.e., private corporations, research centers, and state-affiliated enterprises), by developing the algorithms, managing the databases, and investing in dual-use technologies—like AI and cyber—can also influence the severity of the security dilemma. Social media companies that distribute information which can create and disseminate powerful socially and politically constructed (often false) narratives, especially in times of geopolitical tension and internal strife, can drive security dilemma spiral dynamics between competitive strategic dyads. See Johan Sigholm, "Non-State Actors in Cyberspace Operations," *Journal of Military Studies* 4, no. 1 (2013): pp. 1–37; Buchanan, *The Cybersecurity Dilemma: Hacking, Trust and Fear between Nations*; and James Johnson, "'Catalytic Nuclear War' in the Age of Artificial Intelligence' Autonomy: Emerging Military Technology and Escalation Risk between Nuclear-Armed States," *Journal of Strategic Studies* (2021), doi: 10.1080/01402390.2020.1867541.
87. Herbert A. Simon, "Bandwagon and Underdog Effects and the Possibility of Election Predictions," *Public Opinion Quarterly* 18, no. 3 (1954): pp. 245–253.

88. For cyber, see Diana Saco, "Colonizing Cyberspace: National Security and the Internet," in *Cultures of Insecurity: States, Communities, and the Production of Danger*, edited by Jutta Weldes et al. (Minnesota: University of Minnesota Press, 1999), pp. 261–291. For AI, see Maas "How Viable is International Arms Control for Military Artificial Intelligence?"
89. Greg Allen and Taniel Chan, *Artificial Intelligence and National Security* (Cambridge: Belfer Center for Science and International Affairs, 2017).
90. Both the US and its NATO allies have placed a renewed emphasis on understanding the civilian and military applications of technological advancements in AI technology. D.F. Reading and J. Eaton, *Science & Technology Trends: 2020–2040: Exploring the S&T Edge* (Brussels: NATO Science and Technology Organization, March 2020), p. 57, https://www.nato.int/nato_static_fl2014/assets/pdf/2020/4/pdf/190422-ST_Tech_Trends_Report_2020-2040.pdf.
91. Experts view 5G as a cornerstone technology to increase the speed, stability of data-loads, reduce the latency (i.e., accelerate network response times), and enhance mobile digital communications.
92. The Chinese concept of "cyber sovereignty" is understood to refer to both domestic state control over information and information security and preventing foreign intervention or manipulation of this information. See Rogier Creemers, "China's Conception of Cyber Sovereignty: Rhetoric and Realization," in *Governing Cyberspace: Behavior, Power, and Diplomacy*, edited by Dennis Broeders and Bibi van den Berg (New York: Rowman & Littlefield, 2020), pp. 107–142.
93. Gordon Orr, "China in 2020: Trade Technology and the Path Ahead for US-China Relations," *LinkedIn*, December 16, 2019, https://www.linkedin.com/pulse/china-2020-tradetechnology-path-ahead-us-china-relations-gordon-orr/.
94. Creemers, "China's Conception of Cyber Sovereignty: Rhetoric and Realization"; and James Mulvenon, "A World Divided: The Conflict with Chinese Techno-Nationalism Isn't Coming—It's Already Here," *War on the Rocks*, January 28, 2021, https://warontherocks.com/2021/01/a-world-divided-the-conflict-with-chinese-techno-nationalism-isnt-coming-its-already-here/.
95. "Xi Jinping's Report at the 19th Chinese Communist Party National Congress," *Xinhua*, October 27, 2017, http://www.xinhuanet.com/english/special/2017-11/03/c_136725942.htm.
96. Similarly, Beijing assumes that US technology giants' software applications create backdoors that enable the US to spy on China. Recent revelations about a security software company covertly owned by the CIA (Switzerland-based Crypto) reinforced this perception. "Huawei: US Official Warns 'No Safe Level' of Involvement with Tech Giant," *BBC News*, April 29, 2019, https://www.bbc.com/news/uk-48098362; "Crypto Spy Caper: CIA Secretly Owned Swiss Company That Sold Encryption Products to 120 Countries," *South China Morning Post*, February 12, 2020, https://www.scmp.com/news/world/europe/article/3050193/crypto-spy-caper-cia-secretly-owned-swiss-company-sold-encryption.
97. For example, Germany applied the same security standards to all its 5G equipment providers.
98. Several experts have refuted the genuine (i.e., malign) threat posed by Huawei (and ZTE) to US security and intimated that bans on Chinese information equipment, instead, serve US geopolitical interests—i.e., US military primacy and global hegemony. Anthony

Rutkowski, "The Costs of Trump's 5Gwall," *CircleID*, May 28, 2020, http://www.circleid.com/posts/20200528-the-costs-of-trumps-5g-wall/.
99. Jervis, *Perception and Misperception*, p. 146.
100. White House, "Executive Order on Securing the Information and Communications Technology and Services Supply Chain," May 15, 2019, https://www.whitehouse.gov/presidential-actions/executive-order-securing-informationcommunications-technology-services-supply-chain/.
101. *National Security Strategy of the United States* (Washington, DC: White House, 2017), p. 22.
102. In 2018, the US Congressional Research Service noted: "In 1960, the United States accounted for 69% of global R&D, with US defense-related R&D alone accounting for more than one-third of global R&D (36%). However, from 1960 to 2016, the US share of global R&D fell to 28%, and the federal government's share of total US R&D fell from 65% to 24%, while business's share more than doubled from 33% to 67%. As a result of these global, national, and federal trends, federal defense R&D's share of total global R&D fell to 3.7% in 2016." John F. Sargent and Marcy E. Gallo, "The Global Research and Development Landscape and Implications for the Department of Defense," *Congressional Research Service*, November 8, 2018, https://crsreports.congress.gov/product/pdf/R/R45403.
103. By 2018, on the advice from the Committee on Foreign Investment in the US (CFIUS), Washington blocked four attempts by Chinese companies to invest in advanced semiconductor technology.
104. Eric Sayers, "Thoughts on the Unfolding US-Chinese Competition: Washington's Policy Towards Beijing Enters Its Next Phase," *War on the Rocks*, February 9, 2021, https://warontherocks.com/2021/02/thoughts-on-the-unfolding-u-s-chinese-competition-washingtons-policy-towards-beijing-enters-its-next-phase/.
105. Tom Simonite, "Moore's Law is Dead. Now What?" *MIT Technology Review*, May 13, 2016, www.technologyreview.com/.
106. Under President Xi, dual-use technology—to support indigenous innovation—has been elevated to support Beijing's techno-strategic ambitions (e.g., the Made in China 2025 and Digital Silk Road initiatives). See Elsa Kania, "Made in China 2025, Explained: A Deep Dive into China's Techno-Strategic Ambitions for 2025 and Beyond," *The Diplomat*, February 1, 2019, https://thediplomat.com/2019/02/made-in-china-2025-explained/; Elsa B. Kania and Lorand Laskai, *Myths and Realities of China's Military-Civil Fusion Strategy* (Washington, DC: Center for a New American Security, 2021).
107. Mulvenon, "A World Divided: The Conflict with Chinese Techno-Nationalism Isn't Coming—It's Already Here."
108. Josh Chin, "China Looks to Close the Technology Gap with the US," *The Wall Street Journal*, April 21, 2018, www.wsj.com/articles/china-looks-to-close-technology-gap-with-u-s–1524316953.
109. Johnson, "The End of Military-Techno Pax Americana?" pp. 14–15.
110. For discussion on these broader questions, see Stuart Russell, *Human Compatible* (New York: Penguin Random House, 2019); George Zarkadakis, *In Our Image* (New York: Pegasus Books, 2015); Christianna Ready, "Kurzweil Claims that Singularity Will Happen by 2045," *Futurism*, October 5, 2017, https://futurism.com/kurzweil-claims-that-the-singularity-will-happen-by–2045; Nick Bostrom, *Superintelligence: Paths, Dangers,*

Strategies (Oxford: Oxford University Press, 2014); and Yuval N. Harari, *Homo Deus: A Brief History of Tomorrow* (New York: Harper Collins, 2018).

111. T.V. Paul, Deborah Larson, and William Wohlforth, eds, *Status in World Politics* (New York: Cambridge University Press, 2014); William Wohlforth, "Unipolarity, Status Competition, and Great Power War," in *International Relations Theory and the Consequences of Unipolarity*, edited by John Ikenberry (Cambridge: Cambridge University Press, 2011), pp. 33–65; and Lilach Gilady, *The Price of Prestige: Conspicuous Consumption in International Relations* (Chicago, IL: Chicago University Press, 2018).
112. It is difficult to predict with any degree of certainty, however, how a particular action or mode of behavior (e.g., refusing to back down on an issue or reverse a policy decision) might influence others' expectations of how a state may behave in the future; nor how regime type (i.e., democratic vs. autocratic) might affect these assessments. Jervis, *How Statesmen Think*, pp. 118–119, and 196.
113. Sputnik-like events are likely because potential US adversaries (i.e., China and Russia) consider the exploitation and diffusion of cutting-edge technology to provide asymmetrical responses to Western capabilities. See Robert A. Divine, *The Sputnik Challenge* (New York and Oxford: Oxford University Press, 1993), and Asif A. Siddiqi, *Sputnik and the Soviet Space Challenge* (Gainesville, FL: The University of Florida Press, 2000).
114. Gregory D. Koblentz, *Council Special Report-Strategic Stability in the Second Nuclear Age* (New York: Council on Foreign Relations Press, 2014).
115. P.W. Singer, *Wired for War: The Robotics Revolution and Conflict in the 21st Century* (New York: Penguin, 2009).
116. Michael Horowitz, "AI and the Diffusion of Global Power," *Center for International Innovation Governance*, November 16, 2020, https://www.cigionline.org/articles/ai-and-diffusion-global-power/.
117. Recent scholarship on "strategic" technologies has posited that the link between civilian and military technologies has historically proven tenuous. Thus, if this trajectory follows this established pattern, the diffusion of AI could take far longer than current predictions, particularly for technically advanced weapon systems. See Andrea Gilli and Mauro Gilli, "Why China Has Not Caught Up Yet: Military-Technological Superiority and the Limits of Imitation, Reverse Engineering, and Cyber Espionage," *International Security* 43, no. 3 (2019): pp. 141–189; and Jeffrey Ding and Allan Dafoe, "The Logic of Strategic Assets: From Oil to AI," *Security Studies* 30, no.2 (2021): pp. 182–212, doi: 10.1080/09636412.2021.1915583.
118. Don Rassler, "A View from the CT Foxhole: 'Jack' Shanahan, Director, Joint Artificial Intelligence Center, Department of Defense," *CTC Sentinel* 12, no. 11 (2019): pp. 10–14.
119. Gilli and Gilli, "Why China Has Not Caught Up Yet."
120. Alina Polyakova, "Weapons of the Weak: Russia and AI-Driven Asymmetric Warfare," Brookings Institution, November 15, 2018, https://www.brookings.edu/research/weapons-of-the-weak-russia-and-ai-driven-asymmetric-warfare/.
121. Johnson, "The End of Military-Techno Pax Americana?" p. 16.
122. Will Roper, "There's No Turning Back on AI in the Military," *Wired*, October 24, 2020, https://www.wired.com/story/opinion-theres-no-turning-back-on-ai-in-the-military/.
123. Kania and Laskai, *Myths and Realities of China's Military-Civil Fusion Strategy*.
124. Heather M. Roff, "The Frame Problem: The AI 'Arms Race' Isn't One," *Bulletin of the Atomic Scientists* 75, no. 3 (2019): pp. 1–5; and Paul Scharre, "Debunking the AI Arms

Race Theory," *Texas National Security Review* 4, no. 3 (2021), https://tnsr.org/2021/06/debunking-the-ai-arms-race-theory/.

125. US Department of Defense, "Lt. Gen. Jack Shanahan Media Briefing on AI-Related Initiatives within the Department of Defense" (Washington, DC, August 30, 2019) www.defense.gov/Newsroom/Transcripts/Transcript/Article/1949362/lt-gen-jack-shanahan-media-briefing-on-ai-related-initiatives-within-the-depart/.
126. Henry Markovits, and Guilaine Nantel, "The Belief-Bias Effect in the Production and Evaluation of Logical Conclusions," *Memory & Cognition* 17 (1989): pp. 11–17.
127. Deb Amlen, *The New York Times*, August 4, 2017, https://www.nytimes.com/2017/08/04/crosswords/daily-puzzle-2017-08-05.html.
128. In this context, "image" refers to the overall reputation and status of a state held by the actors of another (usually rival) state; its value to both sides in the assessments (i.e., official reports and speeches) made of the perception of this image vis-à-vis the other state. "Perception" refers to the cognitive processing of information by actors. Images in IR are believed to be comprehensive and enduring, while perceptions of state rhetoric and behavior are more specific and short-lived. Jervis, *Perception and Misperception*, pp. 146, and 117.
129. See Susan M. Puska, *New Century, Old Thinking: The Dangers of the Perceptual Gap in US-China Relations* (Carlisle Barracks, PA: US Army War College Strategic Studies Institute, 1998); and Scobell, "Perceptions and Misperception in US-China Relations."
130. Ibid., p. 638.
131. Jervis, *Perception and Misperception*, p. 118.
132. On US benevolent hegemony, see John G. Ikenberry, *Liberal Leviathan: The Origins, Crisis, and Transformation of the American World Order* (Princeton, NJ: Princeton University Press, 2012); and Stephen G. Brooks, "Can we Identify a Benevolent Hegemon?" *Cambridge Review of International Affairs* 25, no. 1 (2012): pp. 27–38. On China's peaceful rise mantra, see Yong Deng, "China: The Post-Responsible Power," *The Washington Quarterly* 37, no. 4 (2014): pp. 117–132; and Feng Zhang, "Rethinking China's Grand Strategy: Beijing's Evolving National Interests and Strategic Ideas in the Reform Era," *International Politics* 49 (2012): pp. 318–345.
133. The judgment of "reasonableness" of subjective security and military policies is controversial and remains contested. See Jervis, "Was the Cold War a Security Dilemma?" p. 38.
134. Great powers are generally viewed as rational actors and thus interpret their malevolent words and deeds as a reflection of duplicity, deception, secrecy, and stratagem—as opposed to the presumption of irrationality associated with the actions of asymmetric nuclear-armed adversaries such as North Korea. See Thomas G. Mahnken, *Secrecy and Stratagem: Understanding Chinese Strategic Culture* (Sydney: Lowy Institute for International Policy, 2011); and Andrew J. Nathan and Andrew Scobell, "How China Sees America: The Sum of Beijing's Fears," *Foreign Affairs* 91 (September/October 2012): pp. 32–47.
135. See Ifat Maoz et al., "Reactive Devaluation of an 'Israeli' vs. 'Palestinian' Peace Proposal," *The Journal of Conflict Resolution* 46, no. 4 (2002): pp. 515–546.
136. John Spanier, *The Truman-MacArthur Controversy and the Korean War* (New York: Norton, 1965), p. 97.

137. Derek J. Mitchell, "US Self-Image: Pride and Vulnerability in a Changing World," in *Chinese Images of the United States*, edited by Carola McGiffert (Washington, DC: Center for Strategic and International Studies, 2005), pp. 134–151.
138. The "fundamental attribution error" suggests that people tend to attribute the behavior (or character traits) to intentional action while circumstances detect *their own* behavior. This cognitive bias is thought by some to be important in explaining why states fail to cooperate, escalate situations, engage in arms racing, and ultimately go to war. Because actors fail to appreciate the various constraints imposed on others, they often *overestimate* the threat they pose (in mirror imagining) and *underestimate* the threat they pose to others—actors assuming the worst can reduce cooperation and increase the risk of conflict. See Philip E. Tetlock, "Social Psychology and World Politics," in *The Handbook of Social Psychology*, edited by Daniel Gilbert, Susan Fiske, and Gardner Lindzey (New York: McGraw-Hill, 1998), pp. 868–912; and Edward Jones and Victor Harris, "The Attribution Attitudes," *Journal of Experimental Psychology* 3, no. 1 (1967): pp. 167–174.
139. Georgios Theocharous et al., "Personalizing with Human Cognitive Biases," in *Adjunct Publication of the 27th Conference on User Modeling, Adaptation and Personalization* (New York: Association for Computing Machinery, 2019), pp. 13–17.
140. See Daniel Gilbert, *Stumbling on Happiness* (New York: Knopf, 2006).
141. Jervis, *How Statesmen Think*, p. 224.
142. Fearon, "Domestic Political Audiences and the Escalation of International Disputes," pp. 577–592.
143. For example, China has long been critical of the US DoD's annual reports to Congress—and open policy debates that follow—on Chinese military developments and goals.
144. Valerie S. Folkes, "The Availability Heuristic and Perceived Risk," *Journal of Consumer Research* 15, no. 1 (1988): pp. 13–23.
145. "Chinese Military Calls US Biggest Threat to World Peace," *AP News* September 13, 2020, https://apnews.com/article/beijing-china-archive-9c9cf27d653fccbd34baf23f6ed48661.
146. For example, James Steinberg, "China's Arrival: The Long March to Global Power," speech, Center for a New American Security, September 24, 2009, https://www.cnas.org/publications/video/cnasreport-launch-chinas-arrival-the-long-march-to-global-power-highlight-video; Thomas J. Christensen, "China, the US-Japan Alliance, and the Security Dilemma in East Asia," *International Security* 23, no. 4 (1999): pp. 49–80; and Christensen, "The Contemporary Security Dilemma: Deterring a Taiwan Conflict," pp. 7–21.
147. US Department of Defense, "Military and Security Developments Involving the People's Republic of China 2020," September 1, 2020, https://media.defense.gov/2020/Sep/01/2002488689/-1/-1/1/2020-DOD-CHINA-MILITARY-POWER-REPORT-FINAL.PDF, ii.
148. Edward N. Luttwak, *The Rise of China Versus the Logic of Strategy* (Cambridge, MA: Harvard University Press, 2012).
149. See *National Security Strategy of the United States*, p. 1; and State Council Information Office of the People's Republic of China, *China and the World in the New Era*.
150. There is no consensus about how relative power affects crisis dynamics and escalation. While classical and structural realists posit that as relative power balances decrease the likelihood of conflict, the literature on power transitions posits that power balances

increase the risk of escalation and conflict—others argue that power as a factor can be moderated by other variables such as the offense-defense balance (discussed above). See, for example, Jacek Kugler and Douglas Lemke, eds, *Parity and War* (Ann Arbor MI: University of Michigan Press, 1996); and Jervis, "Cooperation under the Security Dilemma," pp. 169–214.

151. Scobell, "Perceptions and Misperception in US-China relations," pp. 645–646.
152. Jervis, *Perception and Misperception*, p. 319.
153. See Hillel J. Einhorn and Robin M. Hogarth, "Confidence in Judgment: Persistence of the Illusion of Validity," *Psychological Review* 85, no. 5 (1978): pp. 395–416.
154. See Keith L. Shimko, *Images and Arms Control: Perceptions of the Soviet Union in the Reagan Administration* (Ann Arbor, MI: University of Michigan Press, 1992); and Robert Jervis and Jack Snyder, *Dominoes and Bandwagons: Strategic Beliefs and Great Power Competition in the Eurasian Rimland* (New York: Oxford University Press, 1991).
155. During the 1961 Berlin Crisis, US military commanders' use of military force to reassure the local population after the Soviets erected the Berlin Wall was taken without orders from Washington. However, it was attributed by the Soviets to high-level US decision-making. Raymond Garthoff, "Berlin, 1961: The Record Corrected," *Foreign Affairs* 84 (1991): pp. 142–156.
156. Daniel S. Geller, "Power Differentials and War in Rival Dyads," *International Studies Quarterly* 37 (June 1993): pp. 173–193.
157. Evan S. Medeiros, "The Changing Fundamentals of US-China Relations," *Washington Quarterly* 42, no. 3 (2019): pp. 93–119; Johnson, "The End of Military-Techno Pax Americana?"
158. China State Council, "State Council Notice on the New Generation Artificial Intelligence Development Plan," July 8, 2017, http://www.gov.cn/zhengce/content/2017-07/20/content_5211996.htm.
159. Ibid.
160. China has assembled a vast database to train AI algorithms. According to a recent industry report, China is on track to possess twenty percent of the world's share of data by 2020 and the potential to have over thirty percent by 2030. Will Knight, "China's AI Awakening," *MIT Technology Review*, October 10, 2017, https://www.technologyreview.com/2017/10/10/148284/chinas-ai-awakening/.
161. In 2017, for example, a Chinese state-run company, Haiyin Capital, outmaneuvered the US Air Force's efforts to acquire AI software developed by Neurala in 2017. See Paul Mozur and Jane Perlez, "China Bets on Sensitive US Start-Ups, Worrying the Pentagon," *The New York Times*, March 22, 2017, https://www.nytimes.com/2017/03/22/technology/china-defense-start-ups.html.
162. See, for example, Anthony Cordesman, *The Biden Transition and U.S. Competition with China and Russia: The Crisis-Driven Need to Change U.S. Strategy* (Washington, DC: Center for Strategic and International Studies, 2021).
163. Johnson, "The End of Military-Techno Pax Americana?" p. 10.
164. In the US-China dyad, this cognitive dimension has been exacerbated by the US's unchallenged global hegemonic status since the demise of the Soviet Union and China's rise steeped in historical memories of imperial glories and colonial-era humiliation. See Aaron L. Friedberg, *A Contest for Supremacy: US, China, and a Struggle for Mastery in Asia* (New York: W.W. Norton, 2012); Andrew Scobell, "Learning to Rise Peacefully? China

and the Security Dilemma," *Journal of Contemporary China* 21 (2012), pp. 713–721; and Johnson, "Washington's Perceptions and Misperceptions of Beijing's Anti-Access Area-Denial (A2–AD) 'strategy'."

165. Einhorn and Hogarth, "Confidence in Judgment: Persistence of the Illusion of Validity."
166. "Status-quo" in international relations is best viewed in a relative sense (or reference point) vis-à-vis states rather than in absolute terms. Several scholars argue that conflict and war are more likely when both sides in a competition dyad believe they are defending the status quo. See Jervis, *How Statesmen Think*, p. 91.
167. Daniel Kahneman and Amos Tversky, "Availability: A Heuristic for Judging Frequency and Probability," *Cognitive Psychology* 5, no. 2 (September 1973): pp. 207–232; Kahneman, *Thinking, Fast and Slow*; and Rose McDermott, *Risk-Taking in International Politics: Prospect Theory Policy Change* (Ann Arbor, MI: University of Michigan Press, 1998).
168. *National Security Strategy of the United States* (Washington, DC: White House, 2017), https://www.whitehouse.gov/wp-content/uploads/2017/12/NSS-Final-12-18-2017-0905.pdf.
169. "US Threatens China, Not Vice Versa: Chinese FM," *Xinhua News Agency*, February 14, 2020, http://www.xinhuanet.com/english/2020-02/15/c_138786792.htm.
170. State Council Information Office of the People's Republic of China, *China and the World in the New Era* (Beijing, September 2019), p. 31. Harry Harding, "China's Co-operative Behavior," in *Chinese Foreign Policy: Theory and Practice*, edited by Thomas W. Robinson and David Shambaugh (Oxford: Clarendon Press, 1994), p. 398.
171. John Bowden, "China: Pompeo Has Sinister Motive for Pushing 'Wuhan Virus' Language," *The Hill*, March 26, 2020, accessed at https://thehill.com/policy/international/489605-china-pompeo-has-sinistermotive-for-pushing-wuhan-virus-language.
172. See Johnson, "Artificial Intelligence in Nuclear Warfare," pp. 197–211.
173. See Keir A. Lieber and Darryl G. Press, "The New Era of Counterforce: Technological Change and the Future of Nuclear Deterrence," *International Security* 41 no. 4 (2017): pp. 9–49.
174. Tanya Ogilvie-White, *The Logic of Nuclear Deterrence* (Geneva: UNIDIR, 2020).
175. Maas "How Viable is International Arms Control for Military Artificial Intelligence?"
176. Johnson, "'Catalytic Nuclear War' in the Age of Artificial Intelligence & Autonomy."
177. The "AI control problem" refers to the issue that, under certain conditions, AI systems can learn in unexpected and counterintuitive ways, which engineers and operators may not have expected and which do not align with their goals. See Russell, *Human Compatible*, p. 251.
178. Heather Williams, "Asymmetric Arms Control and Strategic Stability: Scenarios for Limiting Hypersonic Glide Vehicles," *Journal of Strategic Studies* 42, no. 6 (2019): pp. 789–813.
179. Ali Wyne and Thomas Wright, "Great-Power Competition in the 21st Century," *Paragraph* February 29, 2020, https://www.pairagraph.com/dialogue/70dc26e845aa448cb9cc978e14c0b282/1.
180. Rythoven, "The Securitization Dilemma," p. 479.
181. On the security dilemma, empathy, and conflict de-escalation, see Joshua Baker, "The Empathic Foundations of Security Dilemma De-escalation," *Political Psychology* 40, no. 6 (2019): pp 1251–1266.

182. See Michael C. Horowitz and Paul Scharre, *AI and International Stability: Risks and Confidence-Building Measures* (Washington, DC: Center for a New American Security, 2021).
183. Early scholarship on regime type and escalation risk—notably the democratic peace literature—found that a dispute between democracies is less likely to escalate or reach higher hostility levels and that democracies are more likely to reciprocate in a dispute when the initiator is non-democratic. See Maoz Zeev and Nasrin Abdolali, "Regime Types and International Conflict, 1816–1976," *Journal of Conflict Resolution* 33, no. 1 (1989): pp. 3–35; and Brandon Prins, "Domestic Politics and Interstate Disputes," *International Interactions* 26, no. (2001): pp. 411–438.

5

Catalytic nuclear war

The new "Nth country problem" in the digital age?

Are existing notions of accidental nuclear war caused by third-party actors still relevant in the age of AI and autonomy? The main concern about "catalytic nuclear war"—and the "Nth country problem"[1] more broadly—in the late 1950s and early 1960s centered on the fear that a small or new nuclear power would deliberately set in motion a major exchange between the US and the Soviet Union.[2] As its name suggests, the concept was inspired by chemical reactions where the catalyzing agent (a third-party actor) would remain unscathed by its initiated process.[3] In his work on "catalytic nuclear war" in 1962, Donald Kobe defined the notion of "catalytic war" as the possibility that a third party's actions induce a nuclear war between two nuclear-armed opponents. A third-party attack on nuclear-power would be misattributed to another state, thus sparking a catalytic "chain reaction of retaliation and counter-retaliation."[4]

A broader definition of "catalytic nuclear war" encompasses situations in which the catalyzing actor (non-state or another third party) initiates a simulated attack (e.g., a false alert) that produces the effect of an imminent attack on one or both of the nuclear-armed adversaries, for which pre-emption is considered the most expedient (or only) strategic recourse.[5] This broadened definition, which is used in this chapter, includes any process by which a third-party actor uses its capabilities to intentionally involve nuclear-armed strategic rivals in a war, or to increase the scope or intensity of a situation.[6] The originality of this research lies in revisiting the "catalytic nuclear war" concept, considered by many as unworkable, and reconceptualizing it in light of technological change, as well as an improved understanding of human psychology and other factors.

Because nuclear powers must have some policy of retaliation (i.e., massive, proportional, finite, or random) to act as a deterrent, this chain reaction may, under certain circumstances, become a self-fulfilling prophecy. For instance, the common fear of pre-emption is caused by one or both states being

AI and the Bomb. James Johnson, Oxford University Press. © James Johnson (2023).
DOI: 10.1093/oso/9780192858184.003.0006

vulnerable to a first strike, or if a state's command and control (C2) and communication structures are compromised following an attack.[7] This situation would be compounded by the presence of higher numbers of nuclear-armed states—or nuclear multipolarity—with varied, and possibly conflicting, retaliation and deterrence postures.[8] Moreover, miscalculation, misperception, or an irrational act, would also increase the risk of catalytic war. At the root of these conditions is human psychology, which strains the credibility of arguments that human rationality, perceptions, and nuanced signaling ensure that deterrence will prevail.

Classical nuclear deterrence policies, premised on the assured threat of a retaliatory second strike (or mutually assured destruction), are a fundamentally psychological-political-technical phenomenon that has worked to reduce the chances of deliberate use of nuclear weapons.[9] This chapter argues that increasingly sophisticated, interdependent, and vulnerable nuclear command, control, and communications (NC3) systems—the bedrock of the mutual deterrence regime between nuclear-armed states—increase the risks associated with unintentional catalytic nuclear war.[10] The chapter is not a treatise on the motives or goals of non-state actors for committing such acts, but rather one on the evolving means at the disposal of actors to actualize their malevolent goals—or "apocalyptic world views"—in the digital age.[11] There is a rich body of literature on the intersection of emerging technology and the nuclear enterprise—and the potential for warfare more broadly defined.[12] However, there is a relative paucity of research on how existing concepts of escalation, nuclear terrorism, and classical deterrence theories might apply (or be tested) in the digital age, which is increasingly defined by developments in AI and autonomy, perfect information, and rational decision-making.[13] Are existing theories of military escalation still applicable in the era of AI and autonomy?

The remainder of the chapter is organized into four sections. The first contextualizes the concept of "catalytic nuclear war" with emerging security technologies, particularly AI. The chapter uses this analytical framework to examine how and under what circumstances the nefarious (mis)use of these technologies by non-state actors (i.e., terrorists, criminals, state-proxy actors) might create or exacerbate escalation pathways, leading to an accidental nuclear confrontation between nuclear-armed powers. Next, the chapter considers three aggravating features: information complexity and asymmetry; greater automation of NC3; and mis/disinformation and information manipulation (explored in Chapter 3). The chapter posits that AI-enhanced capabilities will likely amplify these underlying conditions (or second-order effects), thus increasing the risk of sparking unintentional escalation between strategic nuclear-armed powers—especially in crisis and conflict conditions. The next section uses four counterfactual scenarios—cyber "false-flag" operation;[14] cyber-attack on nuclear early-warning systems; drone swarming targeting intelligence surveillance reconnaissance (ISR) systems; and deepfake disinformation—to demonstrate how AI-enhanced capabilities in the hands of non-state actors might accidentally or inadvertently drag a competitive nuclear dyad into conflict. The final section considers possible ways of reducing catalytic escalation risk caused by non-state actors, particularly how militaries can maintain effective command and control of their nuclear forces in a rapidly evolving and complex conflict environment.

Catalytic nuclear war in the digital age: Evolving nuclear strategies of non-state actors

During the later 1950s and early 1960s, the notion of a "catalytic nuclear war" centered on the possibility that a minor nuclear power would deliberately trigger a full-scale nuclear exchange between the US and the Soviet Union. Donald Kobe argued that the notion of "catalytic war" was inspired by chemical reactions whereby the catalyzing agent remains unchanged by the process it has triggered.[15] Several scholars also raised the specter that a minor nuclear power might be tempted to catalyze a devastating exchange that would inflict maximum damage on the superpowers but leave it relatively unscathed and thus dictate conflict termination outcome.[16]

Some scholars have questioned the plausibility of this hypothesis. Even in the unlikely event that a minor power successfully concealed its role in sparking a global war, the prospect of its plot being discovered would almost certainly guarantee decapitation, making any potential gains outweighed by the overwhelming risks thus illusionary.[17] In the event, improvements to nuclear early-warning systems and communication channels—and broader tactic cooperation—between the US and the Soviets did much to reduce the (albeit remote) possibility of a minor power creating the conditions to inadvertently drag the superpowers into nuclear war.[18] What if the risk and rationality considerations mattered less to non-state actors (especially terrorists) in possession of a nuclear weapon to incite chaos and other disruptive means to fulfill its apocalyptic ends?[19]

The nuclear terrorism threat discourse began in the mid-1970s. The terrorist attacks on the US in 2001, coupled with the 2003 exposure of a secret network (led by Abdul Qadeer Khan) supplying Libya, North Korea, and Iran with the know-how for manufacturing critical nuclear equipment, increased the sense of urgency—in particular for the global nuclear security regimes and approaches concerning non-state actors (especially those motivated by extreme political or religious goals).[20] Experts consider the probability of the successful theft of nuclear grade materials (i.e., highly enriched uranium or plutonium) and detonation of atomic weapons or manufacturing of an improvised device (or a "dirty bomb") by non-state actors as extremely low; but it cannot be entirely discounted.[21] To be sure, experts have continued to warn about the opacity of materials' safety and security in smaller nuclear-armed states such as Iran, Israel, Pakistan, and, most notably, North Korea.[22]

If a non-state group realized that the possession of nuclear weapons could trigger a total catalytic war between two or more major nuclear-armed states, what possible situations might set these dynamics in motion? The potential success of a catalytic nuclear detonation by a terrorist group during a crisis between two nuclear-armed states (i.e., the original target of the nefarious attack and the victim of a subsequent inadvertent retaliatory strike), when tensions are running high, and a heightened state of alert is in place—at a nuclear or sub-nuclear level—would likely be amplified. That is, a broader nuclear exchange would likely occur. Moreover, a non-state actor might exploit a tense situation by making provocative false claims that its attack had been sanctioned by the state with which the victim of the attack was already in a crisis, or, worse still, nuclear brinkmanship.

Against the backdrop of fear, acrimony, and paranoia—when decision-makers are more susceptible to misperception and misunderstanding—any denials or assurances would fall on deaf ears.[23] If a repeat of the 1999 Kargil crisis or the more recent Indian-Pakistani stand-off in 2019 occurred, a non-state nuclear detonation (or the credible threat of one) could spark an accidental or inadvertent nuclear exchange.[24] Thus, any confidence Indian or Pakistani has in their ability to manage a conflict below the nuclear level would likely be unreliable in the event of a non-state nuclear attack. Absent unambiguous and incontrovertible evidence to the contrary, an unprovoked non-state attack may compound mutual suspicions about the other side's intentions and behavior, and, in turn, confirm worst-case scenario assessments.[25]

In the contemporary digitized landscape, the theft of nuclear grade materials, the detonation of atomic weapons, and "dirty bombs" by non-state actors are not the only threats that states must worry about. Emerging technologies—most notably cyber, AI technology, and drones—are rapidly creating new (and exacerbating old) low-cost and relatively easy means for non-state actors to fulfill their nefarious goals; without the need for actual physical contact with or manufacture of nuclear weapons to have the "power to hurt"—or the power *to get others to hurt each other*.[26] One of the critical potential "threat vectors" (i.e., a mechanism or means by which an actor can gain access to a network in order to deliver a malicious payload) is the severe consequences that would result from a non-state actor's deliberate actions to penetrate or manipulate NC3 systems with AI-enhanced conventional capabilities. Above all, cyber operations targeting early-warning satellites and radars is a central focus of this chapter.[27]

The presumption of immunity, or overconfidence in the robustness of NC3 networks to external (or insider) threats, may increase the chances that states underestimate nuclear weapon systems' vulnerabilities to a multitude of possible attack mechanisms, *inter alia*, data manipulation, malware cyber-attacks (or hardware vulnerabilities), "false-flag" cyber operations, social media flooding, or spoofing decision-makers with mis/disinformation. Minuteman missile silos are, for example, considered to be particularly vulnerable to cyber-attacks.[28] Nuclear-powered ballistic missile submarines (SSBNs) once believed to be air-gapped—not connected to the internet and considered "hack-proof"—are, however, connected via various electromagnetic signals that create potential vulnerabilities to cyber-attacks.

A committed non-state or third-party actor may, for example, disrupt (digital jamming), deny (denial of service attacks), and distort or destroy information (spoofing or malware attacks) used by SSBN's C2 networks to sow miscalculation and misperceptions, fan the waves of a crisis, or trigger an accidental nuclear launch.[29] *In extremis*, third-party hackers (or "cyber terrorists") might, during a crisis, use AI-augmented false-flag cyber operations to mislead submarine commanders (or political decision-makers) that an SSBN was—or expected to be in the case of a "launch-on-warning" policy—under-attack.[30] Echoing these concerns, a 2009 study commissioned by the International Commission on Nuclear Non-proliferation and Disarmament warned that non-state actors might penetrate C2 systems (and even launch an unauthorized nuclear attack)—an easier and more plausible "alternative for terrorist groups than building or acquiring a nuclear weapon or dirty bomb themselves."[31]

Recent developments in AI-enabling technology have exacerbated these vulnerabilities and introduced additional threat vectors non-state actors might leverage, which may precipitate a catalytic nuclear war that manipulates the information landscape in which decisions about atomic weapons occur—in particular, social media manipulation and the spreading of misinformation, false memes, and fake news. Fake news that goes viral becomes "real" because people believe that if has gone viral it must be true. The more inflammatory false news on Twitter is, for instance, the faster and broader appeal it has compared to genuine news—making it more far more profitable for social media platforms to propagate.[32] For example, in 2016, a false news story appeared on the AWD News site, claiming that Israel had threatened to attack Pakistan with nuclear weapons if Islamabad interfered in Syria. The report consequently caused a tit-for-tat incendiary rhetorical exchange on Twitter—the Pakistani defense minister Khawaja Muhammad Asif warned that Israel should remember that Pakistan is also a nuclear-armed state, debunked as fictitious by the Israeli Defense Ministry.[33] This incident puts a modern spin on the Cold War concept of catalytic nuclear war—in which third party actions provoke a nuclear war between the two nuclear-armed powers—and demonstrates the potentially severe damage caused by the misinformation and manipulation of information by non-state actors.

During a crisis involving a competitive strategic dyad—such as the conflict-prone South Asia or Northeast Asia—when communication is compromised, nuclear arsenals are on high alert, decision-making timeframes are compressed, or launch authority is pre-delegated (e.g., to nuclear-armed submarine commanders), the consequences could be catastrophic.[34] Moreover, in a high pressured crisis environment, with confusion and paranoia running high, the risk of misperceptions of an adversary's intentions and behavior (e.g., putting nuclear arsenals on high alert status, non-routine troop movement), and the temptation for pre-emptive action increases.[35] That is, the catalyzing non-state actor produces the effect of an imminent attack on one or both of two nuclear-armed states, for which pre-emption is considered the most advantageous strategy.[36] What factors might aggravate these escalation pathways during a crisis?

Pathways to catalytic nuclear war and catalyzing aggravators

This section examines three aggravating variables—information complexity; greater automation of NC3; and mis/disinformation and information manipulation—which coalesce in the synthesis between emerging technology and the nuclear weapon systems. How and why might AI and autonomy compound these dynamics and cause unintentional escalation between nuclear-armed powers to occur? Early-warning systems are the core focus of this study because "missiles cannot be recalled; submarine commanders may be out of touch but able to act on their own; missiles may go off accidentally."[37] These variables do not, however, constitute mutually exclusive risk scenarios. Rather, the interplay between these conditions might allow them to feed into one another with uncertain and potentially self-reinforcing effects—a function of the confusion and uncertainty created by the sociotechnical complexity generated in the digital age described in Chapter 3.

Cognitive psychology is relevant to all four aggravating factors described below. Humans exhibit a range of biases that can influence how they observe, collect, and process information, making them less aware of the reality of a particular situation and thus more inclined to interpret events through the lens of existing desires, preferences, and beliefs.[38] Psychological research has shown that people tend to interpret ambiguous information as consistent with their pre-existing beliefs and values—dismissing information that contradicts these views—and accept information that allows them to avoid unpleasant choices.[39] As briefly described in the opening chapter of the book, human psychology studies also tell us that people are predisposed to making fast, intuitive, reflexive, and heuristic judgments—known by cognitive psychologists as "System 1" thinking—rather than slower, more deliberative, conceptual, and analytical thinking—known as "System 2."[40] Fast and reflexive ("System 1") thinking during a nuclear crisis would be inherently problematic if, for example, the potentially catastrophic risks and second-order effects of nuclear war (i.e., loss of human life and environmental destruction) are actively or unconsciously deprioritized in discussions dominated by national defense heuristic imperatives.[41]

When information overload, complexity, and unfamiliar technologies are prominent decision-making features, cognitive biases (i.e., confirmation bias, anchoring, heuristic thinking, and confidence bias) can also be exacerbated. Moreover, the exponential rise and speed by which data emerges from today's information ecosystem creates novel attack vectors to manipulate and propagate mis/disinformation during crisis times, as seen in Chapter 3. In the case of "anchoring," perceptions of historical lessons from past conflict and crises with minimal relevance may have an outsized influence on decision-makers who seek to ground decisions in precedent and lessons from experience.[42] Humans are considered poorly equipped to intuitively understand probability, which is essential for ranking preferences rationally for crisis decision-making.[43] People also tend to misinterpret (or not recognize) randomness and non-linearity and consider the occurrence of unlikely events as virtually impossible.[44]

Three distinct—but not always separate—mechanisms can lead to nuclear escalation, namely, *deliberate* (or intentional), *inadvertent*, and *accidental* escalation (encompassing mistaken or unauthorized usage).[45] These distinctions are not, however, binary, or mutually exclusive. An escalation mechanism that leads from a crisis or conflict to its outcome can involve one or more of these categories: for example, if an accidental or inadvertent escalation signal or event is triggered by a non-state actor's nefarious actions—such as a false-flag cyber operation against a state's NC3 systems—which in turn leads to a deliberate escalatory response.[46] Moreover, the deliberate use of nuclear weapons that originates from a false, manipulated, or distorted assessment of a situation, or in response to an early-warning system false alarm, can quickly muddy the lines of intentionality.[47]

Accidental nuclear war—a nuclear confrontation without a deliberate and properly informed decision to use nuclear weapons on the part of the nuclear-armed state(s) involved—could be caused by a variety of accidents, most often encompassing a combination of human (and human-machine interaction failure) system errors, and procedural or organizational factors.[48] This definition of accidental escalation will be used in this chapter to conceptualize "catalytic nuclear war"—triggered by the deliberate actions of a non-state or third-party actor that is not part of the adversarial nuclear dyad—with developments in AI and autonomy. Therefore, a crucial distinction is between the risk of unintentional (i.e., accidental or inadvertent) escalation and intentional escalation; the fear of deliberate escalation is generally more destabilizing.[49] Despite paying lip-service to Machiavelli's Fortuna—the role of uncertainty in international affairs—decision-makers often underestimate the importance and frequency of accidents and randomness in these interactions.[50] Ultimately, whether the impact of unintended escalation risk is stabilizing or destabilizing depends on the destabilizing force's relative strengths and the fear it instills.[51] Similar to historical cases, human-machine interactions that caused (or compounded) accidents with complex autonomous weapon systems, AI-enhanced systems operating at higher speeds, sophistication, and compressed decision-making timeframes will likely reduce the scope for de-escalating and contribute to future accidents.[52]

Information complexity

In contrast to previous eras of military revolution—when technologies diffused and improved specific capabilities and domains of military power—the coalescence of emerging technologies today (especially AI-ML) can enhance and enable a diverse range of capabilities and missions.[53] The complex interactions within the tightly enmeshed (or dual-use) and commingled systems that control or inform nuclear weapon systems (e.g., early-warning satellites, ISR, electronic data networks, missile defenses), has become a critical risk vector in the digital age. Thus making accidents, short of a nuclear detonation, almost par for the course.[54] Further, the risks of technical and human errors that arise from this complexity and interdependency in modern nuclear systems are compounded by the prospect of cyber-attacks against early-warning and C2 systems (discussed below).[55]

A key characteristic of operating NC3 systems infused with advanced technology (AI-ML, big-data analytics, and cyber) is the vast quantities of data and information collected to inform decision-making.[56] While AI-machine learning (ML) data-mining augmented cyber-surveillance—together with AI-enabled pattern recognition technology—could significantly enhance states' ISR capabilities, the introduction of multiple data streams with varying levels of confidence could overwhelm the ability of decision-makers to determine the credibility of data, especially if the provenance and validity of the information cannot be easily verified.[57] The US Air Force, for example, has characterized this phenomenon as four "Vs"—higher *volume* (collection of magnitudes more data points), *velocity* (the volume of data is acquired at rapid speed), *variety* (numerous formats of information from diverse sources), and *veracity* (the volume, velocity, and variety of data includes a substantial amount of noise and redundant data).[58] Similarly, the US Navy has reported being overwhelmed by the floods of data generated from its existing information-gathering systems.[59]

The quantity and quality of information generated by the advanced technology that supports NC3 systems can increase escalation risks, *inter alia*, in three ways. First, human decision-makers' dependence on the information produced by complex and enmeshed NC3 systems can exacerbate the degradation of the quality and reliability of decision-making if these systems are compromised, especially through cyber-attacks. Scholarship on the risks of a breakdown in NC3 systems and the limitations of human cognitive processing capacity—particularly situations that combine information overload, confusion, and compressed decision-making timeframes—focuses on the accidental escalation risks associated with the organizational reliability of complex and enmeshed systems.[60] Scholars in the "normal accident theory" school argue that despite efforts to enhance the safety and reliability of nuclear weapons, the risk of accidents as a result of complexity and tightly coupled operations cannot be eliminated or even effectively managed.[61]

Second, while information dominance does not ensure stability, juxtaposed to this, information inadequacy (or "information asymmetry") can prompt decision-makers during a crisis to eschew traditional caution and the acute fear of escalation for pre-emption, thereby increasing inadvertent escalation risks.[62] For instance, the information asymmetries by clandestine cyber operations against NC3 systems can narrow or close the crisis bargaining window, producing a trade-off between possessing counterforce capabilities to blunt an adversary's nuclear deterrent and increasing the risks of provoking nuclear warfare in the first place. In short, any new capability that increases (or is perceived to increase) information asymmetry about the balance of power and resolve can sow the seed of bargaining and deterrence failure, and ultimately war.[63] Misperception of another's capabilities, intentions, and assumptions about what information an adversary possesses (or lacks), and, in turn, how it perceives a situation, is an established cause of deterrence failure in international relations.[64]

The Cold War experience demonstrated that the fear an action or signal misinterpreted by an adversary—in the context of uncertainty and incomplete information associated with modern warfare—might spark nuclear pre-emption is a useful point of departure to consider the potential effects of AI and autonomy.[65] Moreover, the increasing use and development of advanced weapon technology, such as hypersonic glide vehicles, dual-use (nuclear and conventional) missile delivery systems, anti-satellite weapons, and dual-use NC3 systems that blur the distinction between nuclear and conventional (and sub-conventional) warfare can heighten strategic ambiguity during crises—thereby creating first-mover advantage incentives, leading states to overestimate an adversary's capabilities and to strike pre-emptively.[66] In sum, the likelihood of escalatory crises originating in cyberspace will continue to rise because of the inherent speed, scope, and opacity surrounding cyber capabilities, coupled with the existential fear of escalatory conflict in cyberspace, the positive feedback loops associated with states' accumulation of offensive cyber capabilities, and the insensitivity of others' perceptions and intentions.[67]

A cyber-attack—dependent on deception, speed, and covertness—that undermines effective information and communication flow can, therefore, increase the rational incentives for escalation.[68] Besides, irregular or opaque communication flow between adversaries may also increase the risk of misperception and miscalculation and assume the worst of others' intentions.[69] Paradoxically, therefore, new technologies designed to enhance information (e.g., modern NC3 systems supported by 5G networks, AI-ML, cyber, big-data analytics, quantum computing) can also erode reliable information flow and communication, which is critical for effective deterrence.[70] Greater volumes of information emerging from the global information ecosystem, rather than reducing the likelihood of misperceptions arising between adversaries, offer motivated third-party actors unprecedented (low-cost, scalable, and fast) opportunities to manipulate decision-makers' perceptions and shape public opinion about others' intentions, capabilities, and the consequences of responding to an attack that originated in the cyber domain—but which might rapidly and inadvertently escalate.

Third, advanced technologies have the potential to qualitatively improve the "always-never" criteria that NC3 systems must meet (i.e., complex failure modes, greater redundancies, and a larger attack surface).[71] The complexity and uncertainties introduced by sophisticated NC3 mechanisms—especially early-warning satellites and radars—can also cause errors, unexpected interactions, and unintended consequences, thus eroding deterrence and creating rational (or sub-rational) incentives to escalate a situation.[72] The notion that sophisticated technologies, designed to improve safety, ultimately become a source of accidents and errors is a documented phenomenon.[73]

This trade-off is a product of the organizational and strategic-cultural variables in the human decision-making process, creating pressures to escalate a situation, rather than viewing technology as an independent variable per se.[74] The assumption of rationality in a nuclear-armed adversary, or that the adversary has more information than might be the case (or "asymmetric information"), *inter alia*, may stem from misperceptions about the nature of an opponent's technological capabilities, the nature of a crisis, or an opponent's intentions. As a result, these factors could exacerbate escalation pressures on states with less robust NC3 systems and safeguards or view its survival as being at stake—North Korea, Pakistan, or India. Furthermore, the normal (or peacetime) perception of events can shift during crises or geopolitical tension, when decision-makers are more prone to harbor worst-case scenario expectations and see things, they *expect or want to see*—known as "cognitive consistency."[75] This logic—the way people interpret situations under pressure due to anxiety—can also become a self-reinforcing process that directly impacts the number of false alarms that occur in early-warning systems. In particular, the probability of Type II errors—or "false negatives."[76] Moreover, as the volume of information received by NC3 systems increases during a period of tension, so do the chances of perceiving a pattern of adversarial behavior where, under typical circumstances, no pattern would be detected.

Technological-deterministic schools of thought that conceptualize technology as the *sole* driver of strategic stability and the conduct of warfare therefore oversimplify this complicated human endeavor—which can mean that rational decisions entail worse consequences than less rational alternatives. For example, if a new state-of-the-art NC3 capability provides a first-mover advantage (or perception of one) during a crisis for one or both sides, thus striking preemptively or causing irreparable damage, the risks of inadvertent nuclear war increase. Besides, problems have often arisen because states have tended to place "new weapons in old bottles." That is, regardless of how sophisticated or capable a system appears to be, these technologies will invariably be assimilated into military organizations with entrenched norms, cultures, and institutional patterns.[77] A complete understanding of the relationship between advanced technology and strategic stability requires a deeper understanding of human tendencies, not merely technical capabilities per se.

Greater automation of NC3 systems

Increasing automation levels in modern NC3 systems—particularly those augmented and supported by AI technology—together with false warnings and other forms of nefarious interference in cyberspace, will likely increase the risk of accidental nuclear war.[78] Moreover, a declared or de facto launch-on-warning posture held by either side during a crisis involving a Type II error (or "false positive") caused, for example, by a false signal or false flag operation, can trigger "use-them-or-lose-them" pressures and increase accidental escalation risk.[79] Thus, it degrades nuclear forces' usability or reduces the damage limitation window to attack enemy nuclear forces if needed.[80] This skewed assessment in the context of nuclear weapons ready to launch at a moment's notice could precipitate worst-case scenario thinking and trigger inadvertent escalation.[81]

The historical record demonstrates the vulnerabilities of NC3 systems to frequent "false alarms" (military exercises and non-military events that look like preparations for a nuclear attack and human/technical errors and glitches),[82] accidents, "close calls," and other risks associated with increasingly complex, porous, and intertwined systems, which, despite their alleged "closed" nature, may offer non-state actors multiple pathways to cause harm.[83] Besides, intertwined NC3 capabilities (to support conventional and nuclear weapons) would make it even more difficult for adversaries to distinguish nuclear from non-nuclear armed missiles.[84] To fully understand and predict vulnerability of NC3 systems is, for a variety of technical, classification, and operational reasons, almost impossible—that is, only a very few individuals with sufficient security clearance can fully comprehend the structure and composition, let alone identify NC3 vulnerabilities in aggregate.[85] Because of the inherent problem of attribution in cyberspace, a non-state actor might plausibly make *both states* the target of its attack, convincing each side that the other party is responsible—or a "double-sided catalytic attack."[86] This problem-set is compounded by the shortened response timeframes available to decision-makers, particularly where a launch-on-warning posture is present.

According to open sources, operators at the North American Aerospace Defense Command (NORAD) have less than three minutes to assess and confirm initial indications from early-warning systems of an incoming attack and decide whether or not to authorize a nuclear response.[87] This compressed decision-making timeframe—for data collection, assessment, and decision-making—could put political leaders under intense pressure to decide to escalate during a crisis, with incomplete (and possibly false) information about a situation. In the context of advanced technologies, response systems like NORAD raise concerns: missiles cannot be recalled, submarines on deterrence patrols may be out of touch for extended times with a high degree of C2 autonomy, and nuclear weapons may be launched accidentally.[88]

These pressures would likely be more intense in situations where nuclear powers are in close proximity to one another (e.g., India and Pakistan, or China and North Korea), or where one side's NC3 systems are asymmetrically more vulnerable or less sophisticated, for example, North Korea, India, or Pakistan.[89] Further, the increasingly interdependent and commingled (or entangled) nature of states' conventional and nuclear C2 systems might exacerbate the incentives to escalate a situation to a nuclear level once a conventional conflict begins.[90] During the open stages of a conventional conflict, for example, State A targets State B's conventional C2 systems to degrade B's *conventional* capacity. Because B's conventional and nuclear systems operate on the same technical infrastructure, A's attack indirectly undermines B's nuclear NC3, and in turn is viewed by A as a deliberately pre-emptive counterforce strike.[91]

The decision to automate nuclear assets may also be influenced by the political stability and the threat perceptions of a nuclear-armed state. A regime that fears either a domestic-political challenge to its rule or foreign interference may elect to automate its nuclear forces, ensuring only a small number of agents are involved in the nuclear enterprise.[92] For example, China maintains strict controls on its nuclear C2 structures (i.e., separating nuclear warhead and delivery systems). Open-source evidence does not suggest Beijing has pre-delegated launch authority down the chain of command if a first strike decapitates the leadership. In short, to retain centralized C2 structures and strict supervision over the use of nuclear weapons, AI-enabled automation might become an increasingly amenable and intoxicating option to authoritarian regimes such as China.

In the context of AI and autonomy, particularly information complexity, misinformation, and manipulation, rationality-based deterrence logic appears an increasingly untenable proposition. In sum, the combination of increased speed, compressed decision-making timeframes, complex and interdependent C2 mechanisms, and increasing levels of AI-enabled NC3 automation and human-machine interactions increases the possibility of sparking a cascading set of errors (human or machine), reinforcing feedback-loops that could ultimately lead to an accidental nuclear exchange.

Furthermore, because there is so much redundancy built into modern NC3 systems, scenarios involving accidental nuclear war as a result of a single technical error or malfunction is less threatening, compared to the simultaneous coalescence of multiple failures in a rapid and unpredictable and reverberating fashion—where malfunctions in once system are prone to cause alarms in others.[93] In particular, if they are compounded by organizational failures, ambiguous information, misperceptions, or excessive trust in technology, this may lead to human operators cognitively offloading judgment to AI algorithms without fully understanding its limitations—also known as "automation bias" (examined in Chapter 1).[94] Such automation bias—especially in human-machine interactions (as discussed in Chapter 2)—could also mean that *both* false negatives and false positives go unnoticed (or are discarded) because the operators are overconfident and complacent in the capabilities of systems augmented with advanced technology such as AI.[95]

In a recent study of the effects of how people respond to the request of robots exhibiting faulty behavior, researchers found that because of automation bias and overtrust,[96] while the robot's erratic behavior "affected its perceived reliability and trustworthiness this had *no impact on participants' willingness to comply with its instructions*, even in the case of unusual requests," including in situations where emergency signals that ran contrary to these requests were present (emphasis added).[97] This "overtrust-automation bias" problem-set is compounded during a crisis when stress, fatigue, information overload, and commingled (nuclear and conventional) systems encounter a priori situations between asymmetric nuclear rivals, thickening the "fog of war" and resulting in irrevocable actions, when use-or-lose-them becomes the only option.[98]

Manipulation of the strategic information landscape

One rapidly developing and increasingly prominent field of AI-augmented technology that can complement and force multiply existing malicious social manipulation behavior and generate campaigns of manipulation—most notably the spread of mis/disinformation—is the ability to generate audio and video images that fabricate events, create fictitious situations, and propagate falsehoods.[99] Experts expect that it will be possible for non-state actors (or amateurs more generally) to generate photorealistic high definition video, audio, and document forgeries using ML tools—in particular generative adversarial networks (GANs)—at a low-cost and on a large scale.[100] In combination with other tools,[101] GANs have already demonstrated that producing realistic images is possible and will likely improve significantly in the next few years.[102] Examples of disinformation and manipulation, *inter alia*, include distributed audio of private statements and conversations between political or military leaders (both domestically or with allies and adversaries); video of crises or conflict (see below) designed to incite public outrage and prompt pressure for retribution; or audio or video material calibrated to deflect, deceive, or otherwise distract attention from an actor's culpability of an aggressor by producing false alternative versions of events.[103]

As AI technology advances, the quality, cost, and availability of GANs and other tools, especially AI-enhanced audio software, will make it increasingly difficult to discern what is real from what is not, eroding public trust in hitherto trust-worthy information sources.[104] In 2014, for example, thousands of residents at St. Mary Parish in Louisiana received a fake text message alert via a bogus Twitter account warning of a "toxic fume hazard" in the area. Further fanning the flames, a fake YouTube video was also posted showing a masked ISIS fighter standing next to looping footage of an explosion.[105] Thus, it is not difficult to imagine how these AI-enhanced technologies in the hands of non-state actors with nefarious goals (or "apocalyptic" world views) might have dangerous consequences for nuclear security and strategic stability.

Deliberate malevolent information manipulation by non-state actors (terrorists, criminals, or state proxies) could destabilize implications on effective deterrence and military planning, both during peace and war. GANs-generated deepfakes might also exacerbate the escalation risks by manipulating the digital information landscape, where decisions about nuclear weapons are made. It is easy to imagine unprovoked escalation caused by a malicious third-party (or state-proxy) clandestine false-flag operation in the competitive strategic environment.[106] During a crisis, a state's inability to determine an attacker's intent may lead an actor to conclude that an attack—threatened or actual—was intended to undermine its nuclear deterrent.[107]

AI systems will likely come under inexorable stress from nefarious attacks using counter-AI techniques (e.g., data pollution, spoofing, false alarms, or tricking a system to reverse engineer algorithms), which might undermine the confidence in a network, creating new vulnerabilities, errors, and unintentional escalation risks. In the emerging deepfakes arms-race—much like cybersecurity more broadly—detection software will likely lag behind advances in offensive enabling solutions—or offense-dominant ones.[108] According to computer science expert Hany Farid, there are probably 100 to 1,000 times "more people developing the technology to manipulate content than there is to detect [it]."[109] Efforts to counter these technologies and regain the upper hand in the war on digital fakery have been underway for several years (e.g., automated anti-virus software, encryption, and other tools to call-out manipulated images and videos). Today, more sophisticated AI-enhanced techniques, coupled with research that suggests pre-existing cognitive schemes, beliefs, and attitudes rather than credulity or gullibility determines whether the public believes fakery is real or not—meaning fake images and videos can be accepted even though they can be easily debunked.[110]

AI-enhanced fake news, deepfakes, bots, and other malevolent social media campaigns could also influence public opinion—creating false narratives or amplifying false alarms—with destabilizing effects on a mass scale, especially in times of geopolitical tension and internal strife.[111] In 2017, for example, a deepfake video was circulated on Russian social media alleging a US B-52 bomber had accidentally dropped a "dummy nuclear bomb" on a Lithuanian building.[112] A state or non-state actor could, for instance, generate an image or recording of a military commander obtained from open-sources to generate and disseminate a deepfake containing false orders, intelligence, or geospatial imagery that, at best, generates confusion and, in the worst-case scenario, aggravates a tense situation or crisis between rival nuclear powers. This idea is examined below. For example, in the world's first reported AI-enabled theft in 2019, non-state actors used AI voice mimicking software to generate a fake recording of a British energy executive.[113] In short, AI-augmented technology is rapidly becoming another capability in non-state actors' toolkit to wage campaigns of disinformation and deception—one that both sides may have used against them.[114]

The democratization of ever more sophisticated technology will likely amplify documented human pathologies,[115] explaining people's attraction to novel and negative information, memes, falsehoods, and digital echo chambers and filter bubbles (or the "information cascade" phenomenon), which deepfakes are adapt to perpetuate (see Chapter 3).[116] In the case of cognitive availability heuristics, people tend to skew their judgments to more recent (or memorable) information, situations, or experiences, thus making new or novel opinions biased toward those that can be more easily recalled.[117] During times of high-pressure crisis and conflict, decision-makers tend to interpret unusual circumstances as threatening, even if an adversary's behavior has not changed. Thus, routine activities (e.g., troop movements) scrutinized in the context of an early-warning alert can be considered as more menacing than they might otherwise be.[118]

In 2017, for example, South Korean counterintelligence officials received fake mobile and social media alerts with orders for US military and DoD personnel to evacuate the Korean Peninsula.[119] Information attacks such as this suggest that non-state actors, state proxy actors—and perhaps state actors—will inevitably attempt to use social media as a tool of war to provoke nuclear confrontation for political-ideological, religious, or other malevolent goals, and with increasing levels of sophistication, stratagem, and AI-enhanced subterfuge.[120] AI might also enable non-state actors to automate, accelerate, and scale synthetic social media accounts and content to support malevolent disinformation operations.[121]

These potentially catalytic dynamics will likely be compounded by human cognitive bias. A motivated non-state actor would be well-positioned to use AI-augmented tools (e.g., "fake news" or deepfake generated propaganda) to exploit this psychological weakness, ensuring control of the dissemination of false narratives and opinions. Threat assessments of catalytic escalation must, therefore, include both the likelihood that a third-party can execute a particular attack and the probability that the falsehoods or fakery that transpire leads to escalation.[122] Strategic decisions made by isolated (and especially pariah) nuclear-armed authoritarian regimes under these circumstances—particularly those that believe their survival is threatened—could trigger dangerous accidental nuclear escalation dynamics. As the Cold War historical record attests, however, when it comes to commanding nuclear weapons responsibility, both democratic and authoritarian regimes have made dangerous decisions that they were unable to fully control.[123] The fictional scenarios that follow highlight the heightened risk of accidental nuclear use in the digital age.

Counterfactual scenarios

The following section includes four counterfactual scenarios to illustrate how AI-augmented capabilities in the hands of non-state actors might accidentally or inadvertently drag a competitive nuclear dyad into conflict: cyber false-flag operation; cyber-attack on nuclear early-warning systems;

drone swarming targeting ISR systems; and deepfake disinformation. These counterfactual scenarios are premised on three assumptions: (1) the involvement of a nuclear-armed adversarial dyad adversary during a crisis or period of tension; (2) the existence of information asymmetry; and (3) the technical feasibility of the operations described—either currently in existence or considered technically viable shortly. Why these capabilities and not others? And, how realistic are these counterfactuals?

These capabilities are of particular interest because they play an increasing role in how non-state actors leverage the asymmetric and low-cost benefits of AI technologies to realize their malevolent objectives. Although multiple pathways may be activated during an actual conflict or crisis—either simultaneously or sequentially—assessing each of these potential escalatory pathways offers insight into the interplay of strategic AI technologies and stability risks. In the real world, the aggravating factors and events described above would naturally interact with each other in varying degrees and intensities, depending on contextual, temporal, spatial, and other shifting variables connecting antecedents to consequences. Actors' relative capabilities vis-à-vis adversaries would, for instance, vary over time, thus producing a much larger universe of possible scenarios than those outlined here. Cognizant of the criticisms of counterfactuals in world politics (subjectivity, circular logic, non-falsifiable, "epistemological slumming" etc.), and the potential for randomly distributed small causes being amplified into more significant effects reminiscent of chaos theory, the scenarios highlight alternative hypothesized consequences caused by complexity and ambiguity.[124]

Scenario 1: Cyber false-flag operation

Party A (a non-state actor or state proxy third party) launches a false-flag cyber operation[125]—data manipulation, social media flooding with bots, a spoofing attack, or other forms of deception—against State B and C's dual-use early-warning satellites (providing both sides with warning of nuclear attack on their homelands) that is not traced to Party A and appears to both State B and C to come from the opposing side. No evidence exists to disprove these claims and both sides deny responsibility.[126] Convinced that the other side is responsible and about to be attacked by the opposing side of the nuclear-dyad, and without evidence of who is behind the attack, State B retaliates against State C in a pre-emptive attack that C views as unprovoked aggression—sparking a catalytic war. During a crisis between two states, leaders would likely be predisposed to assume the worst about the other's intentions, thus making them less likely to exercise rigorous due diligence to establish high confidence attributions (or false positives) due to a cyber-attack.[127] In short, while false-flag deception has long been a feature of strategic affairs, the introduction of AI technology makes these tactics both more likely to occur and more likely to be successful, while also raising concerns about the likelihood that states' automated responses to manipulation (or "gaming") might spark an accidental or misdirected escalation.

Scenario 2: Cyber-terrorism vs. early-warning systems

During a period of heightened tension or crisis between State A and State B, a third-party actor or terrorist floods social media outlets and open-source crowdsourcing platforms with false information (e.g., satellite imagery, 3D models, geospatial data) about the suspicious movement of State A's nuclear road-mobile transporter erector launchers.[128] Because of State B's inability to determine with confidence the veracity of this information, and with mounting public pressures to respond, State B escalates a situation on the falsehood (or false positive) that it is the target of an unprovoked attack.[129] Asymmetries between adversaries' NC3 systems and military capabilities would likely exacerbate the fictional scenario's escalation mechanisms.[130] While social media platforms have had some success in slowing down users' ability to orchestrate manipulative and dangerous campaigns, once these operations (e.g., deepfakes and bots) go viral, the ability to curtail them is extremely difficult for human operators or machines.[131] Taken together, increased sophistication and accessibility of deepfake technology, the inherently dual-use nature of AI, the problem of attribution in cyberspace, the increasingly complex and interdependent nature of NC3 systems, and a compressed timeframe for strategic decision-making associated with hyper-speed warfare, will continue to lower the threshold for false-flag operations.

Scenario 3: Drone swarming vs. ISR systems

An insurgent proxy of State A uses a swarm of autonomous micro-drones over an adversary's territory as part of a clandestine ISR mission.[132] Because this type of sophisticated technology is considered beyond the reach of a lone wolf or unsponsored terrorist group,[133] State B attributes it to its militarily superior adversary, and thus views this proxy as a precursor for an attack planned by State A. The use of autonomous swarms, especially during periods of tension or insensitive anti-access/area denial (A2/AD) environments, might generate incentives for pre-emption. Further, the sophistication of A2/AD capabilities such as drone swarms—and advanced non-nuclear strategic weapons including hypersonic weapons, cyber, and anti-ship ballistic missiles (ASBMs)—complicates Herman Kahn's notion of "ladders of escalation" to achieve a superior position vis-à-vis an adversary across the range of escalation rungs—or "escalation dominance."[134] Where on the escalation ladder is a swarm attack against the US's nuclear ISR? Would a Chinese cyber-attack targeting a US water supply facility be more escalatory than an ASBM attack on a US aircraft carrier in the West Pacific?[135]

Conversely, the discovery of information by State A (e.g., suspicious troop movement or missiles leaving their garrison) could prompt A to escalate a situation, motivated by the desire to capture the perceived strategic upper hand. Moreover, State A might consider the first-mover advantages of using low-cost and relatively dispensable drones as controllable at a conventional level, which would likely increase their appeal, at least in the opening stages of a conflict. In the event that State A's unarmed proxy drone swarm was attacked by B in a counterterrorist or counterinsurgent strike,[136] State A would be forced to either accept this vulnerable asset's loss or escalate a situation—through its surrogate proxy or more directly.[137]

Scenario 4: Deepfake disinformation

To incite conflict between two nuclear-armed rival states, State A hires proxy hackers to launch a GANs generated deepfake video, depicting senior military commanders of State B conspiring to launch a pre-emptive strike on State C.[138] This footage is then deliberately leaked into C's AI-augmented intelligence collection and analysis systems, provoking C to escalate the situation with strategic consequences.[139] State B, fearful of a decapitating strike and losing the first-mover's advantage, swiftly escalates the situation.[140] To be sure, these dynamics might also be set in train once a crisis has begun. For example, in the aftermath of a high-casualty terrorist attack between nuclear-armed adversaries (e.g., India and Pakistan), which triggers a period of heightened tension, a non-state or state proxy actor launches a propaganda campaign on social media, starting a spiral of escalation. The current underdeveloped state of "counter-AI" capabilities and other fail-safe mechanisms (e.g., circuit breakers) to de-escalate escalation in cyberspace will make the unprovoked and unintentional escalation dynamics depicted in this scenario very challenging to anticipate and thus mitigate.[141]

Conclusion

In the emerging digital era, the catalyzing chain of reaction and counter-retaliation dynamics set in motion by a third-party actor's deliberate action is fast becoming a more plausible scenario. The concept of catalytic nuclear war—considered by many as unlikely, given the low probability of a terrorist group gaining access to nuclear weapons—should be revisited considering recent technological change and improved understanding of human psychology. Specifically, the human propensity for making fast, intuitive, reflexive, and heuristic judgments (especially "System 1" thinking),[142] which is exacerbated when information overload and unfamiliar technologies are more prominent features of decision-making. The research found that emerging technologies—most notably cyber, AI technology, hypersonic technology, and drones—are rapidly creating new (and exacerbating old) low-cost and relatively easy means for state and nonstate actors to fulfill their nefarious goals. This is compounded by the exponential rise in data emerging from today's information ecosystem—and in the speed with which it does so—which will create novel threat pathways and escalation mechanisms to manipulate and propagate misinformation and disinformation during times of crisis.

The chapter argues that recent developments in AI-enabling technology have exacerbated these vulnerabilities and introduced additional tools for third-party actors to exploit. A key risk is that these technologies, by enabling the information landscape in which decisions about atomic weapons occur to be manipulated, precipitate a catalytic nuclear war—in particular social media manipulation and the spreading of misinformation, false memes, and fake news. The consequences could be catastrophic during a crisis involving two nuclear powers if communication systems are compromised, nuclear arsenals are on high alert, decision-making timeframes are compressed, or launch authority is pre-delegated (e.g., to nuclear-armed submarine commanders). Furthermore, in a pressured crisis environment, with confusion and paranoia running high, the risk of misperceptions of an adversary's intentions and behavior (e.g., putting nuclear arsenals on high alert status or nonroutine troop movement) increases, along with the temptations for pre-emptive action. That is, the catalyzing action of a state or nonstate actor could produce the effect of an imminent attack on one or both of two nuclear-armed states, for which pre-emption is considered the most profitable strategy.

The chapter identified three features of the digital age that might aggravate the risk of catalytic nuclear war and make nuclear crisis management more difficult than in the past: information complexity; greater automation of

NC3 systems; and information manipulation disinformation. These variables are not, however, mutually exclusive. Instead, the interplay between these conditions might allow them to feed into one another with uncertain and potentially self-reinforcing effects.[143] In short, these conditions are a function of the confusion and uncertainty created by the sociotechnical complexity generated in the digital age.

How might militaries maintain effective command and control of nuclear forces in during times of uncertainty and complexity? While no amount or combination of controls, procedures, or technical enhancements can eliminate the possibility of catalytic nuclear escalation—and accidental escalation more broadly—specific measures that focus on reducing the possibility of accidental nuclear war may help to reduce some of the risks highlighted in the scenarios in this chapter, especially human and technical errors that occur in cyberspace.[144] Because accidental and inadvertent escalation dynamics are not the result of decision-makers electing to escalate, they cannot be deterred in the traditional sense.[145] Deterrence by denial measures by decision-makers, in advance of a non-state actor's attack, might, however, contribute to escalation management.[146]

Specific risk reduction measures can be broadly categorized into three clusters.[147] First, enhancing the safety of nuclear weapons and hardening NC3 systems and processes (e.g., safeguards, risk analysis to strengthen NC3 systems, and building more resilience into permissive action links (PALs)).[148] NC3 infrastructure security could be strengthened further through more robust cyber defenses (e.g., deepfake detection software), identifying false information and ameliorating the impact of misinformation and disinformation.[149]

Second, improving existing NC3 control protocols, oversight, and mechanisms (e.g., adding redundancies; enhancing procedures and launch protocols; authentication codes; back-ups into ISR operations and data-gathering).[150] This cluster could also include establishing a normative commitment to always keep a "human in the loop" in nuclear command and control. Affirming a commitment by nuclear-armed states not to outsource nuclear targeting and launch decisions to machines, thus ensuring the ultimate ethical and moral burden of using nuclear weapons remains with accountable human decision-makers.[151]

Third, enhancing safeguards to contain the consequences of errors and accidents when they occur (e.g., personal training; independent sources to verify and collaborate threat and crisis assessments; collective monitoring of events; and bilateral and multilateral data exchanges in the aftermath of accidents).[152] Social media could be used, for example, as independent sources to verify and collaborate government threat and crisis assessments, as well as to rapidly distribute and disseminate accurate and time-sensitive information during a crisis or conflict.

Broader measures that may also reduce catalytic nuclear escalation risks—and strategic stability and nuclear risk more generally—include, *inter alia*: exploring expanded intelligence sharing arrangements; taking arsenals off high alert status (or launch-on-warning); separating warheads from delivery systems (or de-matting warheads); shifting to a deterrent-only (or minimum deterrence) force posture; and adopting a no-first-use declaratory policy—as China and India do today.[153] Unlikely as it may be that these agreements could (technically or politically) be verified for formal compliance purposes, a normative framework or understanding would be worthwhile exploring nonetheless.[154] Strategic competition between great powers, to reap the perceived first-mover advantages of AI and trapped in the logic of an intense security dilemma described in Chapter 4, will likely become a negative-sum enterprise without resolving the control and safety issues.[155]

Several experts have suggested that integrating AI-ML and autonomy into NC3 systems by improving situational awareness, reducing information overload, and enhancing the speed and scope of intelligence collection and processing, may strengthen nuclear safety.[156] Others contend that the introduction of AI, and other emerging technology, could create new vulnerabilities and sources of errors, which non-state actors will inevitably seek to exploit.[157] These concerns have prompted some to propose that nuclear states commit to retaining humans' role in decisions relating to NC3 (especially early-warning systems) and the nuclear enterprise more broadly.[158]

In a similar preventative approach, some have called for states to prohibit using cyber capabilities against NC3 systems—a commitment that the Permanent Five (P5) states could reiterate during the Nuclear Non-Proliferation Treaty Review Conference. This is one way that P5 states could achieve the stated goals of creating dialogue among states possessing nuclear weapons to increase understanding, alleviate the concerns of non-nuclear states, and reduce the risk of misinterpretation and miscalculation.[159] This would do little to preclude non-state (or state-sponsored) cyber interference, however.[160] In short, given the complex interactions and interdependencies of modern NC3 systems, technical solutions may well create new risks, vulnerabilities, and uncertainties that compound the existing risks of catalytic escalation.

The extent to which these measures decrease the risk of catalytic war will ultimately depend on how the various actors (state and non-state) use the cacophony of AI-augmented technology to achieve their eclectic—and often conflicting—goals. These factors may have confounding effects on AI and autonomy. For instance, some actors might use AI technology to decrease risk, even as others—especially non-state and third-party actors—take deliberate steps that increase escalation risks. In short, the effects of catalytic escalation risk that emerge from the use of these technologies will depend on the relative strength of the destabilizing factors involved, how both sides perceive these factors, and the fear it instills in others.

As decision-makers ruminate over the ways and means by which the actions of state and nonstate actors might spark a catalytic nuclear exchange, they will need to consider how they would respond to such an attack and how an appropriate response might become a self-fulfilling prophecy—a chain reaction of retaliation and counter-retaliation—that leads unintentionally to nuclear war. Towards this end, security studies scholars, diplomatic historians, policymakers, and the defense communities should be encouraged to debate the value of counterfactual thinking to consider nuclear risk in the light of technological change. For example, revisiting nuclear near misses and accidents to consider the role of luck, cognitive bias, and overconfidence in the ability to control nuclear weapons.[161] How does viewing the AI-nuclear strategic nexus through the lens of classical nuclear theorizing help us manage these emerging risks? And what kinds of risk reduction solutions offer the best chance of success? The final chapter will consider these questions.

Notes

1. The "Nth Country Problem" refers to the Cold War-era diffusion of nuclear weapons to additional states—through indigenous development programs, assistance from the present nuclear powers, or a combination of both—which became a key area of concern in the US, and in particular the implications of this proliferation for US disarmament and nuclear assistance programs. See Fred C. Ikle et al., *The Diffusion of Nuclear Weapons to Additional Countries: The "Nth Country" Problem* (Santa Monica, CA: RAND Corporation, 1960).
2. Sections of this chapter are derived in part from James Johnson, "'Catalytic Nuclear War' in the Age of Artificial Intelligence & Autonomy: Emerging Military Technology and Escalation Risk between Nuclear-Armed States," *Journal of Strategic Studies* (2021). Copyright Taylor & Francis, available online: doi: 10.1080/01402390.2020.1867541.
3. Several scholars have argued that catalytic war is unlikely because the catalyzing actor might be destroyed in the resulting war. If the catalyzing country had previously determined that the war would not damage it, it might view the utility of a catalytic war as advantageous. For example, see Henry A. Kissinger, *The Necessity for Choice* (New York: Harper & Bros., 1960), and Herman Kahn, "The Arms Race and Some of Its Hazards," *Dxdalus* 89 (1960): pp. 744–780. On "catalytic nuclear war" theorizing see also Richard S. Leghorn et al., "The Problem of Accidental War," *Bulletin of the Atomic Scientists* 14

(1958): pp. 205–209. Arthur L. Burns, *The Rationale of Catalytic War* (Princeton, NJ: Center of International Studies, Princeton University, Research Monograph no. 3, 1959).

4. Donald H. Kobe, "A Theory of Catalytic War," *The Journal of Conflict Resolution* 6, no. 2 (1962): pp. 125–142.
5. Fred Charles Ikle, "Nth Countries and Disarmament," *Bulletin of the Atomic Scientists* 16 (1960): pp. 391–394.
6. Kahn, "The Arms Race and Some of Its Hazards."
7. Ibid., p. 129.
8. Richard A. Brody, "Deterrence Strategies: An Annotated Bibliography," *The Journal of Conflict Resolution* 4 (1960): pp. 443–457.
9. Optimal arsenal levels and force structure of nuclear weapons would balance the "positive" role of nuclear weapons (i.e., mutual deterrence) and the "negative" one (i.e., accidental nuclear war), allowing for a stabilizing deterrence regime alongside restructured nuclear forces and their attendant support systems to reduce the chances of accidental escalation. Michael D. Intriligator and Dagobert L. Brito, "Minimizing the Risks of Accidental Nuclear War: An Agenda for Action," in *Inadvertent Nuclear War*, edited by Hakan Wilberg, Ib Damgaard Petersen, and Paul Smoker (New York: Pergammon Press, 1993), p. 228.
10. It is beyond the scope of this chapter to examine all the technical vulnerabilities of NC3 systems or the potential implications of possible countermeasures. See Amy Woolf, *Defense Primer: Command and Control of Nuclear Forces* (Washington, DC: Congressional Research Service, December 11, 2018); and Page O. Stoutland and Samantha Pitts-Kiefer, "Nuclear Weapons in the New Cyber Age: Report off the Cyber-Nuclear Weapons Study Group," (Washington, DC: Nuclear Threat Initiative, September 2018).
11. On the nature and drivers of non-state actors' (especially terrorist groups') interest in nuclear weapons, see Mark Fitzpatrick, *The World After: Proliferation, Deterrence and Disarmament if the Nuclear Taboo is Broken* (Paris: Ifri Security Studies Centre, 2009); Charles C. Ferguson and William C. Potter, *The Four Faces of Nuclear Terrorism* (Monterey, CA: Center for Nonproliferation Studies and Nuclear Threat Initiative, 2004); James Forest, "Framework for Analyzing the Future Threat of WMD Terrorism," *Journal of Strategic Security* 5, no. 4 (2012): pp. 51–68; and Graham Allison, *Nuclear Terrorism: The Ultimate Preventable Catastrophe* (New York: Owl Books, 2004).
12. For scholarship on the intersection of technology with the potential for war, see James D. Fearon, "Cooperation, Conflict, and the Costs of Anarchy," *International Organization* 72, no. 3 (2018): pp. 523–559; for escalation dynamics, see Caitlin Talmadge, "Emerging Technology and Intra-War Escalation Risks: Evidence from the Cold War, Implications for Today," *Journal of Strategic Studies* 42, no. 6 (2019): pp. 864–887; for arms racing and strategic stability, see Todd S. Sechser, Neil Narang, and Caitlin Talmadge, "Emerging Technologies and Strategic Stability in Peacetime, Crisis, and War," *Journal of Strategic Studies* 42, no. 6 (2019): pp. 727–735; and for deterrence and the offense-defense balance, see Ben Garfinkel and Allan Dafoe, "How Does the Offense-Defense Balance Scale?" *Journal of Strategic Studies* 42, no. 6 (2019): pp. 736–763; and Joseph S. Nye, "Deterrence and Dissuasion in Cyberspace," *International Security* 41, no. 3 (2017): pp. 44–71.
13. Notable exceptions include Herbert Lin, "Escalation Dynamics and Conflict Termination in Cyberspace," *Strategic Studies Quarterly* 6, no. 3 (Fall 2012): pp. 46–70; Erik Gartzke and Jon R. Lindsay, "Thermonuclear Cyberwar," *Journal of Cybersecurity* 3, no. 1

(February 2017): pp. 37–48; Edward Geist and Andrew Lohn, *How Might Artificial Intelligence Affect the Risk of Nuclear War?* (Santa Monica, CA: RAND Corporation, 2018), and Matthijs M. Maas, "How Viable is International Arms Control for Artificial Military Intelligence? Three Lessons from Nuclear Weapons," *Contemporary Security Policy* 40, no. 3 (2019): pp. 285–311; and James Johnson, "Artificial Intelligence & Future Warfare: Implications for International Security," *Defense & Security Analysis* 35, no. 2 (2019): pp. 147–169.

14. A "false-flag" cyber operation in this context would occur due to a cyber-attack that convinces one side of a nuclear dyad that the other is about to attack, thus increasing incentives for pre-emption. Because of the attribution challenge in cyberspace, "false-flag" operations are particularly concerning.
15. Kobe, "A Theory of Catalytic War."
16. For example, see Arthur L. Burns, *The Rationale of Catalytic War, Research Monograph no. 3* (Princeton, NJ: Center of International Studies, Princeton University, 1959); and William T.R. Fox, "Political and Diplomatic Prerequisites of Arms Control," *Daedalus* 89 (1960): pp. 1000–1014.
17. Fred C. Ikle, "Nth Countries and Disarmament."
18. Herman Kahn, "The Arms Race and Some of Its Hazards," *Daedalus* 89 (1960): pp. 744–780.
19. For a description of the "chain of events" required for a terrorist nuclear detonation, see Ferguson and Potter, *The Four Faces of Nuclear Terrorism*, p. 6.
20. For example, evidence revealed that Al-Qaida actively sought nuclear weapons and expressed a clear desire to use them. Rolf Mowatt-Larssen, "Al Qaeda's Pursuit of Weapons of Mass Destruction: The Authoritative Timeline," *Foreign Policy*, January 25, 2010, https://foreignpolicy.com/2010/01/25/al-qaedas-pursuit-of-weapons-of-mass-destruction/.
21. For a recent discussion of various nuclear terrorism, scenarios see Ferguson and Potter, *The Four Faces of Nuclear Terrorism*, and Matthew Bunn and Anthony Wier, "Terrorist Nuclear Weapon Construction: How Difficult?" *The Annals of the American Academy* 6, no. 1 (2006): pp. 133–149.
22. Erin D. Dumbacher and Page Soutland, "NTI Nuclear Security Index: Building a Framework for Assurance, Accountability, and Action: Fourth Edition," *Nuclear Threat Initiative*, September 2018, https://media.nti.org/documents/NTI_2018_Index_FINAL.pdf.
23. If the false state accused were able to establish the attacker's identity unambiguously, and its false provocations, military action against them might deliver a clear message of resolve and dis-ownership of the group involved, potentially mitigating nuclear exchange between the nuclear states. Robert Ayson, "After a Terrorist Nuclear Attack: Envisaging Catalytic Effects," *Studies in Conflict & Terrorism* 33, no. 7 (2010): pp. 571–593, pp. 589 and 593.
24. Toby Dalton and Gaurav Kalwani, "Might India Start the Next South Asian Crisis?" *War on the Rocks*, November 1, 2019, https://warontherocks.com/2019/11/might-india-start-the-next-south-asia-crisis/.
25. The broader notion that escalation can occur inadvertently (or unintentionally) is attributed to a failure to appreciate the pressures that one's actions put on an adversary. These pressures generate perceived first-mover advantages and are associated with the

"security dilemma" concept—when defensively motivated actions can appear and be perceived as offensive. See Robert Jervis, "Cooperation under the Security Dilemma," *World Politics* 30, no. 2 (1978): pp. 167–214.

26. Thomas C. Schelling, *Arms and Influence* (New Haven, CT, and London: Yale University Press, 1966), p. 2.
27. A cyber-attack targeting a state's NC3 network would require a relatively high level of technical expertise, including: reconnaissance; collecting intelligence; technical knowledge of the NC3 systems being targeted; testing and rehearsing an attack to ensure its efficacy; extensive due diligence on the targets' protocols, strategic culture and cognitive fallibilities; and knowledge of the system's redundancies and countermeasures. Therefore, such an attack would be beyond the abilities and resources of opportunistic or less sophisticated third-party or non-state actors. Innovations in the digital age, such as AI and autonomy, will likely lower this threshold and create new mechanisms for non-state actors to fulfill their goals—primarily through manipulating and distortion of the information landscape. See Patricia Lewis and Beyza Unal, *Cybersecurity of Nuclear Weapons Systems: Threats, Vulnerabilities, and Consequences* (London: The Royal Institute of International Affairs, Chatham House, 2018).
28. Lewis and Unal, *Cybersecurity of Nuclear Weapons Systems*, pp. 4 and 19; and Peter Hayes, *Non-State Terrorism and Inadvertent Nuclear War*, Nautilus Institute for Security and Sustainability Special Reports, January 18, 2018.
29. While SSBNs are not connected to the internet in any meaningful sense, the submarine's missiles, warheads (i.e., launch controls and targeting systems), and the various support systems rely on networked computers and software. Submarines are particularly vulnerable to malware introduced to the network during the procurement phase while the submarine and its missiles and warheads are being built, or when the submarine is in port for maintenance, refurbishment, and software updates. Andrew Futter, "Is Trident Safe from Cyber Attack?" *European Leadership Network*, February 5, 2016, https://www.europeanleadershipnetwork.org/report/is-trident-safe-from-cyber-attack/.
30. While theoretically hackers could acquire submarine launch codes through cyber-espionage and then transmit these "go codes" to submarine commanders, it is highly unlikely that non-state actors could compromise an SSBN in this way. More likely, however, is that hackers steal highly classified information about SSBNs' stealth or navigation technology or information about their locations and patrol itinerary—which will likely be stored or backed up on the internet. Jason Fritz, "Hacking Nuclear Command and Control," International Commission on Nuclear Nonproliferation and Disarmament, July 2009, www.icnnd.org/Documents/Jason_Fritz_Hacking_NC2.doc.
31. Ibid.
32. Soroush Vosoughi, Deb Roy, and Sinan Aral, "The Spread of True and False News Online," *Science* 359, no. 6380 (March 2018): pp. 1146–1151.
33. Russell Goldman, "Reading Fake News, Pakistani Minister Directs Nuclear Threat at Israel," *New York Times*, December 24, 2016, https://www.nytimes.com/2016/12/24/world/asia/pakistan-israel-khawaja-asif-fake-news-nuclear.html.
34. Hannah E. Haegel and Reema Verma, "The Terrifying Geography of Nuclear and Radiological Insecurity in South Asia," *Bulletin of Atomic Scientists*, January 27, 2017.
35. Robert Jervis *How Statesmen Think: The Psychology of International Politics* (New Jersey, Princeton, NJ: Princeton University Press, 2017), p. 222.

36. Burns, *The Rationale of Catalytic War.*
37. Eric Schlosser, *Command and Control* (New York: Penguin Group, 2014), p. 481.
38. Cordelia Fine, *A Mind of Its Own: How Your Brain Distorts and Deceives* (New York: W.W. Norton & Company, 2008); and Robert Jervis, *Perception and Misperception in International Politics* (Princeton, NJ: Princeton University Press, 1976), pp. 117–202.
39. Michela Del Vicario et al., "Modeling Confirmation Bias and Polarization," *Sci Rep* (2017): pp. 1–9.
40. Daniel Kahneman, *Thinking, Fast and Slow* (New York: Farrar, Straus, and Giroux, 2011).
41. Michael J. Mazarr, *Rethinking Risk in National Security: Lessons from the Financial Crisis for Risk Management* (Basingstoke: Palgrave Macmillan, 2016).
42. Amos Tversky and Daniel Kahneman, "Judgment under Uncertainty: Heuristics and Biases," *Science* 185, no. 4157 (September 1974): pp. 1124–1131.
43. Nassim N. Taleb, *Fooled by Randomness: The Hidden Role of Chance in Life and the Markets*, 2nd ed. (London: Penguin, 2004).
44. For recent research on the influence of specific emotions on decision-making in cyber conflict, see Rose McDermott, "Some Emotional Considerations in Cyber Conflict," *Journal of Cyber Policy* 4, no. 3 (2019): pp. 309–325.
45. Forrest E. Morgan et al., *Dangerous Thresholds: Managing Escalation in the 21st Century* (Santa Monica, CA: RAND Corporation, 2008), pp. 19–29.
46. Lin, "Escalation Dynamics and Conflict Termination in Cyberspace."
47. One recent study advanced a more nuanced range of escalation categories: unauthorized use, unintended use, and intended use based on misperceptions. This treatment does not, however, consider the possibility of deliberate use of nuclear weapons. Global Zero, "Global Zero Commission on Nuclear Risk Reduction: De-Alerting and Stabilizing the World's Nuclear Force Postures," April 2015.
48. Intriligator and Brito, "Minimizing the Risks of Accidental Nuclear War," pp. 227–230.
49. The binary distinction between deliberate and unintentional use of nuclear weapons can be problematic, however. Sico van der Meer, "Reducing Nuclear Weapons Risks: A Menu of 11 Policy Options," *Policy Brief*, Clingendael Netherlands Institute of International Relations, June 2018.
50. Jervis, *Perception and Misperception in International Politics*, Ch. 3.
51. Miles, "The Dynamics of Strategic Stability and Instability," p. 429 and 437.
52. Military autonomous systems and inadvertent risks associated with these systems is not a new phenomenon. Examples include landmines, torpedoes, defense systems (such as *Aegis* and *Patriot*), and close-in weapon systems such as *Phalanx*. See Yuna Huh Wong et al., *Deterrence in the Age of Thinking Machines* (Santa Monica, CA: RAND Corporation, 2020), pp. 71–72.
53. See Steven Metz and James Kievit, *Strategy and the Revolution in Military Affairs: From Theory to Policy* (Carlisle: Strategic Studies Institute, 1995); Stephen Biddle, *Military Power: Explaining Victory & Defeat in Modern Battle* (Princeton, New Jersey: Princeton University Press, 2004); Stephen Peter Rosen, *Winning the Next War: Innovation and the Modern Military* (Ithaca, NY: Cornell University Press, 1991); MacGregor Knox and Williamson Murray, eds., *The Dynamics of Military Revolution, 1300–2050* (New York: Cambridge University Press, 2001); and Michael C. Horowitz, *The Diffusion of Military Power: Causes and Consequences for International Politics* (Princeton, N.J: Princeton University Press, 2010).

54. Pavel Podvig, "Risks of Nuclear Command and Control Accidents," in *Understanding Nuclear Weapon Risks*, edited by John Borrie, Tim Caughley, and Wilfred Wan (Geneva: UNIDIR, 2017).
55. Because of the interdependencies between the various components that comprise NC3 systems, it is impossible to calculate the probability of an accidental nuclear confrontation from the risk of the failure of a particular component or system, in isolation from the system and network. Intriligator and Brito, "Minimizing the Risks of Accidental Nuclear War," p. 230.
56. Illustrative of the data-fueled complexity and the challenge of integrating interdependent informational functions into a coherent military concept, C2 morphed into C4ISR (command, control, communications, computers, and ISR).
57. Robotics miniaturization (e.g., CubeSats and swarmed unmanned vehicles) and improvements in networking are enabling the wide deployment of formerly limited capabilities, such as aerial full-motion video, and the harvesting of open-sources (e.g., crowdsourcing, commercial satellite imagery, and geographic information systems data), all of which adds to the information burden in the decision-making process. Shane P. Hamilton and Michael P. Kreuzer, "The Big Data Imperative: Air Force Intelligence for the Information Age," *Air and Space Power Journal* 32, no. 1 (Spring 2018): pp. 4–20.
58. Ibid.
59. According to a RAND Corporation study, the volume of data collected by the US Navy increased at an exponential rate between 2000 and 2015. Isaac R. Porche III et al., *Data Flood: Helping the Navy Address the Rising Tide of Sensor Information* (Santa Monica, CA: RAND Corporation, 2014).
60. Recent studies have identified a range of cognitive and psychological effects whereby subjects tend to discard complex or conflicting information, settle for sub-optimal conclusions to save time, and experience high-stress levels and other adverse psychological effects. David Bawden and Lyn Robinson, "The Dark Side of Information: Overload, Anxiety and Other Paradoxes and Pathologies," *Journal of Information Science* 35, no. 2 (April 2009): pp. 180–191.
61. Scott Sagan, *The Limits of Safety; Diane Vaughan, The Challenger Launch Decision: Risky Technology, Culture, and Deviance at NASA* (Chicago, IL: University of Chicago Press, 1996); Charles Perrow, *Normal Accidents: Living with High-Risk Technologies*, 2nd ed. (Princeton, NJ: Princeton University Press, 1999); Scott A. Snook, *Friendly Fire: The Accidental Shootdown of U.S. Black Hawks over Northern Iraq* (Princeton, NJ: Princeton University Press, 2000); Karl E. Weick and Kathleen M. Sutcliffe, *Managing the Unexpected: Sustained Performance in a Complex World*, 3rd ed. (Hoboken, New Jersey: John Wiley and Sons, 2015).
62. Jervis, *Perception and Misperception in International Politics.*
63. James D. Fearon, "Rationalist Explanations for War," *International Organization* 49, no. 3 (1995): pp. 379–414; and Erik Gartzke, "War is in the Error Term," *International Organization* 53, no. 3 (1999): pp. 567–587.
64. Robert Jervis, "War and Misperception," *Journal of Interdisciplinary History* 18, no. 4 (1988): pp. 675–700.
65. Stephen J. Cimbala, *The Dead Volcano: The Background and Effects of Nuclear War Complacency* (Westport, CT: Praeger, 2002), p. 147.

66. For example, China, India, and Pakistan possess multiple types of ground-launched missiles suspected of dual-use. China's DF-26 intermediate-range ballistic missile, for example, can deploy conventional and nuclear payloads. James M. Acton, "Is It a Nuke? Pre-Launch Ambiguity and Inadvertent Escalation," *Carnegie Endowment for International Peace*, April 9, 2020, https://carnegieendowment.org/2020/04/09/is-it-nuke-pre-launch-ambiguity-and-inadvertent-escalation-pub–81446.
67. Ben Buchanan, *The Hacker and the State: Cyber Attacks and the New Normal of Geopolitics* (Cambridge, MA: Harvard University Press, 2020).
68. During the formulation of Russia's most recent military doctrine, Russian strategists proposed that an attack on Russia's early-warning systems would be interpreted as a sign of an impending nuclear attack. This proposal was not, however, included in the final version. Ryabikhin Leonid, "Russia's NC3 and Early Warning Systems," *Tech4GS* June 11, 2019, https://www.tech4gs.org/leonid-ryabikhin.html.
69. Robert Jervis and Mira Rapp-Hooper, "Perception and Misperception on the Korean Peninsula," *Foreign Affairs*, April 5, 2018, https://www.foreignaffairs.com/articles/north-korea/2018-04-05/perception-and-misperception-korean-peninsula. For scholarship on reputation in international relations, and assuming the worst of other intentions, see Jonathan Mercer, *Reputation and International Politics* (Ithaca, NY: Cornell University Press, 1996); Robert Jervis, *Perception and Misperception in International Politics*.
70. Gartzke and Lindsay, "Thermonuclear Cyberwar," pp. 37–48.
71. However, these systems must also ensure that states' nuclear arsenals will "always" be available if authorized (to enhance deterrence) and "never" be used without authorization (to enhance safety, assurance, and confidence). In this sense, early-warning systems were not designed as safety mechanisms but rather to enable launch-on-warning.
72. The historical record has shown that most international relations situations involve some sort of "sub-rational" thinking or error, which includes misperception, confusion, and "bounded rationality." These scenarios also incorporate a degree of deviation from the logic of rational utility maximization. See James M. Goldgeier and Philip E. Tetlock, "Psychology and International Relations Theory," *Annual Review of Political Science* 4, no. 1 (2001): pp. 67–92; and Janice Gross Stein, "The Micro-Foundations of International Relations Theory: Psychology and Behavioral Economics," *International Organization* 71, no. S1 (April 2017): pp. 249–263.
73. Perrow, *Normal Accidents: Living with High-Risk Technologies.*
74. Knox and Murray, eds., *The Dynamics of Military Revolution.*
75. Richard N. Lebow, *Nuclear Crisis Management* (Ithaca, NY: Cornell University Press, 1987), pp. 67–70.
76. For example, during the 1956 Suez Crisis, misinterpretations, and the tendency to view unrelated and even unconfirmed events as part of a pattern of an adversary's behavior, may have compounded the anxiety and high level of tension that contributed to the misinterpretation of detection data from early-warning systems and, in turn, created a crisis with a momentum of its own. Paul Bracken, *The Command and Control of Nuclear Forces* (New Haven: Yale University Press, 1983).
77. For example, during the height of the 1962 Cuban Crisis, the US absorbed a false report (and ignored a credible one) of an imminent attack by the Soviets on the US due to insufficient institutional checks and balances; and despite extensive indications and warning systems. Lloyd R. Leavitt and Paul Bracken, "Nuclear Proliferation: Neither Safe nor Stable," in *Inadvertent Nuclear War*, edited by Wilberg, Petersen, and Smoker, pp. 203–217.

78. Open sources indicate that the US DoD already uses several technologies that rely on AI. These include various active programs—either under development or deployed—that could directly affect US NC3 systems and the US nuclear deterrence architecture. These incorporate AI-ML to support decision-making support systems (e.g., Deep Green); early-warning, ISR data collection analysis, and remote sensing (e.g., Project Maven, Perdix UAV swarm, Orca XLUUV); cyber and electronic warfare; air-defense systems that use ATR and voice-controlled support assistants in combat aircraft such as the F-15; AI-supported active radar seekers (e.g., Aegis ballistic missile defense systems); to enable autonomous navigation in unmanned vehicles (e.g., X-47B unmanned aerial vehicle); and ML and autonomy to improve precision missile guidance and maneuverability (e.g., Conventional Prompt Strike Guide Vehicles, and long-range hypersonic weapons). Vincent Boulanin, ed., *Artificial Intelligence, Strategic Stability and Nuclear* (Stockholm: SIPRI Publications: June 2020), pp. 42–43.
79. In the context of statistical decision theory, a Type I error (also known as a "false positive") refers to where a particular action was not taken that should have been taken. A Type II error (also known as a "false negative") refers to where an action was taken that was not warranted. See Intriligator and Brito, "Minimizing the Risks of Accidental Nuclear War." Both Type I and II errors can cause misperceptions and mischaracterizations between adversaries in ways that may amplify the risk of accidental escalation. Lora Saalman, "Fear of False Negatives: AI and China's Nuclear Posture," *Bulletin of the Atomic Scientists*, April 24, 2018, https://thebulletin.org/2018/04/fear-of-false-negatives-ai-and-chinas-nuclear-posture/.
80. See Avery Goldstein, "First Things First: The Pressing Danger of Crisis Instability in U.S.-China Relations," *International Security* 37, no. 4 (2013): pp. 49–89; Caitlin Talmadge, "Would China Go Nuclear? Assessing the Risk of Chinese Nuclear Escalation in a Conventional War with the United States," *International Security* 41, no. 4 (Spring 2017): pp. 50–92; and James Acton, "Escalation through Entanglement: How the Vulnerability of Command-and-Control Systems Raises the Risks of an Inadvertent Nuclear War," *International Security* 43, no. 1 (August 1, 2018): pp. 56–99.
81. The uncertainty and ambiguity of states' commingled nuclear and non-nuclear weapon systems can lead to mischaracterization that causes false positives or false negatives. See Acton, "Is it a Nuke?" p. 3.
82. Historical examples of nuclear "false alarms" include: (1) the 1983 NATO Able Archer exercise that raised alarms in the Soviet Union; (2) the 1995 Norwegian rocket incident or the Florida satellite incident that occurred during the Cuban missile crisis; (3) the 1979 incident in which a NORAD technician mistakenly loaded a training tape containing an attack scenario into an operation computer; and (4) the 1980 US Strategic Air Command incident in which a faulty computer chip triggered alerts, causing bomber crews to prepare for take-off. Seth D. Baum, Robert de Neufville, and Anthony M. Barrett, "A Model for the Probability of Nuclear War," *Global Catastrophic Risk Institute*, Global Catastrophic Risk Institute Working Paper 18-1 (March 2018), pp. 19–20.
83. Unexpected accidental situations involving nuclear weapons (or "broken arrow" incidents) have also included missile explosions, aircraft collisions, and even nuclear weapons release—without an actual nuclear detonation. See Pavel Podvig, "Blurring the Line between Nuclear and Non-Nuclear Weapons: Increasing the Risk of Accidental Nuclear War?" *Bulletin of the Atomic Scientists* 72, no. 3 (2016): pp. 145–149; and Eric Schlosser, *Command and Control* (New York: Penguin Group, 2014).

84. For example, the US DoD established the Joint All-Domain Command and Control to integrate sensors and communications across all warfighting domains, reportedly encompassing NC3 systems. Colin Clark, "Nuclear C3 Goes All Domain: Gen. Hyten," *Breaking Defense*, February 20, 2020, https://breakingdefense.com/2020/02/nuclear-c3-goes-all-domain-gen-hyten/.
85. NC3 systems are inherently complex (and increasingly so with the integration of AI technology). They typically include a mixture of old (legacy) components and modern elements, and these systems have been modified, upgraded, and integrated piecemeal over many years to accommodate evolving requirements of technological changes, patching of vulnerabilities, and modernization.
86. Lin, "Escalation Dynamics and Conflict Termination in Cyberspace," p. 59.
87. "Is Launch Under Attack Feasible?" *Nuclear Threat Initiative*, August 4, 2016.
88. Perrow, *Normal Accidents: Living with High-Risk Technologies*, p. 259.
89. Nautilus Institute, Technology for Global Security, Preventive Defense Project, "Social Media Storms and Nuclear Early Warning Systems: A Deep Dive and Speed Scenario Workshop," *NAPSNet Special Reports*, p. 6.
90. Acton, "Escalation through Entanglement"; and Goldstein, "First Things First: The Pressing Danger of Crisis Instability in U.S.-China Relations," p. 89.
91. See Acton, "Escalation through Entanglement."
92. For example, during the Cold War, the Soviets developed a computer program known as VRYAN (a Russian acronym for "Surprise Nuclear Missile Attack") designed to notify Soviet leaders of a pre-emptive US nuclear strike. However, the data used to feed the system was often biased and thus propelled a feedback loop, which heightened the Kremlin's fear that the US was pursuing first-strike superiority. President's Foreign Intelligence Advisory Board, "The Soviet 'War Scare,'" February 15, 1990, vi, 24 et seq.
93. Most nuclear-armed states possess modern NC3 structures with inbuilt redundancies, coupled with robust internal ballistics, mechanical arming, and other safety-related mechanisms and protocols. In some cases (i.e., North Korea and Pakistan), however, national NC3 systems may not be rigorously and continuously tested and may not function coherently or reliably in functional performance under stress. These NC3 systems are also considered insufficiently redundant and lack clear lines of responsibility and accountability. Peter Hayes et al., Binoy Kampmark, Philip Reiner, and Deborah Gordon, "Synthesis Report, NC3 Systems and Strategic Stability: A Global Overview," *NAPSNet Special Reports*, May 5, 2019, https://nautilus.org/?p=97769.
94. On "automation bias", see Linda J. Skitka, Kathleen L. Mosier, and Mark Burdick, "Does Automation Bias Decision-Making?" *International Journal of Human-Computer Studies* 51, no. 5 (1999): pp. 991–1006; and Mary L. Cummings, "Automation Bias in Intelligent Time-Critical Decision Support Systems," *AIAA 1st Intelligent Systems Technical Conference*, 2004, pp. 557–562. On technology and "organizational failure", see Bent Natvig, "Accidental Nuclear War Considered from the Area of Reliability of Large Technological Systems," in *Inadvertent Nuclear War*, edited by Wilberg, Petersen, and Smoker, pp. 55–69.
95. Michael C. Horowitz, Paul Scharre, and Alexander Velez-Green, "A Stable Nuclear Future? The Impact of Autonomous Systems and Artificial Intelligence," *arXiv*, December 2019.

96. Overtrust is related to automation bias in situations where humans perceive technology is more capable than it is, and complacency when individuals become less vigilant towards the functioning of a system even though attention should be paid to it. See Makoto Itoh, "Toward Overtrust-Free Advanced Driver Assistance Systems," *Cognitive Technology & Work* 14, no. 1 (2012): pp. 51–60.
97. Maha Salem et al., "Would you Trust a (Faulty) Robot?" in Proceedings of the Tenth Annual ACM/IEEE International Conference on Human-Robot Interaction, March 2015, pp. 141–148.
98. Militaries today have already begun developing AI solutions to substitute human prediction to deal with information overload—or the so-called "ISR revolution." See Keith Dear, "Artificial Intelligence and Decision-Making," *The RUSI Journal* 164, nos 5–6 (2019): pp. 18–25; and Yang Feilong and Li Shijiang, "Cognitive Warfare: Dominating the Era of Intelligence," *PLA Daily*, March 19, 2020.
99. Bahar Gholipour, "New AI Tech Can Mimic Any Voice," *Scientific American*, May 2, 2017; Will Knight, "These Incredibly Realistic Fake Faces Show How Algorithms Can Now Mess with Us," *MIT Technology Review*, December 14, 2018.
100. Generative adversarial networks (GANs) operate by integrating a competitive function into software, with one network seeking to generate an item (i.e., an image or video clip), while the other network judges the item to determine whether it looks real. See Weiwei Hu and Ying Tan, "Generating Adversarial Malware Examples for Black-Box Attacks Based on GAN," *arXiv*, February 20, 2017, http://arxiv.org/abs/1702.05983.
101. Researchers at the University of Washington are developing AI algorithms that convert audio clips into a realistic video by training software to generate matching mouth shapes on a speaker. Jennifer Langston, "Lip-Syncing Obama: New Tools Turn Audio Clips into Realistic Video," *UW News*, July 11, 2017.
102. For example, an ML artist at Google successfully generated images of imagined faces with a resolution of 768 pixels aside, more than twice as big as anything previously achieved. Also, AI experts anticipate that amateurs will produce compelling YouTube fake videos within the next few years. "Fake News: You Ain't Seen Nothing Yet."
103. Mazarr et al., *The Emerging Risk of Virtual Societal Warfare: Social Manipulation in a Changing Information Environment* (Santa Monica, CA: RAND Corporation, 2019), p. 85.
104. James Johnson, "Delegating Strategic Decision-Making to Machines: Dr. Strangelove Redux?" Journal of Strategic Studies (2020), doi: 10.1080/01402390.2020.1759038.
105. Adrian Chen, "The Agency," *New York Times Magazine*, June 2, 2015.
106. See Martin C. Libicki, *Cyber Deterrence and Cyberwar* (Santa Monica, CA: RAND, 2009), p. 44.
107. For example, even if the malware detected in an attack was only capable of espionage, a target may fear that it also contained a "kill switch" able to disable an early-warning system after activation.
108. Will Knight, "The US Military is Funding an Effort to Catch Deepfakes and Other AI Trickery," *MIT Technology Review*, May 23, 2018, https://www.technologyreview.com/2018/05/23/142770/the-us-military-is-funding-an-effort-to-catch-deepfakes-and-other-ai-trickery/.

109. Richard Fontaine and Kara Frederick, "The Autocrat's New Tool Kit," *The Wall Street Journal*, March 15, 2019, https://www.wsj.com/articles/the-autocrats-new-tool-kit-11552662637.
110. Hazel K. Kwon et al., "Social Media Rumors as an Improvised Public Opinion: Semantic Network Analyses of Twitter Discourses during Korean Sabre-Rattling 2013," *Asian Journal of Communication* (2016): pp. 1–22.
111. Between August 2017 and January 2018, six social media instances playing a role in nuclear-prone conflicts occurred in the Asia-Pacific region alone. See Nautilus Institute, Technology for Global Security, Preventive Defense Project, "Social Media Storms and Nuclear Early Warning Systems: A Deep Dive and Speed Scenario Workshop," p. 1.
112. DFRLab, "#FakeNews: American Bomber Did Not Drop a Bomb on a House in Lithuania," *Medium.com*, June 16, 2017, https://medium.com/dfrlab/fakenews-american-bomber-didnot-drop-a-bomb-on-a-house-in-lithuania-6ae64241fe9e.
113. Drew Harwell, "An Artificial-Intelligence First: Voice-Mimicking Software Reportedly Used in a Major Theft," *Washington Post*, September 4, 2019, https://www.washingtonpost.com/technology/2019/09/04/an-artificial-intelligence-first-voice-mimicking-software-reportedly-used-major-theft/.
114. Disinformation in deception and misinformation campaigns is a familiar aspect of warfare, perhaps most famously demonstrated by the efforts of the Allies during World War II's *Operation Bodyguard* to mislead the Axis regarding the location of what became the D-Day invasion of 1944. Jamie Rubin, "Deception: The Other 'D' in D-Day," *NBC News*, June 5, 2004.
115. Daniel Kahneman, *Thinking, Fast and Slow* (New York: Farrar, Straus and Giroux, 2011).
116. See Judith Möller, "Filter Bubbles and Digital Echo Chambers," in *The Routledge Companion to Media Disinformation and Populism*, edited by Howard Tumber and Silvio Waisbord (London: Routledge, 2021), Ch. 8; and Danielle K. Citron and Robert Chesney, "Deep Fakes: A Looming Challenge for Privacy, Democracy, and National Security," *107 California Law Review* 1753 (2019).
117. Daniel Kahneman and Amos Tversky, "Availability: A Heuristic for Judging Frequency and Probability," *Cognitive Psychology* 5, no. 2 (September 1973): pp. 207–232.
118. Jervis, *How Statesmen Think*, p. 222.
119. Dan Lamothe, "U.S. Families got Fake Orders to Leave South Korea. Now Counterintelligence is Involved," *Washington Post* September 22, 2017.
120. AI systems can track individuals' or groups' online habits, knowledge, and preferences to calibrate specific messages (i.e., propaganda) to maximize the impact on that individual (or group) and minimize the risk of the information being conveyed questioned. This information can then be used by AI systems in real-time to determine the messages' influence, thus learning to become more effective in its task.
121. Alina Polyakova, "Weapons of the Weak: Russia and AI-Driven Asymmetric Warfare," Brookings Institution, November 15, 2018.
122. A non-state third party could relatively easily insert a false narrative into social media or a crowdsourcing platform, with a low probability of increasing the risks of catalytic escalation. For example, non-credible sources or counter-factual information, counter-detection tools, inauspicious timing, or simply that the ML algorithms failed to detect the fakery in time.

123. Scott D. Sagan and Jeremi Suri, "The Madman Nuclear Alert: Secrecy, Signaling, and Safety in October 1969," *International Security* 27, no. 4 (2003): pp. 150–183.
124. Philip E. Tetlock and Aaron Belkin eds., *Counterfactual Thought Experiments in World Politics* (Princeton, NJ: Princeton University Press, 1996), p. 32.
125. Assumptions are needed to explain what an actor might hope to achieve by sparking nuclear confrontation—which would be a sub-rational act. Gartzke and Lindsay, "Thermonuclear Cyberwar," pp. 37–48.
126. While a cyber false-flag operation would not require particularly sophisticated technical expertise—at least compared to the level of scientific and military infrastructure required to develop nuclear weapons—it would need the organizational know-how to collect and parse intelligence and conduct clandestine activities. See C. Shao, et al., "The Spread of Low-Credibility Content by Social Bots," *Nature Communications* 9, no. 4787 (2018): pp. 1–8.
127. Lin, "Escalation Dynamics and Conflict Termination in Cyberspace," p. 59.
128. From an intelligence standpoint, nuclear solid-fuel missiles and tracked transporter erector launchers reduce ISR systems' ability to detect signs of launch preparation. Solid fuel also increases the speed at which missiles can be launched and reduces the number of support vehicles to support an operation.
129. Alternative outcomes from this fictional scenario are, of course, possible. For example, counter-AI systems might uncover the leak's source or false nature before it can do severe damage. State A might also be able to assure State B through backchannel or formal diplomatic communications of this falsehood. While social media platforms have had some success in slowing down users' ability to orchestrate manipulative and dangerous campaigns, once these operations (e.g., deepfakes and bots) go viral, the ability to curtail them becomes inexorably problematic—for human operators or machines.
130. Other information related to weapons of mass destruction that would likely be considered equally escalatory might include intelligence about the movement of support vehicles delivering liquid fuel to prepare a liquid-fueled missile for launch, spikes in radiation levels, or the detection of chemical warfare agents, to name but a few.
131. Facebook, for example, uses its algorithms to anticipate human behavior to create "prediction products" that make people easier to manipulate—that is, profile and micro-target their users to sell more advertising space. This capability was allegedly used to manipulate public perceptions during the 2016 US presidential election and the United Kingdom's referendum on membership of the European Union. See Shoshana Zuboff, *The Age of Surveillance Capitalism: The Fight for a Human Future at the New Frontier of Power* (New York: PublicAffairs, 2019), pp. 3–17.
132. Micro-drone UAVs range from small flying insects to palm-sized devices equivalent to small birds and can carry payloads, including small guns and explosives and unconventional munitions such as poisons and nerve agents. See Berenice Baker, "Dogfighting Drones—Swarms of Unmanned Battle-Bots Take to the Skies," Airforce-technology.com, July 23, 2013, www.airforcetechnology.com/features/featuredogfight-drones-unmanned-battle-bot-swarms/; and Juanjuan Lin, Yuantao Zhang, and Wei Wang, "Military Intelligence is Profoundly Affecting Future Operations," *Ministry of National Defense of the People's Republic of China*, September 10, 2019, http://www.mod.gov.cn/jmsd/2019-09/10/content_4850148.htm.

133. Because of the rapid diffusion of technologies such as virtual reality goggles, 3D printing, modular payloads, and AI, experts anticipate that within two decades terrorist and insurgent forces may be able to field their own autonomous UAVs without state sponsorship. See Zachary Kallenborn and Philipp C. Bleek, "Swarming Destruction: Drone Swarms and Chemical, Biological, Radiological, and Nuclear Weapons," *The Nonproliferation Review*, 25, nos 5–6 (2019): pp. 523–543.
134. As outlined earlier, Herman Kahn defined forty-four "rungs" on a complex metaphorical ladder of escalating conflict, ranging from "sub-crisis maneuvering" to civilization-destroying nuclear exchanges. Herman Kahn, *On Escalation: Metaphors and Scenarios* (New York: Praeger, 1965). Recently, scholars have questioned the usefulness of Kahn's "escalation ladder" in the context of the rapid diffusion of asymmetric technologies to signal intent and attack adversaries. For example, see Rebecca Hersman, "Wormhole Escalation in the New Nuclear Age," *Texas National Security Review* 3, no. 3 (Autumn 2020): pp. 99–110.
135. James Johnson, "China's 'Guam Express' and 'Carrier Killers': The Anti-Ship Asymmetric Challenge to the U.S. in the Western Pacific," *Comparative Strategy* 36, no. 4 (2017): pp. 319–332.
136. Scholars remain divided, however, about the effectiveness of drone strikes in counterterrorism and counterinsurgency activities. Michael J. Boyle, "The Costs and Consequences of Drone Warfare," *International Affairs* 89, no. 1 (2013): pp. 1–29; and Megan Smith and James I. Walsh, "Do Drone Strikes Degrade Al Qaeda? Evidence from Propaganda Output," *Terrorism Political Violence* 25, no. 2 (2012): pp. 311–327.
137. The US Office of Naval Research, for example, envisions currently experimenting with swarms of unmanned surface vehicles to form a defensive perimeter around larger ships and surround enemy ships as part of the counterterrorism effort. Thomas Claburn, "Navy Tests Swarming Autonomous Boats," *InformationWeek*, November 7, 2014.
138. Generative adversarial networks (GANs) are a new approach involving two artificial neural network systems that spar to create an original image, audio, or video content, which machines have never been able to do properly before. See Tim Hwang, "Deepfakes: A Grounded Threat Assessment," *Center for Security and Emerging Technology*, July 2020, https://cset.georgetown.edu/wp-content/uploads/CSET-Deepfakes-Report.pdf; and Cristian Vaccari and Andrew Chadwick, "Deepfakes and Disinformation: Exploring the Impact of Synthetic Political Video on Deception, Uncertainty, and Trust in News," *Social Media + Society*, February 19, 2020, https://journals.sagepub.com/doi/full/10.1177/2056305120903408.
139. Additional assumptions are needed to explain why the non-state actor expects to benefit from starting a nuclear war, which may not be entirely rational or appear tactically and strategically optimal.
140. Like the previous fictional scenario, alternatives to this worse-case scenario outcome can also be conceived. At the most basic level, at the rung of the escalation ladder that led to catastrophe (i.e., the source of the GANs), subsequent subversive tactics could have been detected and, thus, crisis averted. The case study also assumes the victims of this offensive operation viewed each other as adversaries, making them more suspicious of the other's intentions in the event of a crisis. Were *either* side able to demonstrate empathy or display restraint, this escalatory outcome may be avoided.

141. Scott Rosenberg, "Firewalls Don't Stop Hackers, AI Might," *Wired*, August 27, 2017, https://www.wired.com/story/firewalls-dont-stop-hackers-ai-might/.
142. Kahneman, *Thinking, Fast and Slow*.
143. Robert Chesney and Max Smeets, "Policy Roundtable: Cyber Conflict as an Intelligence Contest," *Texas National Security Review*, September 17, 2020, https://tnsr.org/roundtable/policy-roundtable-cyber-conflict-as-an-intelligence-contest/.
144. Andrew Futter, *Hacking the Bomb: Cyber Threats and Nuclear Weapons* (Washington DC: Georgetown University Press, 2018), p. 10.
145. Traditional deterrence theory maintains that for deterrence to succeed, all parties in a conflict or crisis must be interested in limiting their costs and credibly threaten their opponents with punishment should they attempt to escalate—a condition that seldom exists in situations between states and non-state adversaries. Morgan et al., *Dangerous Thresholds: Managing Escalation in the 21st Century*, pp. 173–174.
146. For example, states can deter terrorists from attacking specific targets such as NC3 systems by protecting those targets more aggressively, thus *denying* adversaries the ability to escalate along certain lines. Ibid.
147. The approach used to control or mitigate escalation risk will vary according to the class of escalatory mechanisms involved (i.e., intention vs. unintentional); these approaches might also conflict with each other. Podvig, "Risks of Nuclear Command and Control Accidents"; Wilfred Wan, *Nuclear Risk Reduction: A Framework for Analysis* (Geneva: UNIDIR, 2019).
148. Budget constraints, design modifications, and bureaucratic inertia have hampered previous PAL devices and other safety components to nuclear management systems. Eric Schlosser, *Command and Control* (New York: Penguin Group, 2014).
149. For example, in 2017, NATO established a Strategic Communications Center of Excellence, which supports the development of best practices designed to raise awareness of the risks of disinformation posed by the nefarious dissemination of misinformation. "NATO Takes Aim at Disinformation Campaigns," *NPR Morning Edition*, May 10, 2017.
150. Peter Maurer, *Nuclear Weapons: Averting a Global Catastrophe, ICRC Statement*, April 23, 2018.
151. Michael Quinlan, *Thinking About Nuclear Weapons: Principles, Problems, Prospects* (Oxford: Oxford University Press, 2009).
152. For example, the implementation of best practices to reduce the risk of human fallibilities and measures to extend the decision-making, allowing for clarification of early-warning readings that turn out to be erroneous. US Office of the Director of National Intelligence, *A Guide to Cyber Attribution*, September 14, 2018; Matthew Bunn, Nickolas Roth, and William H. Tobey, "Revitalizing Nuclear Security in an Era of Uncertainty," *Project on Managing the Atom*, January 2019.
153. National Security Commission on Artificial Intelligence Interim Report to Congress, November 2019, p. 46.
154. For a recent multilateral (non-binding) effort, see the UN General Assembly, "Group of Governmental Experts on Developments in the Field of Information and Telecommunications in the Context of International Security: Note by the Secretary-General," A/68/98, June 24, 2013.

155. James Johnson, "The End of Military-Techno Pax Americana? Washington's Strategic Responses to Chinese AI-Enabled Military Technology," *The Pacific Review* 34, no. 3 (2021): pp. 351–378.
156. Vincent Boulanin, ed., *The Impact of Artificial Intelligence on Strategic Stability and Nuclear Risk Vol. I Euro-Atlantic Perspectives* (Stockholm: SIPRI Publications: May 2019).
157. Perrow, *Normal Accidents: Living with High-Risk Technologies*; and Unal and Lewis, *Cybersecurity of Nuclear Weapons Systems*.
158. van der Meer, "Reducing Nuclear Weapons Risks: A Menu of 11 Policy Options."
159. UK HM Government, "Global Britain in a Competitive Age: The Integrated Review of Security, Defence, Development and Foreign Policy," March 2021, https://assets.publishing.service.gov.uk/government/uploads/system/uploads/attachment_data/file/975077/Global_Britain_in_a_Competitive_Age_the_Integrated_Review_of_Security__Defence__Development_and_Foreign_Policy.pdf.
160. For example, the "Memorandum of Agreement Between the United States of America and The Russian Federation on the Establishment of a Joint Center for the Exchange of Data from Early Warning Systems and Notifications of Missile Launches," *The U.S. Department of State*, June 4, 2000.
161. Examples include: Patricia Lewis et al., *Too Close for Comfort: Cases of Near Nuclear Use and Options for Policy* (London: Chatham House, 2014); Gordon Barrass, "Able Archer 83: What Were the Soviets Thinking?" *Survival* 58, no. 6 (2016): pp. 7–30; and Benoît Pelopidas, "The Unbearable Lightness of Luck: Three Sources of Overconfidence in the Manageability of Nuclear Crises," *European Journal of International Security* 2, no. 2 (2017): pp. 240–262.

Conclusion
Managing the AI-nuclear strategic nexus

> Plans are useless, but planning is indispensable
>
> —**Dwight D. Eisenhower**[1]

The book provides a coherent multidisciplinary and much-needed sober examination of the AI-nuclear strategic nexus—the potential effects of AI technology on nuclear strategy and risk. Doing so addresses an important gap in the literature, which considers how AI might impact global security and future warfare but says astonishingly little about the strategic effects of this phenomenon. Revisiting the theoretical foundations of nuclear strategy, the book expounds on a novel theoretical and empirically rich explanation of A's potential effects on nuclear strategic thinking, emphasizing human psychological and other non-technical facets of the rapidly unfolding and omnipresent age of intelligent machines. I believe the book's findings in the intersection of technological change, strategic thinking, and nuclear risk help square the circle between nuclear strategic scholarship and policy.

The road so far . . .

Before we consider some tangible risk reduction solutions, I think it would be helpful at this point to draw out the book's core insights from its examination of the AI-nuclear strategic nexus through the lens of classical nuclear theorizing.

The human factor will for the foreseeable future remain critical to strategic stability

Chapter 1 explained that data limitations and constraints on the ability of AI algorithms to capture the nuanced, dynamic, subjective, and changeable nature of human commanders accurately (or "theory-of-the-mind"

AI and the Bomb. James Johnson, Oxford University Press. © James Johnson (2023).
DOI: 10.1093/oso/9780192858184.003.0007

functions) would mean that, for the foreseeable future, strategic decision-making will remain a fundamentally human endeavor—albeit imbued with increasing degrees of interaction between humans and machines. In other words, for now AI will continue to include some human agency (especially in collaboration with machines) in managing the challenges associated with technological complexity and uncertainty described in this book.

That said, whereas human agency should, in theory, limit the role of AI in the nuclear domain ostensibly to a support one, through the discharge of its "support role" it may still influence strategic decisions that involve nuclear weapons. Therefore, the distinction between AI's impact at a tactical and strategic level is blurred and graduated rather than binary. Irrespective of whether human operators remain in or out of the loop, therefore, the complexities—and concomitant risks—inherent with human-to-machine interactions must be carefully considered. In sum, escalation that begins at the *tactical level* can have *strategic effects*—generating inadvertent escalation risk caused by accidents, disinformation, or misperception.

Technology has long been used throughout history to augment, automate, and enhance human behavior and decision-making in a military context; so far, the "human factor" has eclipsed the impact of technology on strategic stability. As shown in Chapter 1, the underlying forces behind fundamental shifts in strategic stability are generally less concerned with (quantitative or qualitative) net assessments of military capabilities but, rather, are more focused on how nuanced institutional, cognitive, and strategic variables impact strategic decisions and can lead to misperceiving others' intentions. The relationship between AI—and technological change more generally—and strategic stability can be understood, therefore, as part of a complex interaction of disruptive forces, which during periods of heightened geopolitical rivalry, great-power transitions, and strategic surprise, can upend strategic stability and make nuclear use more likely.

In general, technological-deterministic military theories—that conceptualize technology as the *sole* driver of strategic stability and conflict—oversimplify this complicated and fundamentally human endeavor. Lawrence Freedman warned against the trend during the 1970s of strategic theorists being too preoccupied with technical details at the expense of political and other non-technical considerations.[2] Technocratic approaches also underplay the role of chance and friction inherent to strategic affairs that we examined in Chapter 3 and the role of rational and sub-rational state and non-state actors seeking to leverage new tools of disruption to sow confusion to destabilize the strategic balance.

In the final analysis, politics and the breakdown of *human* bargaining ultimately leads to conflict. AI and other emerging technologies (drones, deepfakes, hypersonic, and cyber weapons) can be viewed as independent variables that can complicate and exacerbate escalatory outcomes determined by human choices, biases, emotions, in their human-to-human and human-to-machine interactions. As Colin Gray astutely noted, strategic affairs—and the role of human agency in it—is far more durable than some modern military theorists suggest. Thus, identifying any meaningful change to our understanding of strategic stability—and strategic core logic more generally—requires accepting this continuity as a premise before judging whether the latest whiz-bang technology represents a departure—or revolution in military affairs.[3]

Rational-based deterrence assumptions are becoming obsolete in the age of AI

As shown in Chapter 2, because of the potentially transformative effects of AI and autonomy augmentation on a range of (nuclear and non-nuclear) strategic technologies, we need to urgently rethink existing assumptions, theories, and permutations of deterrence, premised on perceptions, nuanced signaling, and human rationality on both sides to ensure threats will suffice to shape the behavior of others.[4] To make matters worse, AI algorithms to power (semi)autonomous systems today are based on game and decision-theoretic "rational" decision-making—to predict an adversary's intentions and behavior more accurately in simulated battlefield situations. Replacing human agency and rationality with "black box" algorithms, the technology might obscure and erode established deterrence mechanisms, norms, and policies.[5] Will planners become too reliant on AI technology, believing it to be imbued with superhuman intelligence, or distrust its recommendations because of AI's non-human "black box" fuzzy logic? To be sure, in human-machine interactions trust in automation and algorithmic systems includes not only factors related to performance but also those necessary for individuals to interact with them—such as the purpose and goals of algorithms, and the personality, cultural traits, and intentions of human operators.[6]

Notwithstanding the possible reduced risk to human life, the total absence of a normative deterrence framework—particularly to signal to resolve to an adversary while simultaneously seeking to de-escalate a situation—may compress or remove entirely the various rungs of the escalation ladder described in Chapter 3. In sum, an over-reliance on AI to inform strategic decisions by demoting the role of human agents in reviewing, scrutinizing, and invalidating the assumptions about an adversary's intentions and behavior (critical to effective deterrence) increases the risk of inadvertent escalation—that is, complicating the delicate balance between an actor's willingness to escalate a situation as a last resort and keeping the option open to step back from the brink.

Chapter 3 found that traditional nuclear deterrence, crisis stability, and strategic stability theorizing work off the questionable premise that actors are rational, and presume therefore that they can understand their environment and coordinating policy instruments to cut through the "fog of war." In the real world, misperception, cognitive biases, and other human emotional impulses can quickly get out of hand and lead to accidents and unintended and potentially escalatory consequences.[7] Because of the confluence of speed, uncertainty, complexity, and cognitive strains associated with war-faring on the digitized battlefield, the ability for commanders to cut through the "fog" using experience and sound judgment will be severely impaired.

The chapter also found that the agenda-setting framing narrative of decision-makers during a crisis is increasingly influenced by a fragmented and politicized social media information ecosystem and thus more likely to stimulate emotionally visceral reactions than "rational" thought assumed by deterrence theorists. To sum, in a multipolar nuclear world, with imperfect information about the balance of power and resolve, incentives, and new tools to misrepresent and manipulate others' perceptions and emotions of the information landscape, the risk of nuclear use increases. Risk reduction solutions (see below) are not, therefore, reducible to structural processes and assumptions of "rational" actors alone—human agency still matters.[8] A viable solution to the emerging AI-nuclear strategic challenge needs to be as much a political as a technical one.

AI is creating new and disrupting old rungs on the escalation ladder

A key finding from Chapter 3 was that today's global security environment, characterized by great power and regional strategic imbalances in capabilities, domains, and stakes, is creating new rungs and disrupting existing firebreaks (especially between nuclear and non-nuclear warfare) on the psychological escalation ladder—shifting existing thresholds altering shared assumptions and notions of deterrence, strategic stability, and escalation along the "relatively continuous" pathways to war.[9] This chapter underscores the need for more granularity and discussion on the specific characteristics of AI technology that create new and disrupt old rungs, thereby increasing the risk of inadvertently transitioning a regional crisis between nuclear-armed states from a conventional to a nuclear level of conflict.

What characteristics of AI technology may create new rungs on the escalation ladder that increases inadvertent nuclear risk? The chapter highlighted three features of the AI-nuclear strategic nexus—contextualized in the broader digital information ecosystem—that increase inadvertent escalation risk. First, while nuclear states have a mutual interest in avoiding nuclear war—assumed by the "rational deterrence theory" examined in Chapter 2—they tend to place a high value on their nuclear forces, which AI-enhanced conventional counterforce weapons increasingly threaten.

Second, escalatory rhetoric and other aggressive behavior amplified by the digital information ecosystem cause actors to misperceive others' actions as unprovoked malign intent—not, for example, as a response to the initiator's behavior action and reaction spirals of potentially irrevocable inadvertent escalation. Moreover, new AI-enhanced tools enable a more diverse range of state and non-state actors with asymmetric techniques to improve their bargaining hand by delegitimizing (or legitimizing) the use of nuclear weapons during an escalating situation—or to sue for peace or de-escalating a situation.

Third, and related, the state of heightened tension, uncertainty, complexity, and compressed decision-making associated with modern digitized warfare is becoming dramatically increased by AI technology. This emergent trend restricts the options and time available for de-escalation, compounding the problem of offense-defense distinguishability and increasing the risks of horizontal and vertical inadvertent escalation.

The quintessential dual-use nature of AI is intensifying security dilemmas between nuclear rivals

How might dual-use AI technology influence security dilemma dynamics? Chapter 4 demonstrated that specific *structural* (dual-use technology, offense-defense balance, asymmetric distribution of power, and regime type), and *non-structural* (social constructed dynamics and cognitive-psychological) characteristics of dual-use AI technology interact with the security dilemma in ways that intensify this dynamic—or "AI-security dilemma dynamics." The diffusion and synthesis of dual-use AI—predominately commercially derived—increases the danger of spirals of insecurity and distrust that lead to military preparations and doctrinal shifts—to achieve the first-mover advantage and hedge against the uncertainty associated with the indistinguishability of dual-use technology—that risk rivals stumbling into a nuclear conflict.

The backdrop of great-power competition—most notably US-China—predisposes actors to securitize dual-use technology and frame others' actions in ways that fuel uncertainties and insecurities. Viewed this way, AI technology compounds the existing challenges posed by dual-use technology as an aggravator of the security dilemma in competitive dyads, especially when a rising challenger's ambitions are pitted against a declining status-quo hegemon. In short, mutual distrust and an atmosphere of uncertainty surrounding dual-use technology between competitors—viewed through the lens of securitization—means that any additional or new advances will deepen the security dilemma.

The chapter's reworking of the traditional notion of offense-defense distinguishability to view AI technology—that is, dual-use distinguishability uncertainty—generally increases when the development, utilization, and distribution is weak or widely dispersed, particularly to the private sector. This conceptual departure (or "civilian-military distinguishability") can help us to understand contemporary security dilemma spiraling dynamics in the context of rapidly increased commercial investment and political interest in AI—further complicating the perennial knotty conundrum of inferring an adversary's intentions from its capabilities.

Furthermore, because of the commercial impetus and scalability driving advances in AI—compared to other dual-use technologies such as space, cyber, and nuclear—state control over the development and use of AI technology has, thus far, been far more sanguine. I hope that this chapter (and the book more generally) adds to the voice of those who advocate the need for policymakers to be aware of the potential dangers posed by their (deliberate and unintentional) choices and decisions, as well as the importance of developing empathy with others—to determine whether a conflict of interest is genuine or illusory, thus reducing the risk of the unnecessary resort to military force.

AI is creating new tools and novel pathways for third-party actors to spark accidental nuclear war

Chapter 5 reconceptualized the concept of "catalytic nuclear war" in the light of developments in AI and concluded that the concept of catalytic nuclear war should be revisited considering recent technological change and improved understanding of human psychology—especially humans' propensity for making fast, intuitive, reflexive, and heuristic judgments when faced with information overload and unfamiliar technologies. Specifically, the research found that emerging technologies—most notably cyber, AI technology, hypersonic technology, and drones—are rapidly creating new and exacerbating old cheap, scalable, and relatively easy means for state and non-state actors to fulfill their nefarious goals. This trend creates novel threat pathways and escalation mechanisms to manipulate and propagate mis/disinformation during crisis times.

The chapter identified three features of the emerging digital age that are aggravating the risk of catalytic nuclear war (information complexity; greater automation of nuclear command, control, and communication systems; and information manipulation disinformation), complicating nuclear crisis management. These variables are not mutually exclusive. Instead, the interplay between these conditions may allow them to feed into one another with uncertain and potentially self-reinforcing effects. In sum, these conditions are a function of the confusion and uncertainty created by the sociotechnical complexity generated in the digital age as shown in Chapter 3.

This chapter showed that recent developments in AI-enabling technology have exacerbated these vulnerabilities and introduced additional tools for third-party actors to exploit. A critical nuclear risk highlighted in the chapter is: new AI tools make it easier, faster, and more scalable to manipulate the information landscape in which decisions about atomic weapons take place, which in turn increases catalytic nuclear risks. For instance, during a crisis involving two nuclear powers, when communication systems are compromised, nuclear arsenals are on high alert, decision-making timeframes are compressed, or launch authority is pre-delegated, the consequences of accidental nuclear use could be catastrophic. In other words, the catalyzing action of a state or non-state actor could produce the effect of an imminent attack on one or both of two nuclear-armed states, for which pre-emption is considered the most profitable strategy.

The long road home ...

> The only way of discovering the limits of the possible is to venture a little way past them into the impossible
>
> —**Arthur C. Clark's 2nd Law**[10]

This section proposes five broad clusters of potential AI-nuclear nexus risk reduction measures, with relevance to both nuclear and non-nuclear armed states: arms control and verification, using AI and other emerging technologies to support nuclear risk reduction, normative and confidence-building measures to increase mutual trust, bilateral and multilateral strategic stability dialogue, and cooperation and engagement with broader stakeholders.

Arms control: Non-proliferation in the age of AI

Arms control has traditionally been a mechanism for managing certain types of (usually kinetic/tangible) weapons or their applications, and it is one of several tools available to limit competition and increase stability between nuclear-armed adversaries.[11] The long-standing Nuclear Non-Proliferation Treaty (NPT) provides a successful case study in global governance that minimized the threat posed by the weaponization of new (i.e., atomic) technologies while enabling the mutual benefits of sharing nuclear technology to strengthen strategic stability. AI is much more broadly diffused than nuclear technology, however, with the private sector's heavy involvement in the R&D and application of inherently dual-use AI technology.[12] Because of these features, AI technology (like cyberspace) is expected to challenge traditional arms control thinking and raise questions about compliance with international law, which, in turn, will require new tools, competencies, and other means (or "rules of the road") to build trust and confidence between nuclear and non-nuclear armed states.[13]

When the lines between dual-use capabilities and nuclear and non-nuclear are blurred, arms control is much more challenging, and strategic competition and arms racing are more likely to emerge.[14] Moreover, technologies in the digital domain cannot be easily quantified or tracked in the same way that tangible assets as missiles can be counted, monitored, and verified. The intangibility and dual-use nature of emerging technologies like AI, cyberspace, and space capabilities have prompted experts to move away from regulating capabilities and towards regulating behaviors, codes of conduct, and establishing norms—that is, "soft" or informal rather than "hard" and legally binding arrangements like the 1967 Outer Space Treaty, which prohibits the stationing of nuclear weapons in space. In short, legacy state-centric centralized arms control frameworks—and even the notion of strategic stability itself—will increasingly struggle to assimilate, govern, and respond to the decentralized, dynamic, and diversity of actors (state and non-state) that inhabit dual-use emerging technologies like AI.[15]

Another complicating issue is that today there are no standard definitions—or engineering methodology—to formulate new regulatory or arms control frameworks to ensure the safety and robustness of AI—the so-called "AI control problem."[16] For example, some AI experts believe that existing tools (e.g., reinforcement learning techniques) cannot resolve humans' risks from AI-augmented autonomous systems.[17] Thus, a different kind of algorithm than currently exists will probably be needed, prioritizing the safety, robustness, and interpretability of complex AI-infused Nuclear Command, Control, and Communications (NC3) systems. As decision-makers ruminate over the ways and means in which the actions of a non-state actor might spark a catalytic nuclear exchange, they will need to consider how they would respond to such an attack and how an appropriate response might become a self-fulfilling prophecy—a chain of reaction and counter-retaliation—that leads unintentionally to nuclear war. As militaries modify operational concepts and strategic doctrines to manage escalation risk in the digital age, it therefore behooves leaders to account for the rapidly evolving nature of the emerging threat environment, characterized by complexity, uncertainty, deception, misperception, and vulnerability

For now, at least, AI technology is not currently integrated into states' nuclear targeting, C2, or launch systems. A narrow window exists, therefore, for nuclear powers (the P5 members as well as India, Pakistan, Israel, and NATO states) to agree on new principles, practices, and norms (e.g., banning attacks on nuclear-armed states' NC3 systems), and enshrine these into international law—for instance, within the United Nations Convention on Certain Conventional Weapons framework and coordinated by the UN Conference on Disarmament.[18] Identifying individual AI-enhanced weapon systems that nuclear states might develop which would violate existing norms—i.e., cause unnecessary suffering or superfluous injury, are directed at combatants with military objectives, and have effects that cannot be controlled, will likely yield greater success than banning AI-enhanced weapons entirely. Specific measures might include prohibiting or imposing limits on AI technology fusion in nuclear C2 systems, autonomous nuclear-armed missiles, and nuclear weapons launch decisions.[19]

Ultimately, however, until the potential for AI-infused nuclear weapons to cause unintended human consequences and their tactical and strategic potential are better understood, it is unlikely that any self-interested nuclear-power—tempered by a desire to retain the ability to fight effectively to support national security interests—would seriously consider banning AI-enhancements to their nuclear deterrence capabilities.[20] To be sure, there is little historical precedent for prohibiting weapons before they have been fielded.[21] How might AI technology be used to support stability and reduce nuclear risk?

AI as a "solution" to support nuclear risk reduction

To date, the debate about AI technologies, nuclear risk, and non-proliferation tends to focus on ways to manage, restrict (i.e., "keeping humans in the loop" and "meaningful human control"), or ban the technology (i.e., for targeting and using nuclear weapons). High-profile examples include the Campaign to Stop Killer Robots, the UN Convention on Certain Conventional Weapons, and the Group of Government Experts on emerging technology and lethal autonomous weapon systems (LAWS). This campaign, broadly modeled on previous efforts to prohibit "inhuman weapons" (e.g., cluster munitions, anti-personnel mines, and laser weapons), have struggled to gain significant traction at a policy-level on account of the more abstract and nuanced legal, ethical, normative, and security issues surrounding AI technology and autonomous weapons.[22]

Because many observers envision a future in which states will deploy fully autonomous robots to identify and attack combatants, various parties have argued that humans must maintain "humans in the loop" and thus "meaningful human control."[23] Others argue, counter-intuitively, that an outright ban on autonomous weapon systems based solely on speculation as to their future capabilities and usage might forfeit potential uses of them to minimize harm to non-combatants when compared to other systems in military arsenals.[24] Tragedies such as the 1988 USS Vincennes incident—when human error led to the mistaken downing of an Iranian airliner—demonstrate that a "human in the loop" is not necessarily a panacea in situations where it may be difficult to distinguish civilians and civilian objects from combatants and military objectives—or the problem of the "fog of war" examined in Chapter 3.

Much less discussion has explored how AI-enabled and enhanced technologies (e.g., distributed ledgers, additive 3D manufacturing, remote sensing, image-recognition software, and blockchain technology) might, and in some cases already are, creating novel approaches to support nuclear non-proliferation and disarmament efforts.[25] AI technology (i.e., remote-sensing, machine learning (ML), big-data analytics, and image recognition) might also offer innovative solutions to develop new and revise legacy arms control frameworks and contribute to non-interference mechanisms for arms control verification. Thus, reducing the need for "boots on the ground" inspectors in sensitive facilities, for example, authoritarian and "closed" regimes such as Iran, Russia, and North Korea.[26]

AI object identification applications (to augment satellite imagery of missile production facilities or test ranges) and pattern recognition tools (to filter out noise from large datasets and detect anomalies) may be used together with open-source geospatial and intelligence data (e.g., the Open Nuclear Network of One Earth Future, the Geo4Nonpro Project, and Bellingcat) to support arms verification efforts, and call out cheaters under arms control agreements. For example, in the environmental security domain, fish producers can use data on ship AI-enabled geolocations to track fish types and catch sizes to report to regulators and consumers that they are complying with sustainability standards.[27]

Within the arms control context, for instance, intelligence gathered from automatic detection with mounted sensors or ML pattern-recognition tools about preparations, construction, and communication could assess suspicious military movements and missile tests—either in accordance with or in violation of arms control agreements[28]—thus enhancing nuclear safeguards mechanisms and the credibility of future strategic arms control agreements.

The CIA, for example, uses AI technology to support a range of pattern-recognition tasks such as geolocating images without the associated metadata, creating 3D models from satellite imagery, and inferring a building's function based on "pattern-of-life" analysis.[29] AI algorithms can also be used to enhance non-image-based detection, such as identifying small nuclear tests from the vast seismic datasets collected as part of the Comprehensive Nuclear Test Ban Treaty monitoring system.[30] Towards this end, the US National Geospatial Intelligence Agency has recently invested in AI research projects to assist analysts with routine tasks, streamline analysis, identify biases, and work with large and complex datasets.[31] In addition to the US, other states, including the UK, France, Israel, and China are also actively pursuing AI tools to automate and streamline intelligence gathering and analysis processes.[32]

Furthermore, and subject to the reliability and quality of their outputs, algorithms could also be used to identify shifts in news coverage and social media open-sources in authoritarian regimes during a crisis, which may offer critical intelligence to states that consider an attack to be imminent.[33] According to the US DoD, "such a capability may soon be achievable ... new insights into social dynamics offer [the US] the possibility that we could leverage these massive data sources to make actionable predictions about future world events."[34]

AI-augmentation to enhance states' early-warning and detection systems to improve target identification might reduce false positives (or nuclear "close calls"),[35] and bias and ambiguity about an adversary's actions, thus lowering the risk of inadvertent escalation during a crisis.[36] Besides, AI-supported big-data analytics could be used to collate, synthesize, and process electronic data (i.e., signals, imagery, and open-source information) to identify patterns of behavior and launch profiles unique to specific types of capabilities—removing any ambiguity that may exist in datasets viewed individually.[37] For example, AI-enhanced satellites might be used to differentiate between the launch of a demonstration civilian-use rocket and an intercontinental ballistic missile—which have similar radar signatures—or a conventional from a nuclear missile launch. National Security Commissioner and AI expert Jason Matheny notes that this is particularly important for "different scenarios where you may not have a definitive warning from a single sensor," and because "missile designs are meant to evade certain classes of missile warning systems, we're going to have to use some form of sensor fusion that's AI-enabled in order to alert the human operator that a missile has likely been launched."[38]

In addition, fusing AI with early-warning systems may be particularly stabilizing for countries that do not have the advanced satellites, sensors, and forward-deployed radar systems that the US and Russia have developed to ensure that missile launches are detected and assessed for threat potential. For instance, AI-augmented early-warning and detection systems may offer China improved transparency and confidence about US military operations in the Indo-Pacific (i.e., to discern and discriminate between nuclear and conventional weapon systems in an incoming attack), thus reducing inadvertent escalation dynamics caused by miscalculation, false positives, or surprise. AI technology—notably AI-enhanced cybersecurity tools—might also improve the security of NC3 systems (i.e., patch undetected vulnerabilities in nuclear systems), thus potentially strengthening deterrence and stability.[39] The US Defense Advanced Research Projects Agency (DARPA) has, for example, already begun to study the ways AI may be used to identify vulnerabilities in conventional military systems.[40]

AI could also support defense planners' design and manage war-gaming and other virtual training exercises to refine operational concepts, test various conflict scenarios, and identify areas and technologies for potential development.[41] Specifically, AI-ML techniques—modeling, simulation, and analysis—could complement counterfactuals and low-tech table-top war-gaming simulations to identify contingencies under which "perfect storms" might form, not to predict them but rather to challenge conventional wisdom, highlight bias and inertia, to highlight these conditions, thus offering a more accurate tool for assessing the ebb and flow of interdependent global events.[42]

Cognizant of the shortcomings of traditional war-gaming methodologies—particularly the confirmation biases of game developers, defense planners, and related organizations—the US DoD has begun to experiment with AI technology to test a range of scenarios.[43] Thus, enabling participants to better prepare against adversaries in unpredictable dynamics, fast-moving environments, and cross-domain, where unpredictable and counterintuitive (i.e., to Western-centric strategic thinking) human-machine and machine-machine interactions take place.[44] In nuclear deterrence, where understanding an adversary's intentions, and accurately interpreting and sending clear signals (i.e., resolve, reassurance, and credibility), is critical to control nuclear risk, real-world war-gaming and other education tools can help operators, defense planners, allies, and adversaries to consider diverse and unpredictable scenarios from various perspectives.[45]

One promising line of effort is a collaborative initiative that has developed a prototype platform, "SLAFKA,"[46] using blockchain technology—a distributed ledger technology—to safeguard nuclear materials effectively.[47] This initiative demonstrates the potential of blockchain technology to provide data confidentially and securely to the International Atomic Energy Agency: data transmitted by a nuclear member state would only be visible to other member states, maintaining the anonymity of the participants. One analyst notes that these features of blockchain make it uniquely suited to "help to build technical capacity among [non-nuclear weapons states] and habits of cooperation among NPT parties while protecting proliferation-sensitive data."[48]

However, emerging technologies like blockchain and AI are not panaceas for the perennial compliance and global governance challenges surrounding nuclear weapons. If intelligence data was manipulated or otherwise compromised (either deliberately or by design fault), the credibility of this type of information as a transparency tool would be undermined. Tainted evidence would also place an additional burden on intelligence agencies to verify compliance and increase the risk that inaccurate intelligence (i.e., states' self-reporting disclosures) might be inserted into arms control processes.[49] Moreover, as scholars Scott Sagan and Kenneth Waltz remind us, despite advances in verification and safety technology, "there will remain the problem of what to do if an erstwhile nuclear nation is caught secretly preparing to rearm"—a disarmament, deterrence, and, potentially, escalation management problem.[50] In sum, while emerging technology such as AI will unlikely eliminate the complex challenges associated with nuclear verification and safeguards verification, these technologies can play an essential part in strengthening existing measures and creating innovative new ones.

Normative, behavioral, and confidence-building measures to increase mutual trust

The many technical, political, geopolitical, and dual-use challenges surrounding AI technology described in this book make legally binding treaties and formal arms control agreements highly unfeasible. States should, therefore, consider alternative arrangements and policies such as norms, codes of conduct, confidence-building measures, and agreed standards of behavior as the basis for more formal commitments later.[51] Because of the pace of change occurring in AI, even if an agreed framework for limiting and controlling this technology was established, the development and deployment trajectory of AI would likely render it obsolete before the ink had dried. It is noteworthy that some precedent has been established on the principle that the rules of International Humanitarian Law (IHL) regarding the legality of weapon systems can apply to weapons that did not exist when a particular norm was agreed, or customary law enshrined.[52]

Recent initiatives towards this end include, *inter alia*, the following: The UN General Assembly's Secretary-General's "Roadmap for Digital Cooperation" report;[53] a NATO-sponsored initiative that encourages its members to agree on responsible governance principles for new technologies;[54] the US DoD's adoption of "Ethical principles for AI" and "Responsible AI" memorandum;[55] and the Czech Republic, Finland, Germany, the Netherlands, and Sweden calling on the European Union to start a process on the responsible military use of new technologies, including AI.[56] A key takeaway from these efforts is that even though a human might not be in full control of a particular weapon system, it does not diminish human responsibility and accountability for the actions of AI-powered autonomous weapons.

Additional normative and behavioral measures might also include:

- Nuclear states committing not to engage in escalatory attacks or intrusions on each other's (nuclear and non-nuclear) C2 systems.[57]
- Extending a cyber "strategic no-first-use" policy to discourage destabilizing behavior (e.g., cyber-attacks, left-of-launch operations, or data-poisoning) on critical NC3 infrastructure to shore up deterrence and reduce inadvertent escalation risk.[58]

- International dialogue—through channels like the Convention on Certain Conventional Weapons—and investment in AI safety and reliably research.[59]
- Due diligence measures such as testing and impact assessments, exercising oversight, and evaluation, validation, and verification processes.[60]
- Incorporating the risks (and potential stabilizing benefits) of emerging technologies such as AI as a standing topic on the P5's annual agenda.
- Expanding existing verification partnerships (and forging new ones), continuing dialogue.[61]
- Building new skills and capacities between nuclear and non-nuclear powers to identify challenges associated with nuclear risk, disarmament verification, and other broader strategic challenges in the digital age.[62]

Taken together, these measures should, at a minimum, help mitigate some of the inherent "black box" features of AI technology described in the book's opening chapter, and, at best, ensure that the parameters underlying AI algorithms are reliable and accurate, its limitations understood, and whether a particular application is designed to *inform* or *make decisions*.

Bilateral and multilateral strategic stability dialogue

An additional risk reduction channel to help mitigate emerging technologies like AI is a high-level strategic dialogue between nuclear-armed adversaries. States come to the table prepared to discuss strategic stability, deterrence, escalation, and broader nuclear security, focusing on the potential risks created by AI technologies in a strategic context, where AI intersects directly with nuclear weapons and indirectly through non-nuclear strategic weapons. Incorporating emerging technologies like AI into disarmament discussions—for example, a P5/NPT setting—risk reduction dialogue might become an effective means to build broader collaborative pathways among diverse actors who might otherwise have "fundamental ideological differences about pathways to disarmament." The mutual recognition harvested from emerging technologies in NPT exchanges is necessary and may also help encourage cooperation among nuclear powers on other issues.[63]

During US President Joe Biden's first summit meeting with Russian President Vladimir Putin, and after a decade of deadlock, the two presidents agreed to a new round of strategic stability dialogue. At a press conference after the summit, Biden noted that the dialogue would "work on a mechanism that can lead to control of new and dangerous and sophisticated weapons that are coming on the scene now that reduce the times of response, that raise the prospects of accidental war."[64] Biden did not elaborate on the specific weapon systems he had in mind, however. The highest-level meeting between the incoming Biden administration and Chinese officials in March 2021 was accompanied by diplomatic posturing and confusion over whether the exchange was a "strategic dialogue"—a reference to regular talks that fell apart under former President Donald Trump.[65] While positive for the world's most important diplomatic relationship, the willingness to engage on global security issues such as climate change eschewed much-needed discussion on strategic topics such as emerging technology, nuclear risk, and other non-proliferation efforts.

Future high-level bilateral and multilateral dialogue should include issues raised in this book, including the novel non-kinetic escalatory risks associated with complexity in the digital information ecosystem (e.g., dis/misinformation, deepfakes, information sabotage, and social media weaponization) during conventional crises and conflict involving nuclear-armed states. In the case of US-China relations, dialogue and information sharing could be explored to foster mutual understanding—of the potential risks, benefits, and intentions of AI-enabled capabilities—in new and existing bilateral forums, including: the Diplomatic & Security Dialogue, memoranda of understanding for militaries, US-China Joint Staff Dialogue (notably about digital threats posed to NC3 systems), existing hot-lines between China's Ministry of Defense and the US DoD, and the US-China coordination mechanism to manage computer emergency response teams for possibly broader threats with strategic ramifications. Without the confidence that can flow from a mutual understanding between strategic rivals, perceived threats (e.g., cyber-attacks on NC3 systems) could make crisis and conflict harder to manage, leading to an escalation that neither side intend.[66] Sustaining strategic dialogue between adversaries over the long term and implementing nuclear risk reduction and mitigation measures will require political will on both sides, currently in very short supply.

Cooperation and engagement with broader stakeholders

> When we all agree only one of us is doing the thinking
>
> —**Lyndon B. Johnson**[67]

Like other emerging technologies, the quintessential dual-use nature of AI technologies—primarily developed in a bottom-up model by the private sector—is fundamentally distinct from the traditional top-down defense industrial complex model, comprising a relatively small number of defense contractors developing technologies for the defense community.[68] This new bottom-up model requires additional lines of effort between the private and public sectors to ensure that: (a) dual-use technology developed in the private sector is not unwittingly sold to, and misused by, a third party or non-state actor targeting nuclear-armed states;[69] (b) defense planners are aware of AI technology's risks and potential vulnerabilities; and (c) best practices are implemented to mitigate the external costs of actors' misuse of dual-use technology—notably, off-the-shelf commercial technology such as drones, facial recognition, malware, and deepfake software.

Towards this end, partnerships should be forged between commercial AI developers and researchers to explore risk reduction measures in the nuclear enterprise (i.e., NC3 systems). Cooperation might help bridge the gaps between technical domains and policymaking, fostering multidimensional scholarship to support the non-proliferation initiatives.[70] It is noteworthy that the World Economic Forum has acknowledged that bridging this gap is critical to avoid the pitfalls of technologies like AI, while benefiting from their promise.[71] Moreover, the nuclear community should redouble its efforts to educate existing talent about the potential implications—both stabilizing and destabilizing—of the AI-nuclear nexus and recruit new talent with technical expertise in the development of human-centered AI design and the ethical use of these technologies—in security and defense and at a broader societal level.[72]

For instance, in the case of deepfakes, a dearth of high-quality data to train deepfake detection tools and low barriers to generate fake digital content has created a perpetual "cat and mouse game" in detecting manipulated information.[73] To date, the emphasis has been on the development of ML automated detection software to identify questionable content—for example, automated classifiers to predict who might spread disinformation online and AI-enabled tools to detect clickbait, false rumors, and social bots). Though these tools can filter vast amounts of online content, the algorithms used for this purpose are prohibitively time and resource intensive. Previous attempts at using algorithms to moderate mis/disinformation have counter-intuitively increased the prevalence of polarizing topics caused by bias, errors, or algorithms being gamed by malevolent actors.[74] Besides, the limited access to social media data, coupled with the rapid improvements in the automation, speed, and adaptability of bots, makes it increasingly challenging to both identify bots and distinguish authentic from fake accounts.[75]

Public-private sector cooperation should leverage the expertise of existing open-source intelligence and media forensics communities, together with

AI researchers, cognitive scientists, and the nuclear policy community, to rebuild public confidence and trust in manipulated information—notably, mis/disinformation.[76] An example of cooperation in the commercial sector to address this challenge is the Coalition for Content Provenance and Authenticity. Established by Adobe, Arm, the British Broadcasting Corporation, Intel, Microsoft, and Truepic, this emphasizes developing open, global technical standards to support content provenance efforts, including specifications and standards for social media platforms.[77] The open-source intelligence community needs to codify and institutionalize best practices, establish quality standards and shared ethical norms, and improve the collection and analysis skills of (private and public sector) operators to reduce the risk of errors, bias, and myopia. A series of international workshops on open-source intelligence, hosted by the Stanley Center for Peace and Security, a US non-profit, with geospatial and open-source intelligence leaders to examine ethical challenges, is a promising step in this direction.[78]

In the final analysis, the best way to prepare for the AI-nuclear future may be to adhere to a few basic principles to guide the management of nuclear weapons in their interactions with emerging technology.[79] First, nuclear weapon systems should avoid being unduly complex or overcomplicated. Second, these systems must be fortified and robust enough to withstand traditional threats and increasing new threats emerging in the digital domain. Third, nuclear weapons must be disentangled and, where possible, distinctly separate (both physically and doctrinally) from non-nuclear capabilities and command, control, communications, and intelligence systems.

Though achieving consensus on the risk reduction measures outlined in this chapter will not be easy, working towards this goal is critical for the global nuclear and defense communities and broader stakeholders. This goal would represent a positive step towards managing the uncertainties, complexities, and trade-offs associated with the AI-nuclear strategic nexus to: (a) buy more time for decision-makers during crisis and conflict—a precious commodity in the digital age; (b) reduce the perception that AI-enhanced capabilities, in the broader context of nuclear modernization, may be used to support destabilizing nuclear superiority or warfighting postures; and relatedly, (c) helping to ensure that nuclear weapons are not perceived as more usable—that is, reaffirming the taboos, de-escalation options, and deterrence rationale that undergird the ultimate weapon of mass destruction. Serious reflection on these basic principles and the measures outlined in the chapter would represent a positive step towards managing the uncertainties, complexities, and trade-offs associated with the AI-nuclear strategic nexus. As

Alan Turing wrote in 1950: "We can only see a short distance ahead, but we can see plenty there that needs to be done."[80]

Notes

1. Susan Ratcliffe, ed., *Oxford Essential Quotations*, 4th ed. (Oxford University Press, 2016), https://www.oxfordreference.com/view/10.1093/acref/9780191826719.001.0001/q-oro-ed4-00004005.
2. Lawrence Freedman, "Has Strategy Reached a Dead-End?" *Futures* 11, no. 2 (1979): pp. 122–131.
3. Colin S. Gray, *The Future of Strategy* (Malden, MA: Polity Press, 2015), pp. 63–79.
4. See James Fearon, "Rationalist Explanations for War," *International Organization* 49, no. 3 (1995): pp. 379–414; and Bradley A. Thayer, "Thinking about Nuclear Deterrence Theory: Why Evolutionary Psychology Undermines Its Rational Actor Assumptions," *Comparative Strategy* 26, no. 4 (2007): pp. 311–323.
5. See Peter Vrancx, Anne Nowe, and Yann- Michael De Hauwere, eds., *Reinforcement Learning: Adaptation, Learning, and Optimization* (Berlin: Springer, 2012), pp. 441–470.
6. See Alexander Aroyo et al., "Overtrusting Robots: Setting a Research Agenda to Mitigate Overtrust in Automation," *Journal of Behavioral Robotics* 12, no. 1 (2021): pp. 423–443.
7. Robert Jervis, "Arms Control, Stability, and Causes of War," *Political Science Quarterly* 108, no. 2 (1988): pp. 81–110.
8. See Andrew Futter and Benjamin Zala, "Strategic Non-Nuclear Weapons and the Onset of a Third Nuclear Age," *European Journal of International Security* (2021): pp. 18–19.
9. Herman Kahn, *On Escalation: Metaphors and Scenarios* (New York: Praeger, 1965), p. 38.
10. Ratcliffe, ed., *Oxford Essential Quotations*.
11. See, for example, Barry Buzan and Eric Herring, *The Arms Dynamic in World Politics* (London, Boulder, CO: Lynne Reinner, 1998); Thomas Schelling, *Arms and Influence* (New Haven and London: Yale University Press, 1966; and Robert Jervis, "Arms Control, Stability, and Causes of War," *Political Science Quarterly* 108, no. 2 (1988): pp. 81–110.
12. Very few statistics are available on the extent of civilian-sourced technologies in advanced weapon systems, making it challenging to measure civilian breakthroughs' indirect (or intangible) influence on military-use capabilities. The growing literature on "spin-off" (military-to-civilian transfer) and "spin-in" (civil-to-military transfer) can help analysts consider to what extent civilian innovation enables the development of military AI and what this implies for safety, reliability, and regulation. See Matthijs M. Maas, "How Viable is International Arms Control for Military Artificial Intelligence?"; and Maaike Verbruggen, "The Role of Civilian Innovation in the Development of Lethal Autonomous Weapon Systems," *Global Policy* 10, no. 3 (2019): pp. 338–342.
13. Amandeep Singh Gill, "Artificial Intelligence and International Security: The Long View," *Ethics and International Affairs* 33, no. 2 (2019): pp. 169–179.
14. See Heather Williams, "Asymmetric Arms Control and Strategic Stability: Scenarios for Limiting Hypersonic Glide Vehicles," *Journal of Strategic Studies* 42, no. 6 (2019): pp. 789–813.
15. See Futter and Zala, "Strategic Non-Nuclear Weapons and the Onset of a Third Nuclear Age," pp. 1–21.

16. The "AI control problem" refers to the issue that, under certain conditions, AI systems can learn unexpected and counterintuitive behaviors that may run contrary to the goals of engineers and operators. See Stuart Russell, *Human Compatible* (New York: Penguin Random House, 2019), p. 251.
17. Dylan Hadfield-Menell, Anca Dragan, Pieter Abbeel, and Stuart Russell, "Cooperative Inverse Reinforcement Learning," 30th Conference on Neural Information Processing Systems (NIPS 2016), Barcelona, Spain.
18. Whether and how military AI and autonomous weapons can and should be regulated is a highly contested issue among policymakers, scholars, and campaigning activists. See Frank Sauer, "How (Not) to Stop the Killer Robots: A Comparative Analysis of Humanitarian Disarmament Campaign Strategies," *Contemporary Security Policy* 42, no. 1 (2021): pp. 4–29; and Michael Schmitt, "Autonomous Weapon Systems and International Humanitarian Law: A Reply to the Critics," *Harvard National Security Journal* 4 (2013): pp. 1–37.
19. For a recent study on AI and arms control, see Maas, "How Viable is International Arms Control for Military Artificial Intelligence?"
20. Schmitt, "Autonomous Weapon Systems and International Humanitarian Law," p. 35.
21. Ibid. p. 36.
22. These issues include: (1) the contested legal grounds for delegating IHL decision-making authority to a machine—LAWS do not necessarily violate legally binding IHL principles; (2) framing of LAWS is problematic regarding the terminology and definition (e.g., autonomy, AI, LAWS, and "meaningful human control")—degrees of autonomy vary temporally, and dependent on what tasks are performed by the same human-machine-system; (3) technological progress could make LAWS at least as IHL-compliant as existing semi-autonomous and remotely operated weapons. Robert and Sauer, "How (Not) to Stop the Killer Robots: A Comparative Analysis of Humanitarian Disarmament Campaign Strategies"; and Vincent Boulanin, Netta Goussac, and Laura Bruun, *Autonomous Weapon Systems and International Humanitarian Law: Identifying Limits and the Required Type and Degree of Human–Machine Interaction* (SIPRI, Stockholm: June 2021).
23. See UNIDIR, "The Weaponization of Increasingly Autonomous Technologies: Considering How Meaningful Human Control Might Move the Discussion Forward," 2014, https://unidir.org/publication/weaponization-increasingly-autonomous-technologies-considering-how-meaningful-human.
24. See Michael C. Horowitz, "The Ethics & Morality of Robotic Warfare: Assessing the Debate over Autonomous Weapons," *Daedalus* 145, no. 4 (2016): pp. 25–36; and Schmitt, "Autonomous Weapon Systems and International Humanitarian Law," pp. 1–37.
25. Notable exceptions include Jessica Cox and Heather Williams, "The Unavoidable Technology: How Artificial Intelligence Can Strengthen Nuclear Stability," *The Washington Quarterly* 44, no. 1 (2021): pp. 69–85; Michal Onderco and Madeline Zutt, "Emerging Technology and Nuclear Security: What Does the Wisdom of the Crowd Tell Us?" *Contemporary Security Policy* 42, no. 3 (2021): pp. 286–311, doi: 10.1080/13523260.2021.1928963; and Cindy Vestergaard, "Better Than a Floppy: The Potential of Distributed Ledger Technology for Nuclear Safeguards Information Management," *Stanley Center for Peace and Security* (2018), https://stanleycenter.org/wp-content/uploads/2019/09/NPAIa-BetterFloppyVestergaard1018.pdf.

26. Roland LaJoie and Greg Govan, "Don't Let On-Site Inspections Go Extinct," Defense One, March 5, 2020, https://www.defenseone.com/ideas/2020/03/dontlet-site-inspections-adversaries-weapons-sites-go-extinct/163556/; Geo4Nonpro Project, *The James Martin Center for Non-proliferation*, June 25, 2020, https://nonproliferation.org/geo4nonpro-project/.
27. Melina Kourantidou, "Artificial Intelligence Makes Fishing More Sustainable by Tracking Illegal Activity," *Phys.org*, July 12, 2019, https://phys.org/news/2019-07-artificial-intelligence-fishing-sustainable-tracking.html.
28. Reports indicate that the United Nations, the US Central Intelligence Agency, and the US Department of Defense is already using AI technology to parse public open-source information (i.e., scraping and analyzing details from social media, blogs, and other sources) to anticipate when and where conflict may arise. Sheldon Himelfarb, "Can Big Data Stop Wars Before They Happen?" *Foreign Policy*, April 25, 2014, https://foreignpolicy.com/2014/04/25/can-big-data-stop-wars-before-they-happen/.
29. Kelley M. Sayler, "Artificial Intelligence and National Security," *Congressional Research Service*, January 30, 2019, 8, https://crsreports.congress.gov/product/details?prodcode=R45178.
30. National Academies of Sciences, Engineering, and Medicine, Nuclear Proliferation and Arms Control Monitoring, Detection, and Verification: A National Security Priority: Interim Report (Washington, DC: National Academies Press, 2021).
31. See Loren Blinde, "NGA Awards Four Contracts to Enhance Artificial Intelligence and Automation," *Intelligence Community News*, February 16, 2017, https://intelligencecommunitynews.com/nga-awards-four-contracts-to-enhance-artificial-intelligence-and-automation/.
32. Jane Vaynman, "Better Monitoring and Better Spying: The Implications of Emerging Technology for Arms Control," *Texas National Security Review* 4, no. 4 (2021), https://tnsr.org/2021/09/better-monitoring-and-better-spying-the-implications-of-emerging-technology-for-arms-control/#_ftnref55.
33. Elena Labzina and Mark D. Nieman, *Understanding the Use of State-Controlled Media: A Machine-Learning Approach* (2017) (unpublished manuscript), https://cpb-us-w2.wpmucdn.com/sites.wustl.edu/dist/c/549/files/2017/01/labzina_neuman2016_authoritarian_media-252h33v.pdf.
34. Defense Science Board, "Report of the Defense Science Broad: Summer Study on Autonomy," *US Department of Defense* (June 2016), https://www.hsdl.org/?view&did=794641, p. 79.
35. See Patricia Lewis et al., Too Close for Comfort: Cases of Near Nuclear Use and Options for Policy (Chatham House, London: 2014).
36. Cox and Williams, "The Unavoidable Technology: How Artificial Intelligence Can Strengthen Nuclear Stability."
37. Ibid. pp. 73–75.
38. Craig S. Smith, "AI War Games and the Challenge of China," *Forbes*, June 12, 2020, https://www.forbes.com/sites/craigsmith/2020/06/12/ai-war-games-and-the-challenge-ofchina/#603bc8946c74.
39. Most experts agree that AI-ML can offer significant benefits to cybersecurity; however, whether these benefits will amount to a "transformation"—i.e., meaningfully alter cybersecurity practice in cybersecurity—remains contested. See Tim G.J. Rudner and

Helen Toner, "Key Concepts in AI Safety: Robustness and Adversarial Examples," *Center for Security and Emerging Technology*, March 2021. For a competing view, see David Brumley, "Why I'm Not Sold on Machine Learning in Autonomous Security," *CSO Online*, August 27, 2019, https://www.csoonline.com/article/3434081/why-im-not-sold-onmachine-learning-in-autonomous-security.html.

40. Johanna Curiel, "DARPA Cyber Grand Challenge Recap: How Bots Will Help Lift the Security Game," *TechBeacon*, https://techbeacon.com/security/darpa-cyber-grand-challenge-recap-how-bots-will-help-lift-security-game.
41. Craig S. Smith, "AI War Games and the Challenge of China," *Forbes*, June 12, 2020, https://www.forbes.com/sites/craigsmith/2020/06/12/ai-war-games-and-the-challenge-ofchina/#603bc8946c74.
42. Paul K. Davis and Paul Bracken, "Artificial Intelligence for Wargaming and Modeling," *The Journal of Defense Modeling and Simulation* (2022), doi: 10.1177/15485129211073126; and Benjamin Jensen, Scott Cuomo, and Chris Whyte, "Wargaming with Athena: How to Make Militaries Smarter, Faster, and More Efficient with Artificial Intelligence," *War on the Rocks*, June 5, 2018, https://warontherocks.com/2018/06/wargaming-with-athena-how-to-make-militaries-smarter-faster-and-more-efficient-with-artificial-intelligence/.
43. For example, the DoD will often fund a game "with a prior idea about what they want the outcome of the game to be … to justify new authorities or new doctrine." Kendrick Foster, "Games without Frontiers: War-Gaming in US Grand Strategy," *Harvard Political Review*, August 4, 2020, https://harvardpolitics.com/features/gameswithout-frontiers-war-gaming-in-us-grand-strategy/.
44. See Paul K. Davis, *Applying Artificial Intelligence Techniques to Strategic-Level Gaming and Simulation* (Washington, DC: RAND Corporation Paper Series, P-7120, November 1985), https://www.rand.org/content/dam/rand/pubs/papers/2008/P7120.pdf.
45. See Patrick Tucker, "It's Time to Wargame Against an AI-Enabled China," *Defense One*, June 23, 2021, https://www.defenseone.com/technology/2021/06/its-time-wargame-against-ai-enabled-china/174922/; and Niklas Masuhr, "AI in Military Enabling Applications," *CSS Analyses in Security Policy*, no. 251 (October 2019), https://css.ethz.ch/content/dam/ethz/special-interest/gess/cis/center-for-securities-studies/pdfs/CSSAnalyse251-EN.pdf.
46. Participants of the "SLAFKA" project included the nuclear regulator in Finland (STUK), the University of New South Wales in Australia, and the Stimson Center in the US. See "Blockchain Prototype for Safeguarding Nuclear Material Unveiled & Demonstrated," *Stimson Center*, March 10, 2020, https://www.stimson.org/2020/blockchain-prototypefor-safeguarding-nuclear-material-unveiled-demonstrated/.
47. A distributed ledger technology platform is a system of electronic records that enables independent entities to establish consensus around a "ledger"—without relying on a central coordinator to provide the authoritative version of the records. Rauchs, Michel and Glidden, Andrew and Gordon, Brian and Pieters, Gina C. and Recanatini, Martino and Rostand, François and Vagneur, Kathryn and Zhang, Bryan Zheng, "Distributed Ledger Technology Systems: A Conceptual Framework," August 13, 2018, *SSRN* http://dx.doi.org/10.2139/ssrn.3230013.

48. Lyndon Burford, "The Trust Machine: Blockchain in Nuclear Disarmament and Arms Control Verification," *Centre for Science and Security Studies, King's College London* (October 2020), https://www.kcl.ac.uk/csss/assets/the-trust-machine-report.pdf.
49. Jane Vaynman, "Better Monitoring and Better Spying: The Implications of Emerging Technology for Arms Control," *Texas National Security Review* 4, no. 4 (2021) https://tnsr.org/2021/09/better-monitoring-and-better-spying-the-implications-of-emerging-technology-for-arms-control/#_ftnref55.
50. Scott Sagan and Kenneth Waltz, "The Great Debate: Is Nuclear Zero the Best Option?" *The National Interest* (October/November 2010): p. 90.
51. See Michael C. Horowitz and Paul Scharre, "AI and International Stability: Risks and Confidence-Building Measures," *Center for a New American Security*, January 12, 2021, https://www.cnas.org/publications/reports/ai-and-international-stability-risks-and-confidence-building-measures; and Michael C. Horowitz and Lauren Kahn, "How Joe Biden Can Use Confidence-Building Measures for Military Uses of AI," *Bulletin of the Atomic Scientists* 77, no. 1 (2021): pp. 33–35.
52. For example, the US Air Force's 2011 revised guidance on cyber capabilities extends its substantive requirements to mandate compliance with the Hague Convention (IV) on the conduct of war. Schmitt, "Autonomous Weapon Systems and International Humanitarian Law," pp. 30–31.
53. Report of the Secretary-General, "Roadmap for Digital Cooperation," United Nations, June 2020, https://www.un.org/en/content/digital-cooperation-roadmap/.
54. Edward Hunter Christie, "Artificial Intelligence at NATO: Dynamic Adoption, Responsible Use," *NATO Review*, November 24, 2020, https://www.nato.int/docu/review/articles/2020/11/24/artificialintelligence-at-nato-dynamic-adoption-responsible-use/index.html.
55. US Department of Defense, "DOD Adopts Ethical Principles for Artificial Intelligence"; "Responsible AI Champions Pilot," Joint Artificial Intelligence Center, Department of Defense, press release, November 12, 2020, https://www.ai.mil/docs/08_21_20_responsible_ai_champions_pilot.pdf; and "AI Partnership for Defense," Joint Artificial Intelligence Center, Department of Defense, press release, November 12, 2020, https://www.ai.mil/docs/AI_PfD_Joint_Statement_09_16_20.pdf.
56. "Minister's Declaration at the occasion of the Conference," Capturing Technology, Rethinking Arms Control, November 6, 2020, https://rethinkingarmscontrol.de/wp-content/uploads/2020/11/Ministerial-Declaration-RAC2020.pdf.
57. Nuclear weapons states (NWS) continue to actively debate the potential threats posed by attacks on NC3 systems to inform deterrence policies within the NPT regime. See Dmitry Stefanovich, "Russia's Basic Principles and the Cyber-Nuclear Nexus," *European Leadership Network*, July 14, 2020, https://www.europeanleadershipnetwork.org/commentary/russiasbasic-principles-and-the-cyber-nuclear-nexus; Herb Lin, "Cyber Risk Across the Nuclear Enterprise," *Texas National Security Review*, Summer 2021, https://tnsr.org/2021/06/cyber-risk-across-the-u-s-nuclear-enterprise/; and Ariel E. Levite et al., *Carnegie Endowment for International Peace*, "China-U.S. Cyber-Nuclear C3 Stability," April 8, 2021, https://carnegieendowment.org/files/Levite_et_all_C3_Stability.pdf.
58. Jacquelyn Schneider, "A Strategic Cyber No-First-Use Policy? Addressing the US Cyber Strategy Problem," *The Washington Quarterly* 43, no. 2 (2020): pp. 159–175.

59. "Final Report," National Security Commission on Artificial Intelligence, March 2021, https://www.nscai.gov/wp-content/uploads/2021/03/Full-Report-Digital-1.pdf, p. 99.
60. Ibid.
61. Examples of existing collaboration include: The Quad Verification Partnership with membership from the UK, Norway, Sweden, and the US, https://quad-nvp.info; United Kingdom–Norway Initiative, https://ukni.info/#:~:text=United%20Kingdom%20%2D%20Norway%20Initiative,arms%20control%20and%20disarmament%20verificationd; and "AI Partnership for Defense," Joint Artificial Intelligence Center, Department of Defense, Press release, November 12, 2020, https://www.ai.mil/docs/AI_PfD_Joint_Statement_09_16_20.pdf.
62. Marion Messmer, "Strategic Risk Reduction in the European Context," *BASIC*, June 2020, https://basicint.org/wp-content/uploads/2020/06/Strategic-Risk-Reduction-in-the-European-Context-WEB-1.pdf.
63. The relative exclusion of emerging technologies from NPT discussions thus far can be explained in part by: (1) a reticence from both nuclear weapons states and non-nuclear weapons states to include conventional weapons in NPT discussions; and (2) that as an institution, the NPT lacks the flexibility to adapt to rapidly changing geopolitical and technological developments. Heather Williams, "Remaining Relevant: Why the NPT Must Address Emerging Technologies," *Centre for Science and Security Studies, King's College London*, August 2020, https://www.kcl.ac.uk/csss/assets/remaining-relevant-new-technologies.pdf. pp. 7 and 13.
64. Kingston Reif and Shannon Bugos, "US, Russia Agree to Strategic Stability Dialogue," *Arms Control Association*, July/August 2021, https://www.armscontrol.org/act/2021-07/news/us-russia-agree-strategic-stability-dialogue.
65. Philip Glamann and Jing Li, "US, China Clash Over How to Describe Crucial First Meeting," *Bloomberg News*, March 11, 2021, https://www.bloomberg.com/news/articles/2021-03-11/u-s-china-clash-over-how-to-describe-crucial-first-meeting.
66. See Ariel E. Levite et al., "China-US Cyber-Nuclear C3 Stability," *Carnegie Endowment for International Peace*, April 9, 2020, https://carnegieendowment.org/files/Levite_et_all_C3_Stability.pdf.
67. Quotes by Lyndon Johnson, *The Best Quotations*, https://best-quotations.com/authquotes.php?auth=442.
68. See Antonio Missiroli, "Game of Drones? How New Technologies Alter Deterrence, Defence, and Security," *NATO Review*, May 5, 2020, https://www.nato.int/docu/review/articles/2020/05/05/game-ofdrones-how-new-technologies-affect-deterrence-defence-and-security/.index.html.
69. See Doug Britton, "The CANOPY WING Vulnerability: Weaponizing the Weakness," *IIoT World*, October 24, 2018, https://iiotworld.com/ics-security/cybersecurity/the-canopy-wing-vulnerabilityweaponizing-the-weakness.
70. Cox and Williams, "The Unavoidable Technology: How Artificial Intelligence Can Strengthen Nuclear Stability," p. 81.
71. Bruce Schneier, "We Must Bridge the Gap between Technology and Policymaking: Our Future Depends on It," *World Economic Forum*, November 12, 2019, https://www.weforum.org/agenda/2019/11/we-must-bridge-the-gap-between-technology-andpolicy-our-future-depends-on-it.

72. See Elizabeth Seger et al., "Tackling Threats to Informed Decision-Making in Democratic Societies," *The Alan Turing Institute*, October 14, 2020, https://www.turing.ac.uk/sites/default/files/2020-10/epistemic-security-report_final.pdf.
73. Alex Engler, "Fighting Deepfakes When Detection Fails," Brookings Institution, November 14, 2019, https://www.brookings.edu/research/fighting-deepfakes-when-detection-fails/; and Karen Hao, "Deepfakes Have Got Congress Panicking: This is What it Needs to do," *MIT Technology Review*, June 12, 2019, https://www.technologyreview.com/2019/06/12/134977/deepfakes-aicongress-politics-election-facebook-social/.
74. Sergey Sanovich, and Denis Stukal, "Strategies and Tactics of Spreading Disinformation Through Online Platforms," in *Social Media, Political Polarization, and Political Disinformation: A Review of the Scientific Literature*, edited by J.A. Tucker et al., pp. 30–39 (Menlo Park, CA: William & Flora Hewlett Foundation, 2018).
75. Ibid.
76. See Katarina Kertysova, "Artificial Intelligence and Disinformation: How AI Changes the Way Disinformation is Produced, Disseminated, and Can Be Countered," *Security and Human Rights* 29, no. 1–4 (2018): pp. 55–81; and Rachel Armitage and Cristian Vaccari, "Misinformation and Disinformation," in *The Routledge Companion to Media Disinformation and Populism*, edited by Howard Tumber and Silvio Waisbord (London: Routledge, 2021), pp. 38–49.
77. "Technology and Media Entities Join Forces to Create Standards Group aimed at Building Trust in Online Content," *Microsoft News Center*, February 22, 2021, https://news.microsoft.com/2021/02/22/technology-and-media-entities-join-forces-to-create-standards-group-aimed-at-building-trust-in-online-content/.
78. "The Gray Spectrum: Ethical Decision Making with Geospatial and Open-Source Analysis," *The Stanley Center for Peace & Security*, January 2020, https://stanleycenter.org/publications/the-gray-spectrum/.
79. Andrew Futter, *Hacking the Bomb: Cyber Threats and Nuclear Weapons* (Washington, DC: Georgetown University Press, 2018), p. 158.
80. "What Alan Turing Means to Us," *The Alan Turing Institute*, https://www.turing.ac.uk/blog/what-alan-turing-means-us.

Index

For the benefit of digital users, indexed terms that span two pages (e.g., 52-53) may, on occasion, appear on only one of those pages.

H

I